Evolutionary Economics and Social Complexity Science

Volume 23

The Japan Association for Evolutionary Economics (JAFEE) always has adhered to its original aim of taking an explicit "integrated" approach. This path has been followed steadfastly since the Association's establishment in 1997 and, as well, since the inauguration of our international journal in 2004. We have deployed an agenda encompassing a contemporary array of subjects including but not limited to: foundations of institutional and evolutionary economics, criticism of mainstream views in the social sciences, knowledge and learning in socio-economic life, development and innovation of technologies, transformation of industrial organizations and economic systems, experimental studies in economics, agent-based modeling of socio-economic systems, evolution of the governance structure of firms and other organizations, comparison of dynamically changing institutions of the world, and policy proposals in the transformational process of economic life. In short, our starting point is an "integrative science" of evolutionary and institutional views. Furthermore, we always endeavor to stay abreast of newly established methods such as agent-based modeling, socio/econo-physics, and network analysis as part of our integrative links.

More fundamentally, "evolution" in social science is interpreted as an essential key word, i.e., an integrative and /or communicative link to understand and re-domain various preceding dichotomies in the sciences: ontological or epistemological, subjective or objective, homogeneous or heterogeneous, natural or artificial, selfish or altruistic, individualistic or collective, rational or irrational, axiomatic or psychological-based, causal nexus or cyclic networked, optimal or adaptive, micro- or macroscopic, deterministic or stochastic, historical or theoretical, mathematical or computational, experimental or empirical, agent-based or socio/econo-physical, institutional or evolutionary, regional or global, and so on. The conventional meanings adhering to various traditional dichotomies may be more or less obsolete, to be replaced with more current ones vis-à-vis contemporary academic trends. Thus we are strongly encouraged to integrate some of the conventional dichotomies.

These attempts are not limited to the field of economic sciences, including management sciences, but also include social science in general. In that way, understanding the social profiles of complex science may then be within our reach. In the meantime, contemporary society appears to be evolving into a newly emerging phase, chiefly characterized by an information and communication technology (ICT) mode of production and a service network system replacing the earlier established factory system with a new one that is suited to actual observations. In the face of these changes we are urgently compelled to explore a set of new properties for a new socio/economic system by implementing new ideas. We thus are keen to look for "integrated principles" common to the above-mentioned dichotomies throughout our serial compilation of publications. We are also encouraged to create a new, broader spectrum for establishing a specific method positively integrated in our own original way.

Editors-in-Chief
Takahiro Fujimoto, Tokyo, Japan
Yuji Aruka, Tokyo, Japan

More information about this series at http://www.springer.com/series/11930

Jun Tanimoto

Sociophysics Approach to Epidemics

Springer

Jun Tanimoto
Faculty of Engineering Sciences; Interdisciplinary
Graduate School of Engineering Sciences
Kyushu University
Kasuga, Japan

ISSN 2198-4204 ISSN 2198-4212 (electronic)
Evolutionary Economics and Social Complexity Science
ISBN 978-981-33-6480-6 ISBN 978-981-33-6481-3 (eBook)
https://doi.org/10.1007/978-981-33-6481-3

This Springer imprint is published by the registered company Springer Nature Singapore Pte Ltd.
The registered company address is: 152 Beach Road, #21-01/04 Gateway East, Singapore 189721, Singapore

Preface

The Spanish Flu (1918), SARS (2003), Swine Influenza (2009), and MERS (2013) illustrate the ongoing threat of worldwide pandemics that intermittently arise and jeopardize the safety and security of the modern social system and our daily lives. The COVID-19 pandemic has brought unparalleled experiences: locked-down cities, quarantines and social-isolation on a massive scale, economic downturn, enforced "social distancing," collapse of medical systems, and more. I am writing this preface in the aftermath of COVID-19 without any optimistic forecast of when we may return to normal life. COVID-19 has taught us that the modern social system is vulnerable to unknown viruses with no available vaccines or treatments. The only possible provision for protecting people is the employment of massive quarantines or systematic social isolation. These are quite primitive tools that are essentially the same as those depicted in Giovanni Boccaccio's "Decameron" in the 14th century or in 17th-century London, where William Shakespeare wrote "King Lear" while theaters were closed due to a pandemic. It seems particularly ironic that this is happening in the twenty-first-century world shaped by advanced science and technology and mature engineering practices. We have supercomputers, supersonic airplanes, interplanetary spacecraft, new medical treatments and pharmaceuticals, and abundant and high-quality food. All these advances have enriched our lives but are inadequate for coping with an unknown virus epidemic. COVID-19 is forcing us to redefine our modern social systems and the purpose of science and technology.

However, I am optimistic that science will play an important role in managing this epidemic by analysis, modeling, and prediction, which will guide the implementation of practical social provisions to navigate this situation.

This book delivers readers a powerful tool for combating this epidemic.

Disease spread can be suppressed by medical developments such as new vaccinations and anti-viral drugs. These are critical issues for scientists to tackle. However, social physics, as a new emerging science, can also help control the spread of disease as an epidemic is a unified dynamic phenomenon comprising two different events: the physics of pathogen diffusion through a physical network, and human decision-making and social attitudes shared through a virtual network. The former component can be modeled by mathematical epidemiology, which was established by Kermack and McKendric's SIR model in the early 20th century. Evolutionary game theory underpins the latter component and has become one of the hottest subfields of complex science and statistical physics in the twenty-first century.

This book introduces the new extended concept of the vaccination game, in which the mathematical epidemiology describing the physics of diffusion is merged with evolutionary game theory to model the human decision-making process. This book begins by presenting the fundamental basis of evolutionary game theory and mathematical epidemiology and then introduces various applications based on the extended vaccination game concept.

I would be most honored if this book somehow serves as a guidebook for readers seeking new transdisciplinary areas merging epidemics and human decision-making process.

Kasuga, Japan Jun Tanimoto

Acknowledgments

This book owes its greatest debt to my coworkers, who were my excellent students. Chapter 2 partially relies on the contributions of Mr. Md. Rajib Arefin (Assistant Professor at Department of Mathematics, University of Dhaka, Bangladesh). He and Dr Kazuki Kuga (Assistant Professor at Faculty of Engineering Sciences, Kyushu University) gave great amount of input to the content of Chap. 3. Chapters 4–9 are products from brilliant projects by those two gentlemen above and Dr. K. M. Ariful Kabir (Assistant Professor at Department of Mathematics, Bangladesh University of Engineering and Technology, Bangladesh), Dr. Muntasir Alam (Lecturer at Department of Mathematics, University of Dhaka, Bangladesh), Mr. Masaki Tanaka (SEENET JOUHOU, Ltd.), and Keisuke Nagashima (Money Forward, Inc.).

I greatly appreciate all of their contributions.

Last, but not the least, I am grateful to Dr. Prof. Yuji Aruka; Professor Emeritus of Chuo University for providing me with the opportunity to publish this book.

Contents

About the Author

Jun Tanimoto The author was born in 1965 in Fukuoka but he grew up in Yokohama. He graduated in 1988 from the Department of Architecture, Undergraduate School of Science and Engineering at Waseda University. In 1990, he completed his master's program, and in 1993, he earned his doctoral degree from Waseda University. He started his professional career as a Research Associate at Tokyo Metropolitan University in 1990, moved to Kyushu University and was promoted to Assistant Professor (Senior Lecturer) in 1995, and became Associate Professor in 1998. Since 2003, he has served as Professor and Head of the Laboratory of Complex Social and Environmental Systems. He served as a Visiting Professor at the National Renewable Energy Laboratory (NREL), USA; at the University of New South Wales, Australia; at Eindhoven University of Technology, Netherlands; and at Max Planck Institute for Evolutionary Biology, Germany. Professor Tanimoto has published numerous scientific papers in building physics, urban climatology, statistical physics, and social physics, and is the author of books such as *Evolutionary Games with Sociophysics: Analysis of Traffic Flow and Epidemics* (Springer; ISBN: 978-981-13-2769-8), *Fundamentals of Evolutionary Game Theory and its Applications* (Springer; ISBN: 978-4-431-54961-1), and *Mathematical Analysis of Environmental System* (Springer; ISBN: 978-4-431-54621-4). He was a recipient of the Award of the Society of Heating, Air-Conditioning, and Sanitary Engineers of Japan

(SHASE), the Fosterage Award from the Architectural Institute of Japan (AIJ), the Award of AIJ, and the IEEE CEC2009 Best Paper Award. He is involved in numerous activities worldwide, including editor at several international journals including Applied Mathematics and Computation, PLOS One, and Journal of Building Performance Simulation, among others; committee member for many conferences; and expert at the IEA Solar Heating and Cooling Program Task 23. He is also an active painter and novelist and has been awarded numerous prizes in fine art and literature. He has created many works of art and published several books. He specializes in scenic drawing with watercolors and romantic fiction. For more information, please visit http://ktlabo.cm.kyushu-u.ac.jp/.

Chapter 1
A Social-Physics Approach to Modeling and Analyzing Epidemics

This chapter presents the background and motivation for why this book came about. Needless to say, this book has two key words besides "sociophysics," and these are "evolutionary game" and "mathematical epidemiology." Game theory is a contemporary mathematical concept founded in the middle of the twentieth century by the milestone work of John von Neumann and Oskar Morgenstern; "Theory of Games and Economic Behavior" (Princeton University Press, 1944). Following their work, John Nash played an important role in applying the theory to various domains, including economics, information science, and biology; for his efforts, he was awarded the Nobel Prize in Economics in 1994 due to his establishment of the

J. Tanimoto, *Sociophysics Approach to Epidemics*, Evolutionary Economics and Social Complexity Science 23, https://doi.org/10.1007/978-981-33-6481-3_1

concept of Nash Equilibrium.[1] Briefly, game theory can be described as a mathematical template for modeling human decision-making processes. If one intends to address the time-evolutionary aspect of a dynamical system, this is called evolutionary game theory (EGT), as distinct from classic game theory, which focuses on a static situation in which several game players with several strategies mutually interact to maximize their benefits (called a "payoff") at a certain moment. After the concept of complex science emerged in the 1990s and computational resources surged, EGT was combined with multi-agent simulation (MAS) to open up a new horizon through which it is possible to approach many complex problems related to human social systems that contain physical systems as subordinates. Such problems had previously been considered unsolvable because human behavior was seen as too stochastic to appropriately predict the dynamics of decision-making. A disease spreading in our society is a good example. Epidemiology—meaning the study of how an epidemic spreads on a human network—can be said to be sufficiently predictable because we know that it obeys a simple physics: the principle of diffusion phenomena. But the behavior of each individual is so varied, so prone to stochastic deviation, and so significantly influenced by information from the media that it is hard to predict how a disease will really spread through a complex human social system. In fact, whether one commits to pre-emptive vaccination or not is deeply related to the costs of the illness and the vaccination, and is also significantly influenced by the extent of the current outbreak as it stochastically evolves in time. This is quite a difficult task: but this book aims to address it in the following chapters.

To begin with, this chapter presents the holistic background that has motivated the author to produce the present book. We also introduce the concepts of complex human–physics systems, mathematical epidemiology, and evolutionary game theory.

1.1 Modeling of a Social–Complex System: A Human–Physics System

Most fields in science and engineering are primarily concerned with physical systems. *Science* specializes in understanding the physical phenomena of a natural, or even an artificial system, and building a mathematical model to describe the dynamics and underlying mechanisms of the phenomena. If that particular phenomenon brings profitable outputs like resources and energy to human society, *engineering* picks up where science leaves off so that the output can be maximized by means of careful design and optimized parameterization of the controlling variables. No matter what differences there are between systems, the most important issue in

[1]Any standard textbooks on game theory will carefully explain the concept of Nash Equilibrium. For example, see Tanimoto (2014, 2015, 2019).

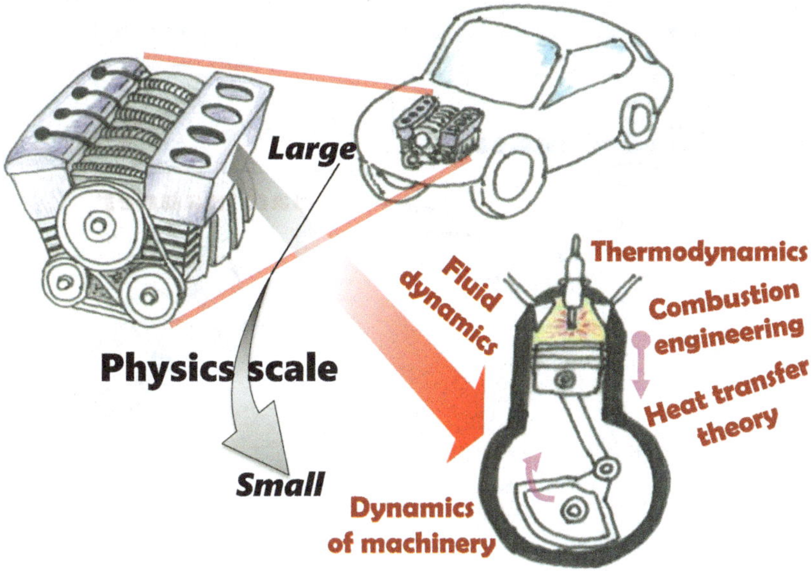

Fig. 1.1 Designing an engine, which can be described fully in terms of physics. Breaking it down into pieces, we expect that a cylinder-level optimization is likely to result in an optimal design for the vehicle as a whole

such an approach is how the focal physical phenomenon can be modeled to ensure dynamical prediction and optimization.

Let us start with a simple and familiar example. We imagine the design process of an engine, as shown in Fig. 1.1. Under the given constraint(s), we seek an optimal design for the whole vehicle system, which requires the best-performing engine, as well as the best-streamlined shape of a body. To optimize the engine performance considering the multi-objective situation, in practice, we consider not only maximizing the fuel efficiency, but also minimizing the exhaust of SOx, NOx, and PM. Numerous engineers have made great efforts towards this end since the nineteenth century, and these efforts continue so as to control what happens in each cylinder of the engine, which is basically a problem of complex physics. At one time, this was very difficult; but now, the foundations for this research have been revealed by several sophisticated sciences, including combustion engineering, heat-transfer theory, thermodynamics, and fluid dynamics. All of this basically lies within the sphere of physics, which is more tangible and controllable than a fully stochastic system like human psychology, even if there are some noise effects resulting from the uncertainty of mechanical design and from material, fuel, and other considerations. This is because most of the phenomena involved in an internal-combustion engine can be traced by a set of laws of physics–mathematical equations.

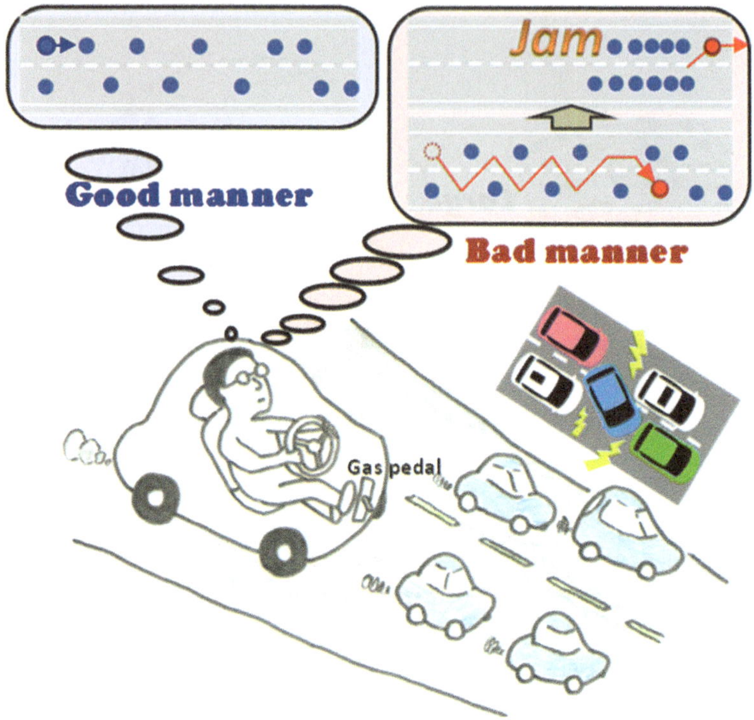

Fig. 1.2 A human decision made on the spot critically determines the system's holistic efficiency. "Efficiency" means the traffic flux, which would be influenced by drivers' actions, rather than inherent considerations such as each vehicle's mechanical performance

Here, we may notice that one crucially important element has been neglected in the discussion above: the human. If human commitment is not considered, you might miss out on the substance of reality. In this case, the subject is a human; that is a driver. By committing, as a driver, to a vehicle system, a human is able to draw a benefit. A driver may intend to maximize the performance of their engine through their control of the gas pedal; but in general, most drivers usually aim not to maximize their fuel efficiency nor to minimize their environmental impact, but to maximize their own benefit by minimizing the travel time to their destination. And this becomes much more complicated when there are many vehicles, with many drivers, each in turn driven by their respective decision-making processes. A society collecting many vehicles, say agents, might suffer a conflict between mutually competing agents individually seeking each local maximization, which consequently triggers a traffic jam. A jam inevitably leaves all agents with a meager level of benefit (see Fig. 1.2). Here, we must note that predicting human decision making is quite a difficult subject, as humans do not obey a set of deterministic equations like those of an engine. Rather, their behavior must be guessed through the empirical findings of experimental psychology with a fully stochastic mathematical framework.

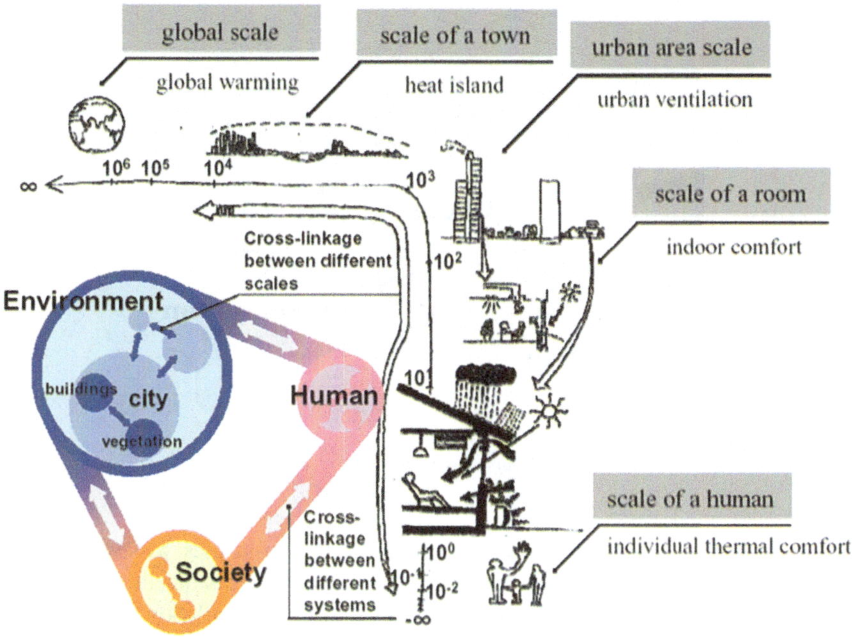

Fig. 1.3 The wide range of spatial scales over which environmental systems act, and the concept of a human–environmental–social system

Therefore, a precise discussion and prediction at the level of each cylinder, engine, or vehicle no longer works. We must account for the existence of a driver, and also consider a multiple-driver situation; this can be called a "society," since the performance of an engine is totally controlled by its driver; and the driver's control is fully dependent upon social context—how other drivers are driving and whether it results in a jam. Thus, we must model not only an independent physical system, but also simultaneously model the effect due to interference by the human who controls the system, in the context of a society that we can regard as an ensemble of individuals. All of these components—physics, human, and society—must be integrated into the same modeling template with a common spatiotemporal structure. This is the concept of a *human–physics system* from the standpoint of social physics (sociophysics).

We once introduced the concept of a human–environmental–social system,[2] where the sphere of the "environment" implies an assembly of many physical sub-systems. For example, an urban climate system contains an urban block-scale model and a building-physics model as its subordinates while it is connected with a regional meso-scale climate model as a supra-structure (see Fig. 1.3). The human–physics system is a more generalized concept than the human–environmental–social

[2]Tanimoto (2015, 2019).

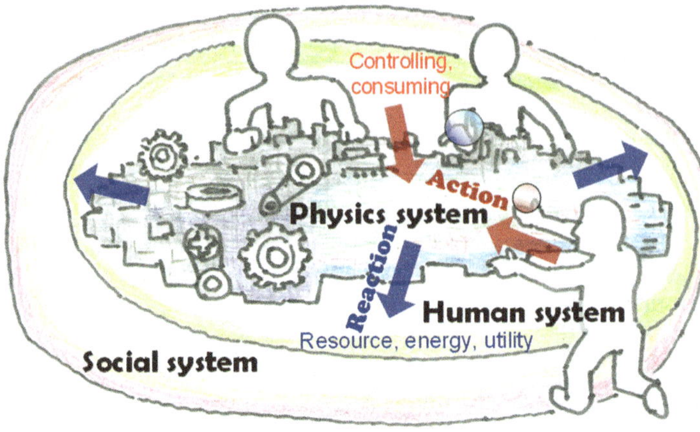

Fig. 1.4 Conceptual schema of a human–physics system

system; the three are, respectively, driven by different principles, as schematically illustrated in Fig. 1.4. The physical, human, and social systems together compose a kind of a layer structure. They are mutually connected by relations between the action–reaction chain. If individuals try to extract much more benefit from a physical system, or if agents take extreme action towards it, the physics system may harshly react to them, depending on the inherent potential capacity that the physical system has. Thus, we must see the three subsystems as an integrated system, and must build up a certain comprehensive model. This is the most important kernel of a human–physics system.

As you might imagine, the dynamical system in which an epidemic spreads and individuals protect themselves based on own decisions is one of the most appropriate applications of the concept of a human–physics system, and is the subject of the present book.

1.2 How the Spread of an Infectious Disease Can be Modeled?—Mathematical Epidemiology

The counterpart to the "physics model" in the previous section in the context of an epidemic is provided by "mathematical epidemiology." A communicable disease mediated by either bacteria or a virus spreads through human contact; typically, the frequency with which a focal agent engages in physical human contact is represented by their degree on the physical network with which they are involved. Here, "degree" means the number of links the agent has; or, say, the number of immediate neighbors around them. Thus, an agent with a higher degree would be exposed to a higher risk of infection. Such agents are called "hubs." Whether the focal agent would be infected when they have at least a single infectious[3] neighbor depends on a certain stochastic process. The so-called transmission rate, usually denoted by β [day^{-1} person^{-1}], which indicates how frequently an infectious agent is communicated to a healthy agent, would differ from one agent to another depending on their specific situation. But speaking approximately, the spread of a disease on a social network connecting all agents in a society can be mathematically formulated by the physics of diffusion phenomena. Such physics are basically represented by a simple linear partial differential equation, $\frac{\partial C}{\partial t} = \kappa \frac{\partial^2 C}{\partial x^2}$, where C is the concentration of the transferred object, κ is so-called diffusion coefficient for the object, and t and x, respectively, indicate time and space. If you have already introduced an underlying network into your model to represent physical contacts, the evolution of a physical system in time can be treated in a much easier way. The state of an agent is plural. Presuming what is known as SIR dynamics—one of the most general mathematical epidemiology models, which will be deliberately introduced in Chap. 3—we can note that there are three distinct subpopulations: *Susceptible* (S), meaning those who are healthy yet vulnerable to infection, *Infected* (I), and *Recovered* (R), meaning those who may never be infected again due to acquired immunity. The main question to be answered is whether or not an S agent gets infected (i.e., whether S transfers to I) if this S agent has an I agent in their neighborhood. The answer primarily depends upon the transmission rate β. Such a binary-state transfer from S to I is ubiquitously observed on a certain underlying network in many natural systems. A good analogy is that of a forest fire (see Fig. 1.5), which has been used as a template for *percolation theory*. Percolation theory is the mathematical model used to describe stochastic diffusion processes to emulate a "fire" (or an "epidemic" in the current discussion) spreading on a network. Inspired by Fig. 1.5, we can draw another schematic explanation; Fig. 1.6, which explains that mathematical epidemiology works as a model for describing the physical dynamics composing a human–physics system, thereby quantifying how an epidemic spreads on a human complex network. A more detailed mathematical framework is presented to readers in Chap. 3.

1.3 How Human Behavior Can be Modeled?—Evolutionary Game Theory

In this section, we should discuss the remaining component in a human–physics system: human commitment. Needless to say, a human is not like a tree in the woods in the case of a forest fire (Fig. 1.5). Trees are passive, and cannot take actions to protect themselves; but humans can.

[3]Not "infected" but "infectious". Those two terms have different meanings, as discussed on Chap. 3.

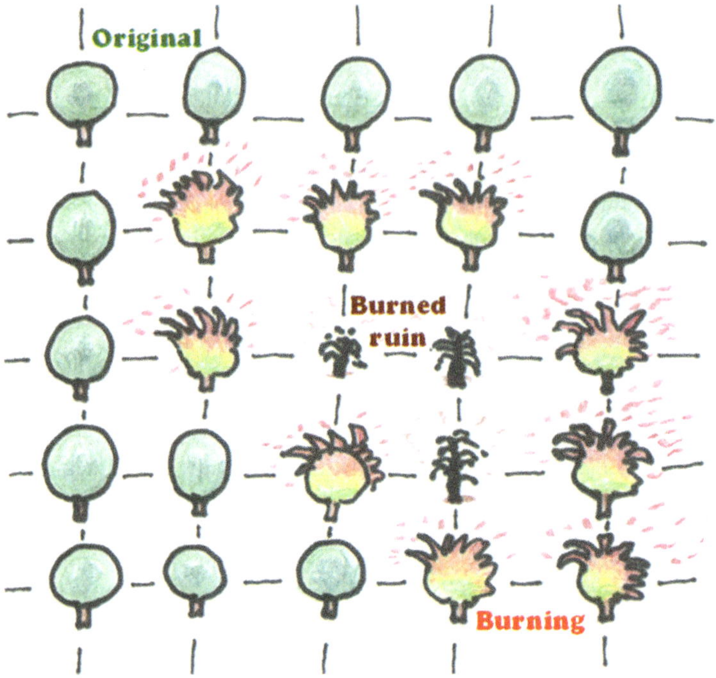

Fig. 1.5 Forest fire modeled by percolation theory. The fire spreads to healthy trees. The front region consists a "wave" of burning, which can be likened to the I state of the SIR model. Sites of burned ruins can be likened to the R state, although in this case, they do not recover

Let us suppose a situation of exposure to the risk of an epidemic breakout. In fact, we have gained practical experience of this in the wake of the worldwide COVID-19 pandemic. Each individual has a different social background (nationality, income, gender, job, religion, etc.) and thus a different decision-making mechanism; however, they have several provisions available for avoiding infection, as schematically shown in Fig. 1.7.

Committing to pre-emptive vaccination has been thought effective against such epidemics as influenza; and a late vaccination can be another alternative to a pre-emptive one, by which an individual commits to a vaccination in the middle of a season, which might be thought more beneficial (information-advantageous, precisely speaking) than pre-emptive vaccination for someone who seeks to free ride on herd immunity (i.e., doing nothing despite others vaccinating). Because such an egocentric person may avoid infection without paying any cost for vaccination due to the shield of herd immunity, which can be regarded as a kind of public good.

As other alternatives, gargling and wearing a mask may be worthy of consideration. In fact, COVID-19 shows that these strategies are not insignificant at all although it is said that some American people disapprove. Such "intermediate defense measures" backed by new lifestyle habits are thought meaningful, since

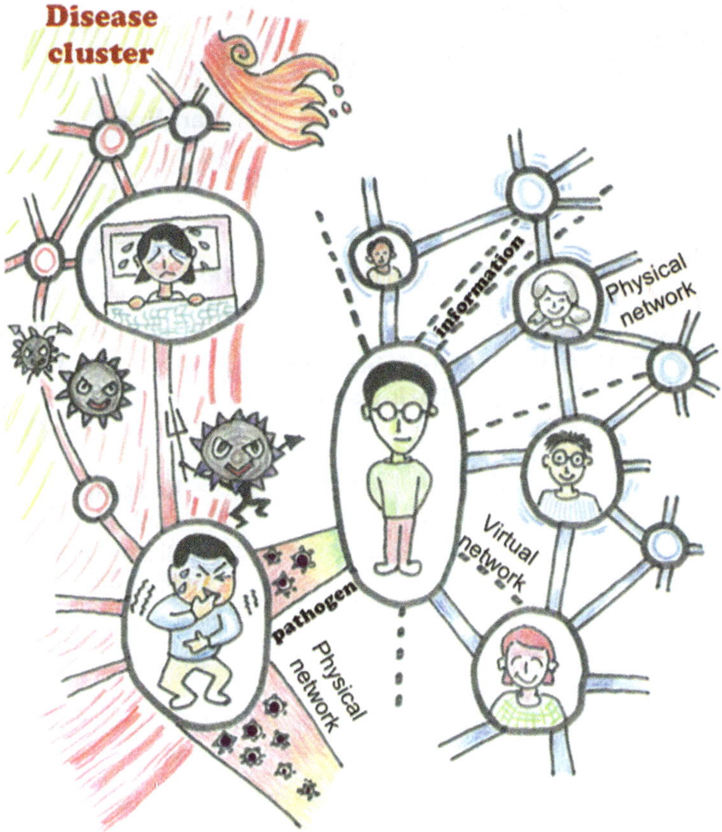

Fig. 1.6 The role that the mathematical epidemiology playing to predict a disease spreading on a human complex network. Agents are connected by both physical network, actual physical contacts and virtual network, emulating a social relation mutually sharing information. Epidemic spreads on the physical network. If an S agent who has an infectious neighbor, depending on transmission rate, β, he may be infected. If he spreads out some useful information about the disease to his neighbors, it would be helpful for them, since those neighbors are able to prepare the disease front comes to their site

the required cost is quite small, even if the expected effect might be smaller than that of medical provisions like vaccination.

Another new lifestyle habit we have learnt from COVID-19 is so-called social distancing. Central and local governments, as well as public-health authorities, have addressed this at length, because maintaining sufficient physical distance from others can directly reduce the transmission rate; β. A recommendation of "keeping social distance" encourages people to refrain from taking public transportation during rush hour, for an example, which may result in the emergence of a new lifestyle in our society.

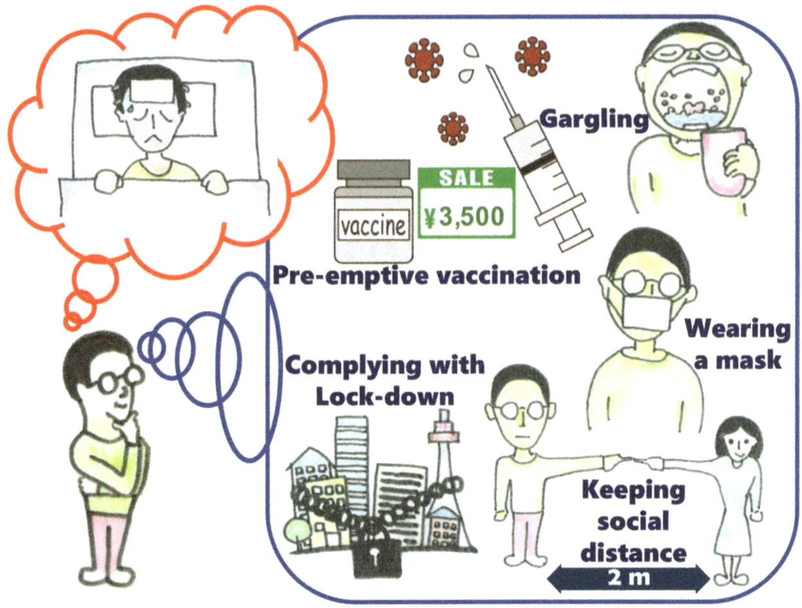

Fig. 1.7 The decision maker elects on a certain strategy to cope with the potential risk of an infectious disease. If he is confident that he will not to be infected, he may do nothing to protect himself. Yet, this may lead him to be infected, which would compel him to pay a huge expense: the disease cost. This gives him an incentive to take provisions. In terms of cost and effectiveness, there are various provisions. Most of them are spontaneous and rely on voluntary action. For example, the cost of an influenza vaccine in some countries like Japan is fully covered by each individual, rather than being covered by public health or insurance. Meanwhile, the government may impose several compulsory provisions on all the people, such as lockdowns, forceful quarantine, and so forth, which we continue to experience in the aftermath of COVID-19

Several measures have been taken by governments and public-health authorities to suppress the spread of COVID-19 that have annoyed individuals and devastated the economy. The biggest of these is the so-called lock-down, whereby most activities in an area are forcefully subdued; thus, people must stay home basically all of the time, which certainly decreases transmission, but also brings down economic indicators. If the spread of COVID-19 could have been successfully controlled in the early stage of breakout in China, such fierce provisions all over the world may have been avoided. Perhaps we missed a chance to cope with the virus.

The key point is that the extent of each alternative intervention used to control the spread of a disease depends on each individual's decision at a certain time and in a certain social context; however, its effect is shared with all people in a society. A situation of "expecting individual dedication but the fruits are given to all" carries with it a social dilemma. This is why we must dovetail the concept of evolutionary game theory with the physical process of the spread of disease.

Game theory provides us with a quite powerful quantifying framework for modeling human decision-making processes, including whether or not each strategy should be taken if an individual was to refer to its expected benefits. Evolutionary game theory adds the concept of dynamics to classic game theory. Chapter 2 presents some of the mathematical fundamentals of evolutionary game theory.

For last several years, there has been a wave of epidemiological studies in which mathematical epidemiology is coupled with the evolutionary game theory. The challenge of modeling such a human–physics system is highlighted in this book. Such studies have been concerned with what is known as the "vaccination game," which is mathematically explained in Chap. 3. In the chapters following Chap. 3, we will deliver several applications considering various provisions besides vaccination, which we name the "intervention game"; this is a more advanced and resilient concept than the vaccination game.

References

Tanimoto J (2014) Mathematical analysis of environmental system. Springer, Tokyo

Tanimoto J (2015) Fundamentals of evolutionary game theory and its applications. Springer, Tokyo

Tanimoto J (2019) Evolutionary games with sociophysics: analysis of traffic flow and epidemics. Springer, Singapore

Chapter 2
Evolutionary Game Theory: Fundamentals and Applications for Epidemiology

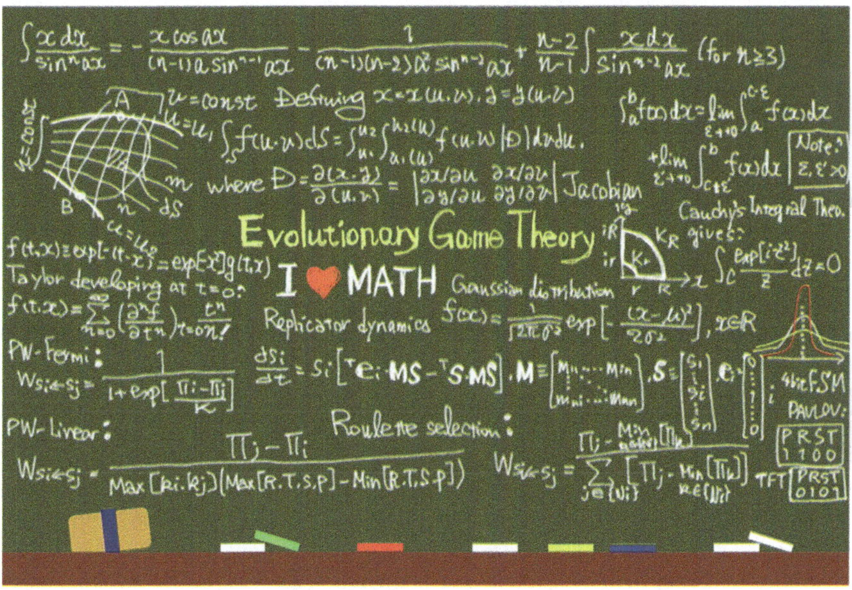

This chapter presents the mathematical fundamentals of evolutionary game theory. Game theory provides the mathematical foundation for quantifying human decision-making for choosing strategies. There are several varieties of "games" in game theory, including zero-sum (constant-sum) games (in which one is either a winner or a loser) or non-zero-sum (non-constant-sum) games; symmetric games (in which both the focal player and their opponent share a common payoff structure) or asymmetric games; 2-player games or a multi-player game; 2-strategy games or multi-strategy games; and so forth. In any case, classical game theory primarily concerns game equilibrium, or game solutions, which can be understood as steady-

J. Tanimoto, *Sociophysics Approach to Epidemics*, Evolutionary Economics and Social Complexity Science 23, https://doi.org/10.1007/978-981-33-6481-3_2

state solutions or static solutions in the field of conventional science and engineering. Conversely, evolutionary game theory rather concerns the time-evolution of a system.

First, let us examine the foundation of a symmetric 2-player and 2-strategy (2×2) game, which is the most important archetype among evolutionary games. Subsequently, multi-player and 2-strategy games are also introduced. In the latter parts of this chapter, we present two important theoretical frameworks that will work as powerful tools to quantify social dilemmas. One is the universal concept of dilemma strength, which enables one to quantify the extent to which a reciprocity mechanism can ameliorate a social-dilemma situation in any 2 by 2 game. Another is the concept of a social efficiency deficit (SED), which also quantifies the existence of a social dilemma and its features, which are applicable to any type of game, unlike the universal concept of dilemma strength. According to the mathematical definition, a social dilemma exists whenever a game's equilibrium deviates from a situation that ensures a socially optimal payoff.

2.1 Two-Player and Two-Strategy Games

Generally, game theory treats games that are mathematical frameworks in which m players exist and the set of strategies that each of them take, of which number is n, is defined. More importantly, the payoff structure, $\pi_i(s_1, s_2, \cdots; s_i; \cdots, s_m)$, indicates player i's payoff when any player j's strategy, $s_j \in \{st_1, \cdots, st_n\}$, is given. If the strategy set is not defined as discrete as above but rather as continuous, this is fine. In such a case, we cannot say the number of strategies. When the strategy set is defined as discrete, the payoff π can be given by a look-up table, whereas π must be defined as a continuous function when the strategy is defined in a continuous form. If $m = 2$ and $n \geq 2$ is presumed, i.e., the 2-player, n-strategy game, π is described by a $n \times n$ (n by n) matrix. In general, $\pi_1(s_1, s_2) \neq \pi_2(s_1, s_2)$ would be "asymmetric," thus, this is known as an asymmetric 2-player and n-strategy game. A game with $\pi_1(s_1, s_2) = \pi_2(s_1, s_2)$ is called a symmetric 2-player and n-strategy game.

In this book, our main concern is how cooperation can evolve in a social context. In fact, as we will discuss in the following chapters, a voluntary commitment to pre-emptive vaccination at one's own expense is obviously regarded as a socially "cooperative" behavior, which helps to ensure the so-called herd immunity. To elucidate a mechanism for such a cooperative behavior, a non-constant-sum symmetric 2-strategy game is the simplest template, but is still suggestive. Therefore, let us discuss symmetric two-player and two-strategy games (abbreviated as 2 by 2 games or 2×2 games) in this section, which have most commonly been accepted as archetypical templates in the field of evolutionary game theory.

2.1.1 Theoretical Foundation

Let us suppose an unlimited population. The individuals (hereafter sometimes called "agents") are well-mixed; in other words, there is no spatial special structure mutually connecting individuals, no iteration of games, and no mechanisms for specifying each individual such as tag, skin color, or mutual communication to add "social viscosity" (Nowak 2006) to the system, as explained in sub-section 2.1.2. This specific situation is called an "infinite and well-mixed population," which implies that an agent never plays with the same opponent again. In a nutshell, the assumption of an infinite and well-mixed population ensures a perfect anonymity amid agents.

From the infinite and well-mixed population, two individuals are selected at random and made to play the game. The game uses two discrete strategies (shown in Fig. 2.1): cooperation (C) and defection (D). The pair of players receives payoffs in each of the four combinations of C and D. A symmetrical structure between the two players is assumed. In Fig. 2.1, the payoff of player 1 (the "row" player) is represented by the entries preceding the commas; that of player 2 (the "column" player) is represented by the entries after the commas. The payoff matrix is denoted by $\begin{bmatrix} R & S \\ T & P \end{bmatrix}$. Depending on the relative magnitudes of the matrix elements elements P, S, and T, the game can be divided into 4 classes: the Trivial game with no dilemma, the Prisoner's Dilemma game (sometimes abbreviated PD)c the Chicken game (also known as the Snowdrift or Hawk–Dove game), and the Stag Hunt game (sometimes abbreviated SH). The system denoted by this payoff matrix has four degrees-of-freedom, since the number of variables is four.

To go beyond classical game theory to evolutionary game theory, let us introduce the concept of "time" here. The strategies (called "hands" in some contexts) adopted by an agent are cooperation (C) or defection (D), expressed by the following state vectors[1]:

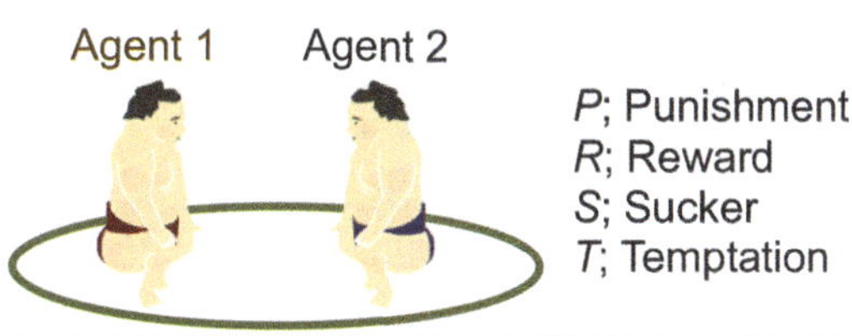

Agent 2 / Agent 1	Cooperation (C)	Defection (D)
Cooperation (C)	R, R	S, T
Defection (D)	T, S	P, P

Fig. 2.1 Payoff matrix of a 2 × 2 game

[1]To precisely learn the following mathematical procedure, you should consult; Tanimoto (2014, 2015).

$$\text{Strategy C} : {}^T\mathbf{e}_1 = (1 \ \ 0), \tag{2.1.1}$$

$$\text{Strategy D} : {}^T\mathbf{e}_2 = (0 \quad 1). \tag{2.1.2}$$

Here, the operator T means taking transpose for either a vector (thus, a row (column) vector being transferred to a column (row) vector) or a matrix. The payoff matrix of the game structure is

$$\begin{bmatrix} R & S \\ T & P \end{bmatrix} \equiv \mathbf{M}. \tag{2.2}$$

Moreover, let the proportions of agents adopting strategies C and D at a given time be defined as s_1 and s_2, respectively. These strategy ratios are expressed as

$${}^T\mathbf{s} = (s_1 \quad s_2). \tag{2.3}$$

From the condition of the simplex, we get

$$s_2 = 1 - s_1, \tag{2.4}$$

which implies that we can actually reduce the number of variables from 2 to 1. The validity of Eqs. (2.2) and (2.3) should be understood from the following vector–matrix equation describing the battle between two agents adopting strategy D, for which the outcome is P:

$$\pi_{\mathrm{DD}} = (0 \quad 1) \cdot \begin{bmatrix} P & S \\ T & P \end{bmatrix} \begin{pmatrix} 0 \\ 1 \end{pmatrix} = P. \tag{2.5}$$

A variant form of Eq. (2.5) also computes the payoff when one strategy plays a game $\mathbf{M}$ against another player with a different strategy. Thus, the expected payoff when an agent using strategy C battles with a randomly sampled agent at the present time expressed as strategy ratio $\mathbf{s}$ is

$${}^T\mathbf{e}_1 \cdot \mathbf{M}\mathbf{s}.$$

Similarly, the expected payoff when an agent using strategy D fights a randomly sampled agent at the present time expressed by strategy ratio $\mathbf{s}$ is

$${}^T\mathbf{e}_2 \cdot \mathbf{M}\,\mathbf{s}.$$

The replicator-dynamicsequation is defined as the strategy-ratio dynamics of strategy i, expressed as

$$\frac{\dot{s}_i}{s_i} = {}^T\mathbf{e}_\mathrm{i} \cdot \mathbf{M}\,\mathbf{s} - {}^T\mathbf{s} \cdot \mathbf{M}\,\mathbf{s}. \tag{2.6}$$

The dimensionless quantity on the left-hand side of Eq. (2.6), which is obtained by dividing $\dot{s}_i$ by the strategy ratio itself, is called the coefficient of deviation in the field of statistics and indicates the extent of strategy change. As the reader should certainly appreciate, this quantity is determined by the extent of how the payoff for strategy i playing against the societal average at a given time differs from the expected societal payoff at that time. Replicator dynamics seem appropriate as a driving rule for the time evolution of a certain dynamical system for the following reason: after a game, the use of successful strategies (those achieving higher payoffs than the average accumulated by the strategic ratio) will increase in the next time step, whereas that of less successful strategies will decrease. The ratio of this extent is thought to be decided through comparison with the aforementioned level of "success." In such a system, good conduct is rewarded whereas bad conduct is punished (a form of survival of the fittest). Selection mechanisms in the natural world (including human social systems) tend to operate in this manner.

Substituting Eqs. (2.1.1)–(2.3) into Eq. (2.6) and explicitly writing the elements, we obtain

$$\begin{cases} \dot{s}_1 = [(R-T)\cdot s_1 - (P-S)\cdot s_2]\cdot s_1 \cdot s_2 \\ \dot{s}_2 = -[(R-T)\cdot s_1 - (P-S)\cdot s_2]\cdot s_1 \cdot s_2 \end{cases}. \tag{2.7}$$

Note that when the right-hand side of (2.7) is equal to 0, the equation becomes cubic in s_1 and s_2; that is, the system contains three equilibrium points. Two of these are trivial:

$$\begin{pmatrix} s_1 & s_2 \end{pmatrix} = \begin{pmatrix} 1 & 0 \end{pmatrix} \equiv \mathbf{s}*|_{\text{C-dominate}}, \tag{2.8.1}$$

$$\begin{pmatrix} s_1 & s_2 \end{pmatrix} = \begin{pmatrix} 0 & 1 \end{pmatrix} \equiv \mathbf{s}*|_{\text{D-dominate}}. \tag{2.8.2}$$

In the former, all individuals ultimately become cooperative (all-cooperators state); the latter leads to the defection state (all-defectors state), denoted as C-dominant and D-dominant, respectively. The remaining equilibrium point is obtained by simultaneously solving Eq. (2.7), setting [...] on the right-hand side to 0, and eliminating s_2 through Eq. (2.4):

$$\begin{pmatrix} s_1 & s_2 \end{pmatrix} = \left(\frac{P-S}{P-T-S+R} \quad \frac{R-T}{P-T-S+R} \right) \equiv \mathbf{s}*\Big|_{\text{Polymorphic}}. \tag{2.8.3}$$

This third equilibrium point lies within [0, 1] and depends on the values of P, R, S, and T. In this case, as we discuss later, the dynamics become polymorphic or bi-stable. Equation (2.8.3) defines an internal-equilibrium point. As we discuss below, by considering the dynamics of a game, an evolutionary path is absorbed by one of those of $\mathbf{s}*|_{\text{C-dominate}}$, $\mathbf{s}*|_{\text{D-dominate}}$, or $\mathbf{s}*|_{\text{Polymorphic}}$, irrespective to an initial state; or either $\mathbf{s}^*|_{\text{C-dominate}}$, $\mathbf{s}^*|_{\text{D-dominate}}$ depending on an initial state, which is called Nash equilibrium (NE)).

Once the three equilibrium points are obtained, the signs of the eigenvalues of the Jacobian matrix at each are determined, and the equilibrium points are assessed to be sinks, sources, or saddles. To this end, we re-write Eq. (2.7) as follows:

$$\dot{s}_1 \equiv f_1(s_1, s_2), \tag{2.9.1}$$

$$\dot{s}_2 \equiv f_2(s_1, s_2). \tag{2.9.2}$$

From Eq. (2.4), we observe that $f_1 = -f_2$.

Meanwhile, let us consider an arbitrary continuous dynamical system in which the system-state equations are expressed by a non-linear function $\mathbf{f}$:

$$\frac{d\mathbf{x}}{dt} = \dot{\mathbf{x}} = \mathbf{f}(\mathbf{x}). \tag{2.10}$$

In general, non-linear functions can be approximated as linear functions at any arbitrary point, $\mathbf{x}^*$, by Taylor expansion. Expanding the right-hand side of Eq. (2.10), we obtain

$$\begin{aligned} &\mathbf{f}(\mathbf{x}) = \mathbf{f}\,(\mathbf{x}^*) + \mathbf{f}'(\mathbf{x}^*)(\mathbf{x} - \mathbf{x}^*) + \frac{\mathbf{f}''(\mathbf{x}^*)}{2!}(\mathbf{x} - \mathbf{x}^*)^2 + \cdots \\ &\Leftrightarrow \mathbf{f}(\mathbf{x}) \cong \mathbf{f}\,(\mathbf{x}^*) + \mathbf{f}'(\mathbf{x}^*)(\mathbf{x} - \mathbf{x}^*). \end{aligned} \tag{2.11}$$

From the definition of an equilibrium point, note that $\mathbf{f}\,(\mathbf{x}^*) = \mathbf{0}$. This should be evident by substituting $\frac{dx}{dt}\big|_{\mathbf{x}=\mathbf{x}*} = \mathbf{0}$ into Eq. (2.10). By ignoring higher-derivative terms, Eq. (2.11) is approximately equal to

$$\mathbf{f}(\mathbf{x}) \cong \mathbf{f}'(\mathbf{x}^*)(\mathbf{x} - \mathbf{x}^*). \tag{2.12}$$

Eq. (2.12) can then be approximated as the following linear equation:

$$\mathbf{f}(\mathbf{x}) = \mathbf{f}'(\mathbf{x}^*)(\mathbf{x} - \mathbf{x}^*) = \mathbf{f}'(\mathbf{x}^*)\mathbf{x} - \mathbf{f}'(\mathbf{x}^*)\mathbf{x}^*. \tag{2.13}$$

The first term on the right-hand side of Eq. (2.13) is first-order in $\mathbf{x}$, while the second term is constant. Following mathematical conventions, $\mathbf{A}$ is called the transition matrix in the case of a linear dynamical system, $\dot{\mathbf{x}} = \mathbf{A}\mathbf{x}$. Observing Eq. (2.13), the original non-linear system, Eq. (2.10), is simplified as the linear system, Eq. (2.13). Thus, the transition matrix is $\mathbf{f}'(\mathbf{x}^*)$. According to the theory of a linear dynamical system, we can say whether each equilibrium point absorbs or expels a specific time-evolutionary episode by identifying whether the slope of the system at each equilibrium is positive or negative, which can be evaluated by the signs of the eigenvalues corresponding to the equilibrium points of this matrix.

In the focal case, the transition matrix happens to be the Jacobian matrix of the tangent gradients of the multi-variable vector function:

$$\mathbf{f}'(\mathbf{x}^*) = \left.\frac{\partial \mathbf{f}(\mathbf{x})}{\partial \mathbf{x}}\right|_{\mathbf{x}=\mathbf{x}*} = \begin{bmatrix} \frac{\partial f_1(\mathbf{x})}{\partial x_1} & \cdots & \frac{\partial f_1(\mathbf{x})}{\partial x_n} \\ \vdots & \ddots & \vdots \\ \frac{\partial f_n(\mathbf{x})}{\partial x_1} & \cdots & \frac{\partial f_n(\mathbf{x})}{\partial x_n} \end{bmatrix}_{\mathbf{x}=\mathbf{x}*}. \tag{2.14}$$

Here, let us apply Eq. (2.14) to the discussion leading up to Eqs. (2.9.1) and (2.9.2). Each element of the Jacobian of Eq. (2.14) is calculated as

$$\begin{cases} \frac{\partial f_1}{\partial s_1} = -\frac{\partial f_2}{\partial s_1} = 3(-R+S+T-P){s_1}^2 \\ \quad +2(R-2S-T+2P)_{S_1} + S - P \end{cases}, \tag{2.15.1}$$

$$\frac{\partial f_1}{\partial s_2} = -\frac{\partial f_2}{\partial s_2} = -3(-R+S+T-P){s_1}^2 \\ \text{-}2(R-2S-T+2P)s_1 - S + P. \tag{2.15.2}$$

The Jacobian matrix $\mathbf{J} = \begin{bmatrix} \frac{\partial f_1}{\partial s_1} & \frac{\partial f_1}{\partial s_2} \\ \frac{\partial f_2}{\partial s_1} & \frac{\partial f_2}{\partial s_2} \end{bmatrix} = \begin{bmatrix} \frac{\partial f_1}{\partial s_1} & \frac{\partial f_1}{\partial s_2} \\ -\frac{\partial f_1}{\partial s_1} & -\frac{\partial f_1}{\partial s_2} \end{bmatrix}$ is 2×2 in this case, so its eigenvalues are easily obtained in an analytical procedure as 0 and $\frac{\partial f_1}{\partial s_1} - \frac{\partial f_1}{\partial s_2}$. Since 0 is unsigned, we need to only obtain the sign of $\frac{\partial f_1}{\partial s_1} - \frac{\partial f_1}{\partial s_2}$ to establish the equilibrium conditions. Explicitly, these eigenvalues are

$$\lambda = \frac{\partial f_1}{\partial s_1} - \frac{\partial f_1}{\partial s_2} = 6(-R+S+T-P){s_1}^2 \\ +4(R-2S-T+2P)s_1 + 2(S-P). \tag{2.16}$$

Note that the following three cases must be checked, because there are three equilibria, $\mathbf{s}^*|_{\text{C-dominate}}$, $\mathbf{s}^*|_{\text{D-dominate}}$, and $\mathbf{s}^*|_{\text{Polymorphic}}$, as obtained in Eqs. (2.8.1)–(2.8.3).

1. The necessary and sufficient condition for the equilibrium point $\mathbf{s}^*|_{\text{C-dominate}}$ to be a sink is $\lambda < 0$ when substituting $(s_1 \quad s_2) = (1 \quad 0)$ into Eq. (2.16). The following conditions are sought:

$$T - R = D_g < 0. \tag{2.17}$$

2. The necessary and sufficient condition for the equilibrium point $\mathbf{s}^*|_{\text{D-dominate}}$ to be a sink is $\lambda < 0$ when substituting $(s_1 \quad s_2) = (0 \quad 1)$ into Eq. (2.16). We now require that:

$$P - S = D_r > 0. \tag{2.18}$$

3. The necessary and sufficient condition for the equilibrium point $\mathbf{s}^*|_{\text{Polymorphic}}$ to be a sink is $\lambda < 0$ with $(s_1 \quad s_2) = \left(\frac{P-S}{P-T-S+R} \quad \frac{R-T}{P-T-S+R} \right)$ substituted into Eq. (2.16). Noting that $\lambda = 2\frac{(R-T)(P-S)}{R-S-T+P}$, we seek the following conditions:

$$P < S \wedge R < T \Leftrightarrow P - S = D_r < 0 \wedge T - R = D_g > 0. \tag{2.19}$$

Here, let us define the gamble-intending dilemma (hereafter referred to as GID or the Chicken-type dilemma), D_g, and the risk-averting dilemma (hereafter referred to as RAD or the Stag Hunt (SH)-type dilemma), D_r, respectively[2]:

$$D_g \equiv T - R, \tag{2.20.1}$$

$$D_r \equiv P - S. \tag{2.20.2}$$

GID can be described in a word as "greed," while RAD can be described as "fear" in a symmetric 2 × 2 game.

The above conditions are summarized in Table 2.1, with the following substitution:

$$\begin{aligned} \mathbf{s}^*|_{\text{Polymorphic}} &= \left(\frac{P-S}{P-T-S+R} \quad \frac{R-T}{P-T-S+R} \right) \\ &= \left(\frac{D_r}{D_r - D_g} \quad \frac{-D_g}{D_r - D_g} \right), \end{aligned} \tag{2.21}$$

which comes from Eq. (2.8.3).

Referring to D_g and D_r in Eqs. (2.20.1) and (2.20.2), the four game classes are transparently established as PD, Chicken, SH, and Trivial, as summarized in Tables 2.1 and 2.2. Here, these divisions are represented by the difference between the signs of the three equilibrium points. More importantly, which class—PD, Chicken, SH, or Trivial—is embedded in a given 2 × 2 game can be fully evaluated

[2]To learn in detail about GID and RAD and D_g and D_r, you should consult: Tanimoto and Sagara (2007).

Table 2.1 2×2 game dynamics derived analytically

Game class	Phase	Nash equilibrium	Sign of D_g	Sign of D_r	Each point sink, source, or saddle		
					(1,0)	(0,1)	$\left(\frac{D_r}{D_g - D_r} \quad \frac{-D_g}{D_r - D_g}\right)$
PD	**D**-dominate	(0,1)	+	+	Source	Sink	Saddle
Chicken	Polymorphic	$\left(\frac{D_r}{D_g - D_r} \quad \frac{-D_g}{D_r - D_g}\right)$	+	−	Source	Source	Sink
SH	Bi-stable	(0,1) or (1,0)	−	+	Sink	Sink	Source
Trivial	**C**-dominate	(1,0)	−	−	Sink	Source	Saddle

Table 2.2 Class type in a 2 × 2 game

Game class	Greed; GID? ($D_g > 0$?)	Fear; RAD? ($D_r > 0$?)	Dilemma as a whole?
Prisoner's dilemma; PD	Yes	Yes	Yes
Chicken(snow drift; hawk–dove)	Yes	No	Yes
Stag hunt; SH	No	Yes	Yes
Trivial	No	No	No

by both signs of D_g and D_r. In other words, we note whether positive or negative signs for D_g and D_r strictly regulate the game class for any arbitrary 2 × 2 game.

Note that this important statement is true as long as an infinite and well-mixed population is presumed. As we discuss later, if there is any mechanism to add social viscosity to the present 2 × 2 game (meaning that the assumption of a well-mixed and infinite population is violated), the classification procedure relying on D_g and D_r does not work. Hence, we need a further theoretical underpinning, which is discussed in Sect. 2.3.

In PD, $\mathbf{s}^*|_{\text{C-dominate}}$ and $\mathbf{s}^*|_{\text{D-dominate}}$ are the source and the sink, respectively; hence, regardless of the initial cooperation proportion in [0, 1] the ultimate state is one of complete defection at $t \to \infty$. In Chicken, $\mathbf{s}^*|_{\text{C-dominate}}$ and $\mathbf{s}^*|_{\text{D-dominate}}$ are both sources. In this case $\mathbf{s}^*|_{\text{Polymorphic}}$ (value in [0, 1]) is a sink, so regardless of the initial cooperation proportion, as $t \to \infty$, the system will settle into the internal-equilibrium point $\mathbf{s}^*|_{\text{Polymorphic}}$. It is worth noting that this state does not imply a specific agent being fixed into either cooperation or defection, but that when the infinitely large group is viewed as a whole, the proportions of cooperative and defective players are (dynamically) steady.

In SH, the internal-equilibrium point $\mathbf{s}^*|_{\text{Polymorphic}}$ is a source, while $\mathbf{s}^*|_{\text{C-dominate}}$ and $\mathbf{s}^*|_{\text{D-dominate}}$ are both sinks. Therefore, if the initial proportion of cooperative players is smaller (or larger) than $\mathbf{s}^*|_{\text{Polymorphic}}$, the ultimate state is pure defection (or pure cooperation), and the system is bi-stable. In the Trivial case, $\mathbf{s}^*|_{\text{C-dominate}}$ is a sink and $\mathbf{s}^*|_{\text{D-dominate}}$ is a source, so regardless of the initial cooperation proportion, the pure-cooperation state is inevitable. Hence, Trivial is a game with no dilemmas.

The above discussion is summarized schematically in Fig. 2.2.

2.1.2 *Social Viscosity*

As long as an infinite and well-mixed population is assumed, the theory correctly predicts the dynamics of any symmetric 2 × 2 game, as we have discussed. Going on this fact alone, if one were exposed to the social dilemma modeled by a PD, there would no way for them to cooperate rather than to defect in the long term. In the natural world, however, cooperative behavior is found ubiquitously—not only in human societies, but also among social insects such as ants and bees. This raises the

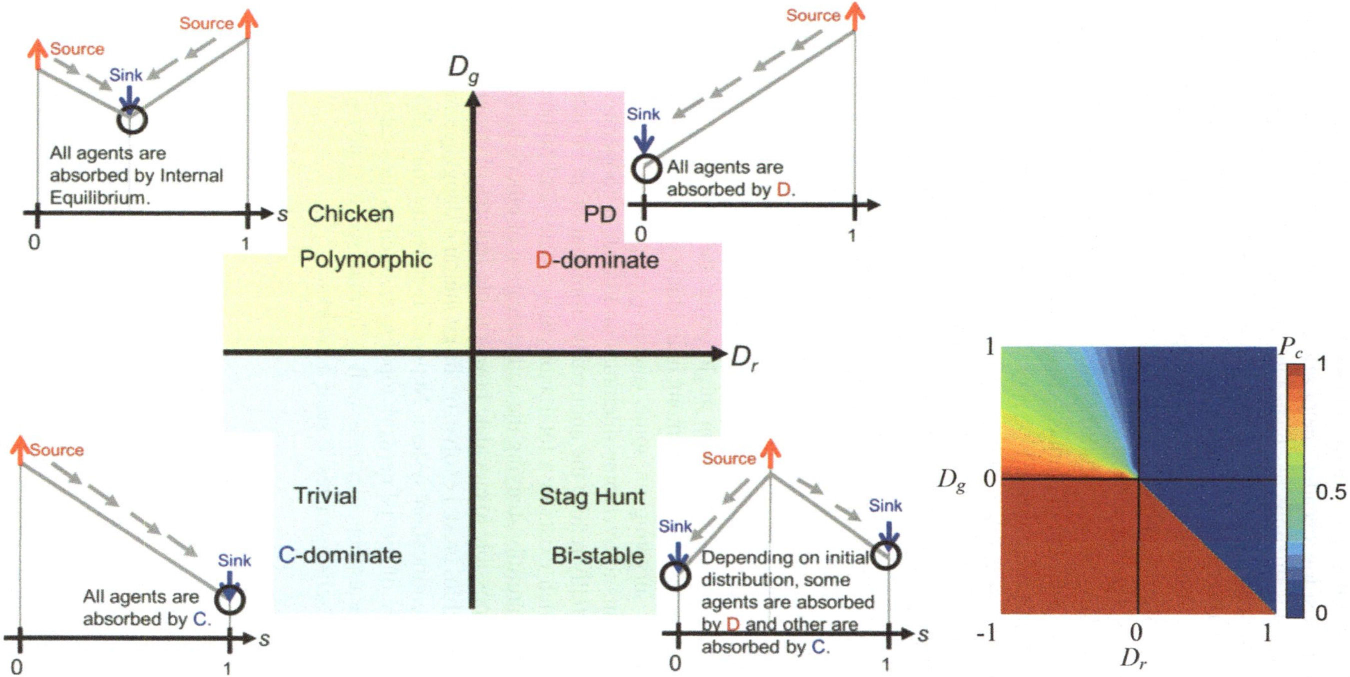

Fig. 2.2 Phase diagram of the dynamics classified by D_g and D_r of a two-by-two game and a summary of the dynamics of each game class (left panel). The right panel shows the cooperation fraction at equilibrium when an infinite and well-mixed population with replicator dynamics is assumed with an initial cooperation fraction, P_c, of 0.5. PD and Trivial are filled out in blue and red, respectively, since D-dominated and C-dominated phases are established. In the Chicken game region, a gradual shift of the cooperation fraction at equilibrium is observed due to the polymorphic phase. In the SH game region, bi-stability shows a twofold phase: either all cooperation or all defection

question of what mechanisms must exist in addition to the original situation of an infinite and well-mixed population to promote cooperation among agents?

Therefore, for the last few decades, the puzzle of what can be the "supplementary framework" to solve this dilemma has been subject to heavy research by biologists, physicists, mathematicians, and information scientists. Many studies have taken a simulation approach, respectively reporting that models introducing a specific new additional framework somehow enable enhanced cooperation beyond the default presumption of an infinite and well-mixed population. However, they have not directly answered the abovementioned question.

Among these studies, Nowak (2006) showed that there are five fundamental protocols to mitigate or cancel dilemmas,[3] as summarized in Fig. 2.3. The mechanisms of these protocols are governed by very ordinary and beautiful mathematical expressions similar to those of kin selection (Hamilton 1963). Nowak refers to these mechanisms as "social viscosity." Assuming a well-mixed and infinite population, each game is played by a single person whose next encounter is unknown (well-mixed). However, in repeated game battles between a pair of individuals (direct reciprocity),[4] or observing the tag of an opponent (indirect reciprocity), the behavior of an opponent, whether cooperative or defective, can be distinguished. When players play games against only their neighboring players on a certain underlying network, information relating to the strategy is obtained through the network (network reciprocity). If agents are distributed on island societies in which games are played and little inter-island competition might occur, the selection process would work in two layers; among islands and among individuals on an island; this may lead to cooperative agents surviving on some islands (group selection). All of these conditions enable agents to overcome dilemmas and create a cooperative society.[5] These processes essentially reduce the anonymity from that of an infinite and well-mixed population (which exists in a total anonymous state) and authenticates the battle opponents. By carefully studying the authentication of others through indirect reciprocity, it may be possible to elucidate how notable features of organisms (such as color differences in bird crests) evolve, or the evolution of language, which is the ultimate third-party-identification system. Network reciprocity may also help us to understand the structure of special network topologies, such as the scale-free graphs observed in many natural phenomena, as well as human social systems—in particular, how cooperation self-organizes in such networks.

---

[3]Strictly speaking, PDs satisfying $D_g = D_r$.

[4]This situation accords with common sense. If a game is played against the same partner each time, rather than against an unknown one, both individuals should accept the cooperation option to avoid strategies leading merely to short term profit. If both individuals take the defection option P, neither will benefit in the long term. Our daily behavior follows the former pattern.

[5]Many of these dynamics can be verified by simulation. Games are repeated between multiple agents in a simulated society; this approach is known as multi-agent simulation.

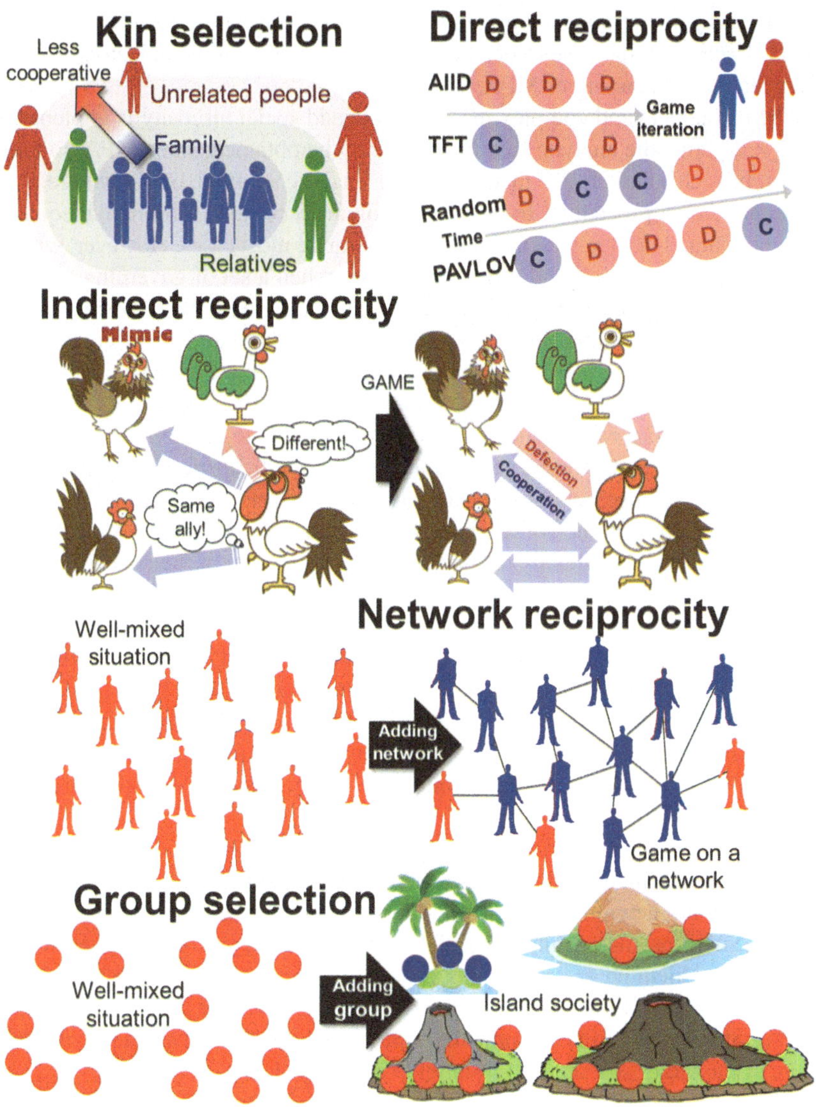

Fig. 2.3 Five basic mechanisms of dilemma resolution and examples of network reciprocity

2.1.3 Multi-Agent-Simulation Approach

The assumption of a well-mixed and infinite population is equivalent to the so-called mean-field approximation (MFA)), which has been heavily applied to various theoretical problems in the field of statistical physics to obtain analytical solutions. In fact, in the procedure of replicator dynamics—solving Eq. (2.6)—assumes a hypothetical game opponent who either cooperates with the frequency of the cooperators' fraction at that time or defects with the frequency of the defectors' fraction; this is the literal meaning of MFA.

When a realistic and complex mechanism to add social viscosity is implemented in a 2 × 2 game, the analytic solution can rarely be obtained, since MFA no longer applies. In such a case, rather than seeking an analytic solution, one should rely on a certain numerical procedure. In the case of solving a complex differential equation, one cannot help but to rely on a numerical solution in most cases. However, what we are discussing here is little bit different, because when a set of governing equations are given, seeking an exact solution can be called "regular", whereas relying on a numerical solution should be called "irregular" or "alternative". By contrast, in a complex but realistic 2 × 2 game presuming some mechanisms to add social viscosity, no set of governing equations can exactly describe its dynamics. Replicator dynamics can be said to result from the ultimate approximation premised on a well-mixed and infinite population, which is unrealistic. Thus, its analytic solution is, in a sense, unrealistic. The only way to obtain a "real" solution is an experimental approach, i.e., observing what really happens to the dynamical system that the game model is literally describing. This is the germ of the concept of multi-agent simulation (MAS)). Taking a MAS approach, one should build an artificial society on a computational environment, i.e., *on silica*, where a sufficiently large number of agents for statistical robustness can be generated, each having their own strategy of either cooperation or defection. Those agents can be exposed to the specific game framework defined by the model by giving them respective payoffs, leading to a selection process that reproduces a population in the next time-step. Because of the digitally computed simulation, time-evolution should be treated as discrete.

Let us touch upon an intelligible example of how a MAS approach can be applied to a 2 × 2 game with an additional mechanism for adding social viscosity. Let us be concerned with network reciprocity, one of the five mechanisms discussed in Fig. 2.3. Such a game is called a spatial version of the 2 × 2 game. If a game has a Prisoner's Dilemma structure, it is called a Spatial Prisoner's Dilemma (SPD) game; such games have been commonly accepted as a metaphor for real social systems. Study of the SPD began with a milestone work by Nowak and May (1992). As we discussed in sub-sect. 2.1.1, any PD games with $D_g > 0$ and $D_r > 0$, a well-mixed and infinite population, and replicator dynamics will always be absorbed by an all-defectors state, which is D-dominated (see Fig. 2.4a) at equilibrium. By contrast, an SPD game exposed to the same D_{g} and D_{r} may be able to allow cooperators to survive at equilibrium (see Fig. 2.4b).

For a spatial 2 × 2 game, we must note (1) that an agent placed on each vertex (node) of the presumed underlying network only plays with his neighbors and (2) that he updates his strategy by copying from one of his neighbors. Namely, the basic concept of SPD regulates how an agent interacts, i.e., with whom he plays game, rather than randomly playing with all agents, and how he updates his strategy other than through replicator dynamics.

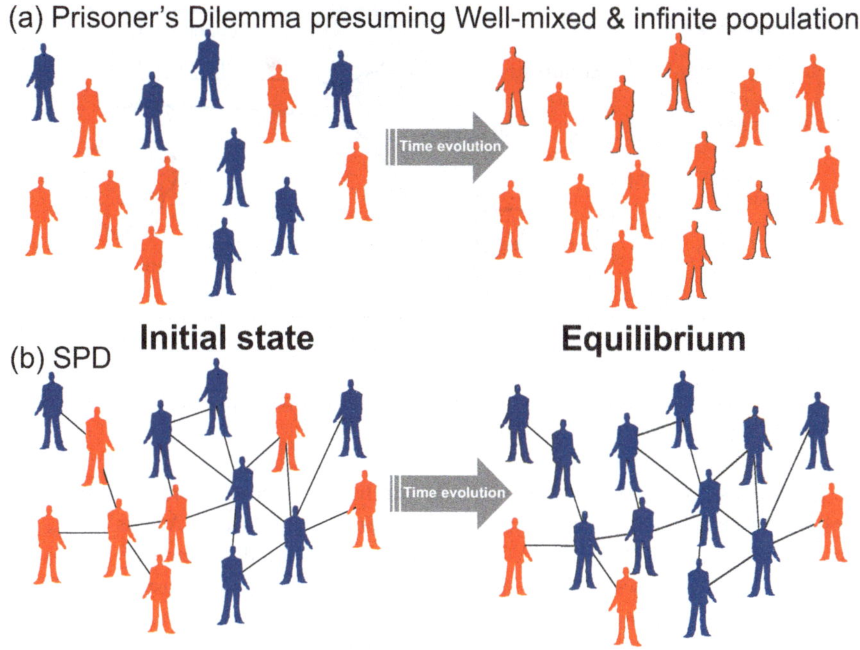

Fig. 2.4 The initial state has an equal fraction of cooperators and defectors. For a well-mixed and infinite population, the equilibrium becomes an all-defectors-state (**a**), while the SPD setting allows cooperators to survive (**b**), this is called network reciprocity

In practice, we should start to generate a finite number of agents on the presumed underlying network in the artificial society. Note that the example shown in Fig. 2.5 presumes a lattice of which the degree (number of links) is four. At the beginning of each simulation episode, $t = 0$, the generated agents are given cooperation or defection strategies with equal probabilities of 1/2, and are randomly placed on the lattice.

At every time-step, each agent plays PD games with their four neighbors; this is the gaming (interaction) phase. Subsequently, each agent updates their strategy by copying one of their neighbors; this is the strategy-updating phase. Two types of copying procedures may be used: deterministic or stochastic. The most representative deterministic procedure is Imitation Max (IM)), by which the focal agent copies the strategy of the agent in their neighborhood (including themself) who obtains the largest payoff. Meanwhile, the most representative stochastic procedure is pairwise Fermi (PW-Fermi), by which the focal agent (i) copies a randomly selected neighbor j's strategy with a probability of $\frac{1}{1+\exp\left[\frac{\pi_i-\pi_j}{\kappa}\right]}$ or otherwise keeps their original strategy. Here, $\pi_i(\pi_j)$ is i's (j's) accumulated payoff over four games with neighbors and κ is the noise parameter of decision making, meaning the thermodynamic temperature (as κ increases, the decision-process becomes approximately random). Completing the strategy-updating process of all agents in the domain, the fractions of cooperators and defectors evolve in time. In this way, the time-step is advanced.

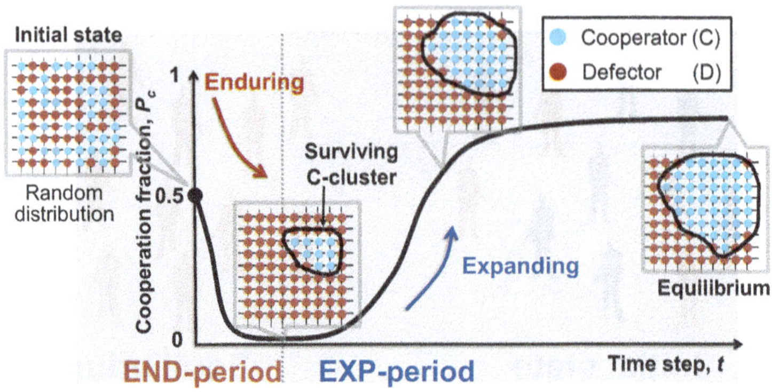

Fig. 2.5 The actual MAS procedure for an SPD game is visually presented, and how the END- and EXP-periods work as the kernel of network reciprocity is explained

The process at each time-step is repeated until the global-cooperation fraction no longer changes, which can be regarded as a pseudo-equilibrium. At this point, this simulation should be terminated. Note that hundreds or thousands of simulation episodes (realizations) should be done to obtain a statistically robust result; this is known as the global-cooperation fraction.

As Fig. 2.5 visually suggests, in a typical SPD under a moderate dilemma strength, cooperation can survive. The mechanism at work is described below. Because of the PD situation, cooperators initially placed in the domain are exploited by neighboring defectors (due to $T > R$). Thus, cooperation fraction decreases rapidly. However, if some cooperators happen to form a cooperators' cluster, they may be able to survive this initial ordeal (which is called the Enduring (END)-period), since boundary cooperators in the cluster are exposed to neighboring defectors' exploitation, but are compelled to maintain cooperation by support from neighboring cooperators in the cluster. By this mechanism, the global-cooperation fraction stops declining and subsequently starts to increase, since neighboring defectors tend to copy cooperation from boundary cooperators in the cluster. This process following the END-period is called the Expanding (EXP)-period. Eventually, cooperation can survive until equilibrium. The concept of the END- and EXP-periods mechanically explains the substance of network reciprocity (Wang et al. 2013; Kabir et al. 2018).

2.2 Multi-Player Games

Though we have so far assumed that there are two game players, a multi-player situation is more typical in a realistic context. It is therefore natural to extend our discussion to multi-player games. In fact, evolutionary game theory can be applied to such real situations as analysis of vaccination games, as will be discussed in the following part of this book. For instance, a social dilemma game of many people regulates how rapidly an influenza spreads. Thus, a fundamental base is needed to quantify multi-player games.

First, we outline the so-called Public Goods Game (PGG)), which has been used most often in the field as a template for multi-player games. This game is based upon a social dilemma around a public good that can only be sustained by a reasonable number of moral-minded cooperators through their donations. The sum of the donations—known as fruits of cooperation—is amplified and equally distributed to all game participants including defectors. This means that a player has an incentive not to donate (cooperate), but definitely to want to get their share of the fruits brought about by the donations of others. As will be discussed in later chapters, in the case of the vaccination game presuming pre-emptive vaccination, herd immunity constitutes a public good, by which a non-vaccinated individual can be protected from infection as well as a vaccinated individual, because a higher vaccination coverage than the threshold prevents stochastic infection.

Suppose that G players participate in a single multi-player game in which a cooperator is requested to donate a cost c to a public pool (see Fig. 2.6). We call G the game size. We assume that the population is infinite and well-mixed; hence, at each time for gaming, G players are randomly selected from the infinitely large mother population. Let the number of cooperators among G be n_c. After collecting donations from all cooperators among the G players, the total pooled donation is multiplied by an amplifying factor, r. Thus, the public good is amplified. The fruits of this public good are distributed equally to all game participants, irrespective of whether they are cooperators or defectors. In this sense, a defector can be called a

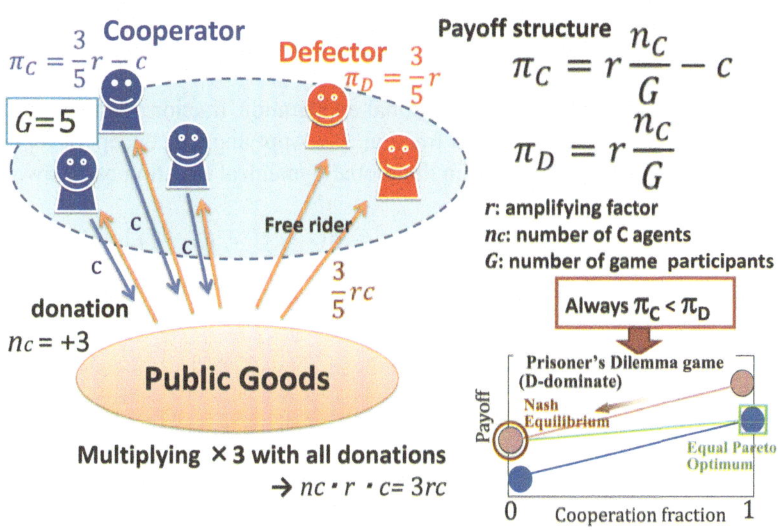

Fig. 2.6 Public Goods Game (PGG); multi-player Prisoner's Dilemma (n-PD) game

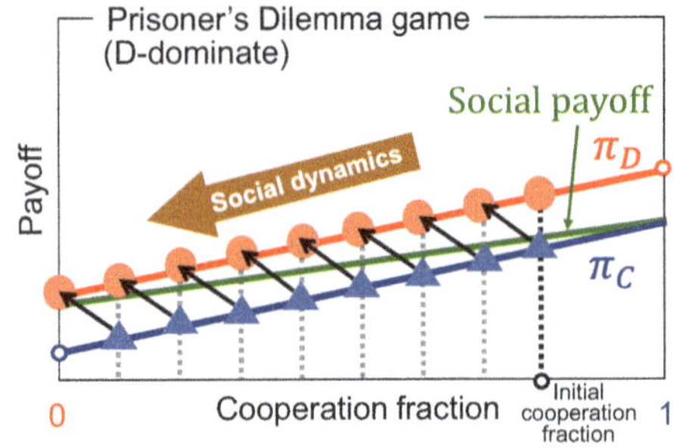

Fig. 2.7 Payoff-structure function of the n-PD

free rider.[6] Here, we can define the payoff-structure functions for both cooperators and defectors as $\pi_C(n_c)$ and $\pi_D(n_c)$, as shown in Fig. 2.6. The functions are defined by the number of cooperators in the group and can also be defined by the local-cooperation fraction, $n_c/G \equiv P_C$. One important thing is that the defectors' payoff is always larger than that of the cooperators at any particular cooperation fraction for a PGG; this schematic relation is redrawn more precisely in Fig. 2.7, where the cooperator and defector plots π_C and π_D indicate the respective payoffs at the local-cooperation fractions, $P_C = n_c/G$. The figure obviously suggests that, as long as $\pi_D(n_c - 1) > \pi_C(n_c)$ is satisfied, a cooperator has no incentive to keep cooperating at any cooperation fraction, and thus the cooperation fraction is always declining regardless of its initial value. This is stronger than the condition in a continuous formulation, $\pi_D(x) > \pi_C(x)$, whereby x is the global-cooperation fraction, of which a stringent definition is given below. Consequently, NE is absorbed by an all-defectors-state, i.e., $P_C = 0$ $(x = 0)$. Conversely, the optimal (i.e., maximum) social payoff appears at the all-cooperators state, $P_C = 1$ $(x = 1)$. This is why we can basically identify PGG as multi-player Prisoner's Dilemma (n-PD) game, featuring a D-dominated equilibrium.

One significant thing is that the global-cooperation fraction, denoted by x, is different from the local-cooperation fraction, P_C. Applying the concept of replicator dynamics (i.e., Eq. (2.6)), we obtain the global dynamical equation as below:

$$\dot{x} = x[f_C - (x \cdot f_C + (1 - x) \cdot f_D)] = x(1 - x) \cdot [f_C - f_D], \tag{2.22.1}$$

$$f_C = \sum_{j=0}^{G-1} \binom{G-1}{j} x^j (1 - x)^{G-1-j} \pi_C(j + 1), \tag{2.22.2}$$

[6] Precisely speaking, we should call it a first-order free rider, because models considering punishment mechanisms also include second-order free riders, who are cooperators who do not punish defectors.

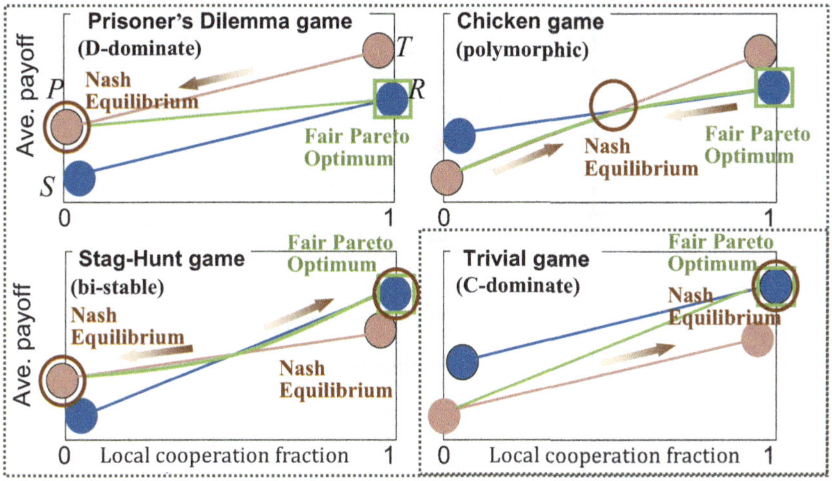

	C	D
Cooperation (C)	*R*, *R*	*S*, *T*
Defection (D)	*T*, *S*	*P*, *P*

Fig. 2.8 Four game classes and payoff-structure functions for multi-player games

$$f_D = \sum_{j=0}^{G-1} \binom{G-1}{j} x^j (1-x)^{G-1-j} \pi_D(j). \tag{2.22.3}$$

As a simple case, let us limit the both payoff functions are featured with monotonic increase as shown in Fig. 2.8. Solving Eqs. (2.22.1)–(2.22.3), we can classify any game as one of the four types, as was the case for 2×2 games: PD, Chicken, SH, and Trivial. Multi-player Chicken (n-Chicken) is featured when the cooperator's payoff function crosses the defector's at a certain local-cooperation fraction. Thus, a multi-player Chicken game has an internal-equilibrium point, as in 2×2 Chicken. Note that internal equilibrium is inconsistent with the crossing points of both payoff functions. Returning to Eqs. (2.22.1)–(2.22.3), the internal equilibrium x_* must be obtained from

$$\sum_{j=0}^{G-1} \binom{G-1}{j} {x_*}^j (1-x_*)^{G-1-j} [\pi_C(j+1) - \pi_D(j)] = 0. \tag{2.23}$$

In general, the number of x_* internal equilibria is not necessarily one but may be more, depending on the forms of both payoff functions. A multi-player Chicken (n-Chicken) game termed "the tragedy of the commons" has been accepted as a typical template model for describing the social dilemma caused by environmental problems. Multi-player Stag Hunt (n-SH) games also have a crossing point between two

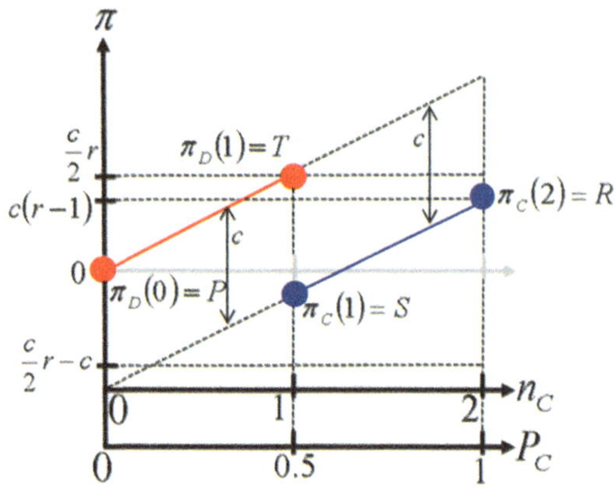

Fig. 2.9 Payoff-structure functions $\pi_C(n_c)$ and $\pi_D(n_c)$, where $G = 2$ in the Public Goods Game

payoff functions. Multi-player Trivial (n-Trivial) has no social dilemma, since cooperation dominates defection, meaning that the cooperator's payoff exceeds the defector's at any local-cooperation fraction, $\pi_D(n_c) < \pi_C(n_c + 1)$.

Let us confirm that Eqs. (2.22.1)–(2.22.3) recovers a 2 × 2 PD game when presuming $G = 2$. In practice, Eqs. (2.22.1)–(2.22.3) becomes

$$\dot{x} = x(1-x)[(1-x)(\pi_C(1) - \pi_D(0)) + x(\pi_C(2) - \pi_D(1))]. \tag{2.24}$$

Substituting $\pi_D(0) = P$, $\pi_D(1)=T$, $\pi_C(1) = S$, and $\pi_C(2)=R$, we get

$$\dot{x} = x(1-x) \cdot [(1-x)(S-P) + x(R-T)]. \tag{2.25}$$

Equation (2.25) is completely consistent with the first equation of Eq. (2.7).

In particular, presuming PD, the payoff functions suggest $\pi_D(1) > \pi_C(2) > \pi_D(0) > \pi_C(1)$, as confirmed in Fig. 2.9, which covers the PD's necessary and sufficient conditions ($D_g > 0$ & $D_r > 0$); $T > R > P > S$.

Let us consider the PGG presuming $G = 2$. Figure 2.9 gives the payoff-structure functions, $\pi_C(n_c)$ and $\pi_D(n_c)$, respectively. According to the definition of PGG and noting that this game recovers a 2 × 2 PD game, we obtain

$$R = \pi_C(2) = \frac{2c}{2}r - c = c(r-1), \tag{2.26.1}$$

$$T = \pi_D(1) = \frac{c}{2}r, \tag{2.26.2}$$

$$S = \pi_C(1) = \frac{c}{2}r - c, \tag{2.26.3}$$

$$P = \pi_D(0) = 0, \tag{2.26.4}$$

where $1 < r < 2$ to make sure $T > R > P > S$. Substituting these into Eqs. (2.20.1) and (2.20.2), we get;

$$D_g = D_r = c\left(1 - \frac{r}{2}\right). \tag{2.27}$$

As long as $1 < r < 2$ is assumed, $D_g = D_r > 0$. One may imagine that this particular game can refer to the Donor and Recipient (D & R) game featuring the same quantity of Chicken-type and SH-type dilemmas, which will be discussed in the next section.

2.3 Social Dilemma and its Mathematical Quantification

We are very much concerned with whether a given game has a "social dilemma" or not. Thus far, we have termed a game without a dilemma as Trivial, in that there is no friction among agents in selecting a dominant strategy.[7] First, we should remind the reader of the stringent mathematical meaning of dilemma: that is, a situation in which the Nash Equilibrium, NE, is not consistent with the socially optimal state. The socially optimal state is the situation in which the average social payoff—that is, the sum of individual payoffs over the population—becomes maximal. This proposition is universal and can be applied not only to 2×2 games, but also to multi-player and even multi-strategy games.

For a symmetric 2×2 game with a well-mixed and infinite population, dilemmas can be simply classified by whether Chicken-type (D_g) or SH-type (D_r) has a positive sign or not as we discussed in Eqs. (2.20.1) and (2.20.2).

One notable point for discussion is what the socially optimal situation would be. One may guess that the socially optimal state for a PD game as well as a Chicken game is always the all-cooperators state, since every agent enjoys obtaining an R that is greater than P. This is not necessarily correct. If met with $S + T < 2R$, the socially optimal state is certainly an all-cooperators state; thus, a PD or Chicken game with $S + T < 2R$ requires mutual cooperation to be socially optimal, which is known as the R-reciprocity game. Conversely, for $S + T > 2R$, mutual cooperation does not realize social optimality. In such a situation, coexistence of certain fractions of cooperators and defectors rather ensures a higher social-average payoff than total cooperation; this is called an ST-reciprocity game.[8]

Although the respective signs of D_g and D_r suggest whether a dilemma exists, they do not quantify how the social dilemma works. If we add some social viscosity by implementing any reciprocity mechanism in an original game, this fact becomes

[7] Presuming $R > P$, which might be an acceptable constraint, the dominant strategy incurring any friction is cooperation. Hence, a Trivial game has a C-dominant equilibrium.

[8] Concerning R-reciprocity and ST-reciprocity games, you should refer to: Wakiyama and Tanimoto (2011).

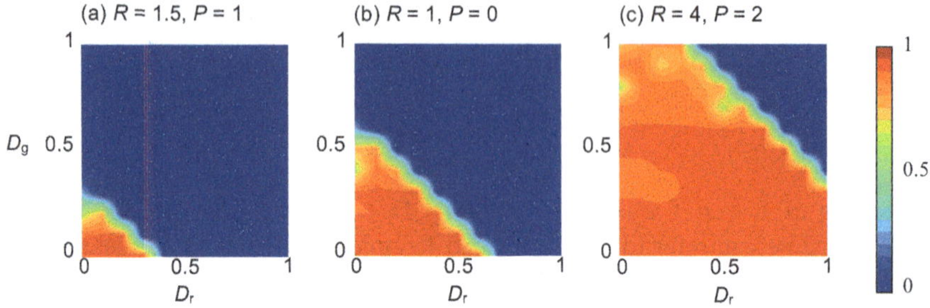

Fig. 2.10 Averaged cooperation fraction D_r–D_g diagrams for (**a**) $R = 1.5$, $P = 1$, (**b**) $R = 1$, $P = 0$, and (**c**) $R = 4$, $P = 2$. Games are played on an 8-neighbor lattice, and the degree of the network is $k = 8$. The population is not infinite but finite; $N = 10^4$ is presumed. In each simulation, an agent plays with their eight immediate neighbors and sums up all payoffs as their accumulated payoff. After the gaming session at each time-step, agents synchronously update their strategy to be either C or D. We adopted Imitation Max (IM) as the strategy-update rule, whereby each agent deterministically copies the best-performing strategy

more tangible. Figure 2.10 displays the MAS result showing the cooperation fraction in the PD region ($D_g > 0$ and $D_r > 0$) when varying R and P. Returning to Fig. 2.2, the theory based on the formulation of replicator dynamics with a well-mixed and infinite population correctly predicts that the entire PD region is always absorbed by an all-defectors-state (see the red region for PD in the left panel of Fig. 2.2). Being concerned with panel (a), we can evidently confirm that the region with smaller D_g and D_r allows the survival of cooperators, of which the cooperation fraction is quite high (red-triangle region). This is because network reciprocity solves the dilemma situation, letting agents cooperate to obtain mutual cooperation with a payoff of R. Moving to panels (b) and (c), we should note that, with the increase of the difference between R and P (i.e., $R - P$), the red-triangle region significantly expands. This fact proves that D_g and D_r fail to appropriately quantify the dilemma strength when a certain mechanism for adding social viscosity is implemented. It would be quantitatively acceptable when we compare two different PD games having same D_g and D_r; say, for example, $\begin{bmatrix} R & S \\ T & P \end{bmatrix} = \begin{bmatrix} 1 & -1 \\ 2 & 0 \end{bmatrix}$ and $\begin{bmatrix} 100 & -1 \\ 101 & 0 \end{bmatrix}$. Those two games have $D_g = D_r = 1$, and thus both belong to the D & R (Donor and Recipient) game, a sub-class of PD. Suppose that there is a certain reciprocity mechanism, which implies that an agent has a non-zero possibility of playing with the same opponent more than once. This is possible because the social viscosity added by the reciprocity mechanism breaks the perfectly anonymous situation. If that is the case, cooperation may become a meaningful strategy for the focal player expecting an opponent's cooperation in return. Such a tendency becomes much more significant if there is a huge gap between R and P, since it ensures a sufficiently higher payoff to any cooperator, regardless of whether their opponent cooperates or defects (the payoff difference between 101 and 100 might be negligible as compared with $1 = D_g = D_r$). Now it should be apparent why such a thing takes place and why the set of D_g and D_r is inappropriate as an index of dilemma strength.

A question arising to us is whether an alternative measure can quantify a dilemma strength in the context of a symmetric 2 × 2 game, even if a reciprocity mechanism is presumed to be implemented.

The answer is "yes," as shown below.

2.3.1 Concept of the Universal Scaling for Dilemma Strength

Let us jump directly to the conclusion: the answer lies in the simple idea that both D_g and D_r should be divided by $R - P$ to normalize the influence of the difference. Thus, we define a new set of GID and RAD as D_g' and D_r', respectively:

$$D'_g \equiv \frac{T-R}{R-P} = \frac{D_g}{R-P}, \tag{2.28.1}$$

$$D'_r \equiv \frac{P-S}{R-P} = \frac{D_r}{R-P}. \tag{2.28.2}$$

This definition conveys the payoff structure of a game, as represented by **M** in Eq. (2.29):

$$\mathbf{M} \equiv [a_{ij}] = \begin{pmatrix} R & P-(R-P)D'_r \\ R+(R-P)D'_g & P \end{pmatrix}. \tag{2.29}$$

Here, let us rely upon the splendid result of Taylor and Nowak (2007), who successfully deduced that any of the Nowak's five reciprocity mechanisms can be expressed by each transformation applied to an original 2 × 2 game payoff given by Eq. (2.2). This allows us to obtain the equivalent payoff matrix when each of the five reciprocity mechanisms is applied. More importantly, their theory also presumes a well-mixed and infinite population, allowing us to apply the replicator dynamics to quantify an equilibrium. In a word, by slightly altering the equivalent payoff matrix, each of the five reciprocity mechanisms can be evaluated, quantified, and discussed based on the same template of a 2 × 2 game payoff presuming a well-mixed and infinite population.

Let us discuss each mechanism in turn.

2.3.1.1 Direct Reciprocity

In repeated games with a pair composed of the same two agents or when the same pair of agents plays a different game in another round, direct reciprocity stimulates cooperation (Trivers 1971, 1985). In each round, the two agents must choose either cooperation or defection; with a probability of w, these same two agents play another round. We assume that defectors, denoted by D, always choose defection and cooperators, C, play tit-for-tat (TFT); they start with cooperation and then follow what the other player has done in the previous move. According to Taylor and Nowak's work, the equivalent payoff matrix is

$$\begin{array}{cc} & \begin{array}{cc} C & D \end{array} \\ \begin{array}{c} C \\ D \end{array} & \begin{pmatrix} \frac{R}{1-w} & S+\frac{wP}{1-w} \\ T+\frac{wP}{1-w} & \frac{P}{1-w} \end{pmatrix} \end{array}. \tag{2.30}$$

Hence, we can revise D'_g and D'_r to $D'_{g\ \mathrm{DR}}$ and $D'_{r\ \mathrm{DR}}$,

$$D'_{g\mathrm{DR}} = \frac{\left(T+\frac{wP}{1-\mathrm{w}}\right)-\left(\frac{R}{1-\mathrm{w}}\right)}{\left(\frac{R}{1-\mathrm{w}}\right)-\left(\frac{P}{1-\mathrm{w}}\right)}, \tag{2.31.1}$$

$$D'_{r\mathrm{DR}} = \frac{\left(\frac{P}{1-\mathrm{w}}\right)-\left(S+\frac{wP}{1-\mathrm{w}}\right)}{\left(\frac{R}{1-\mathrm{w}}\right)-\left(\frac{P}{1-\mathrm{w}}\right)}. \tag{2.31.2}$$

Assuming that $w = 0.1$, we obtain Figs. 2.11, wherein the D_g–D_r diagrams (upper panels) as well as the D_g'–D_r' diagrams (lower panels) of equilibria for different $R - P$ values are shown. To draw Fig. 2.11 (as well as Figs. 2.11, 2.12, 2.13, 2.14, and 2.15), we assumed the initial cooperation fraction to be 0.5.

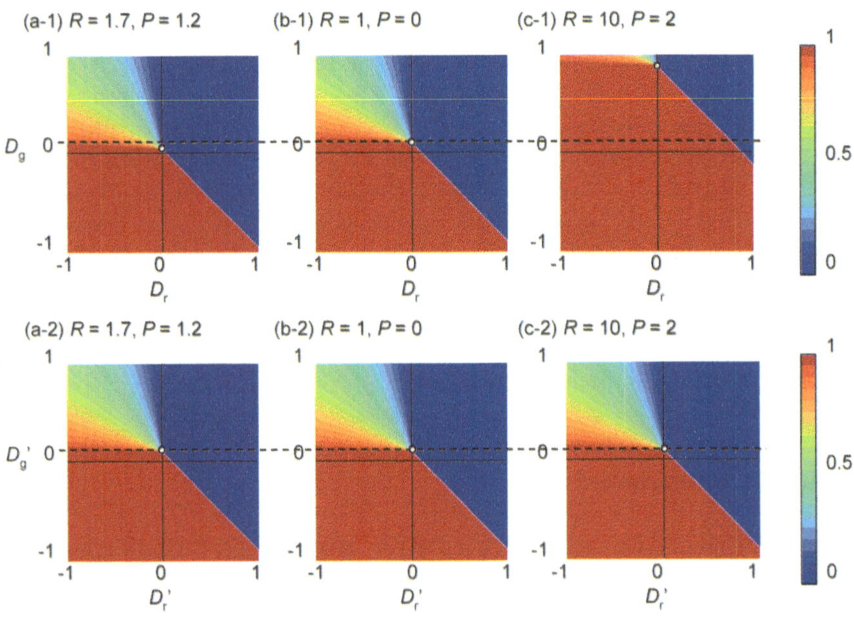

Fig. 2.11 Direct-reciprocity diagrams for equilibrium cooperation fractions of D_r–D_g (upper line) and D_r'–D_g' (lower line) for (**a**) $R = 1.7$, $P = 1.2$, (**b**) $R = 1$, $P = 0$, and (**c**) $R = 10$, $P = 2$, with the probability of another round being $w = 0.1$

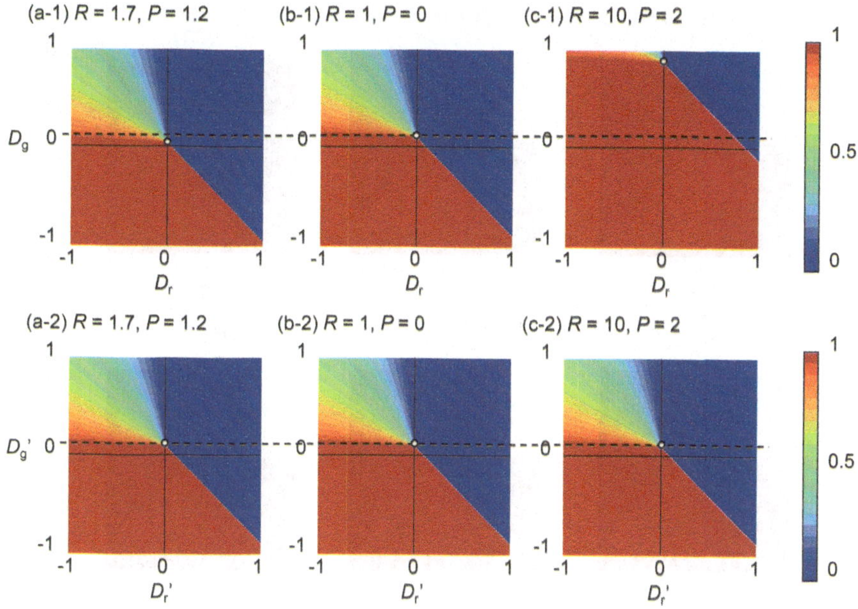

Fig. 2.12 Equilibrium-cooperation fraction D_r–D_g (in the upper line) and D_r'–D_g' (in the lower line) diagrams of indirect reciprocity for (**a**) $R = 1.7$, $P = 1.2$, (**b**) $R = 1$, $P = 0$, and (**c**) $R = 10$, $P = 2$ with a $q = 0.1$ probability of knowing the reputation of another individual

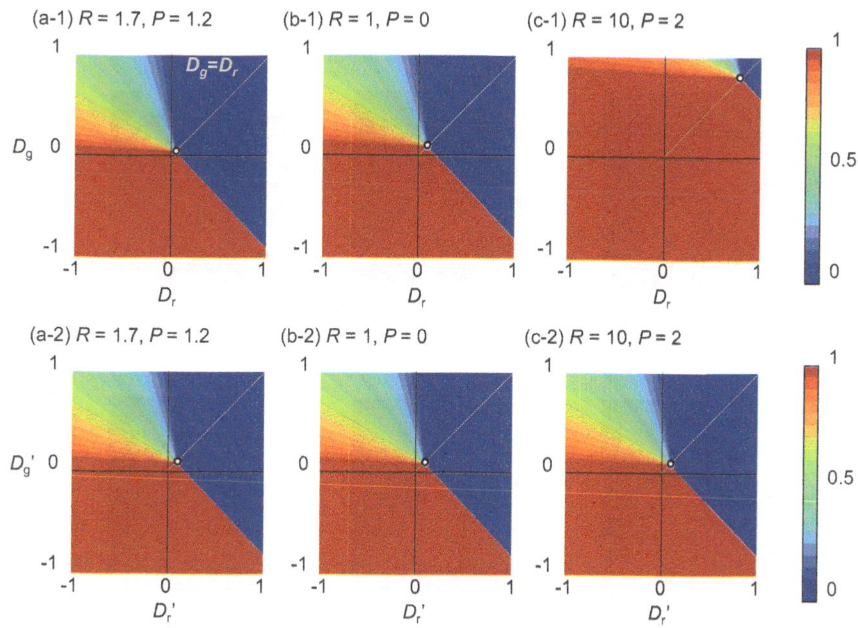

Fig. 2.13 Equilibrium-cooperation fractions D_r–D_g (in the upper line) and D_r'–D_g' (in the lower line); diagrams of kin selection for (**a**) $R = 1.7$, $P = 1.2$, (**b**) $R = 1$, $P = 0$, and (**c**) $R = 10$, $P = 2$, with the average relatedness between interacting individuals given by $r = 0.1$

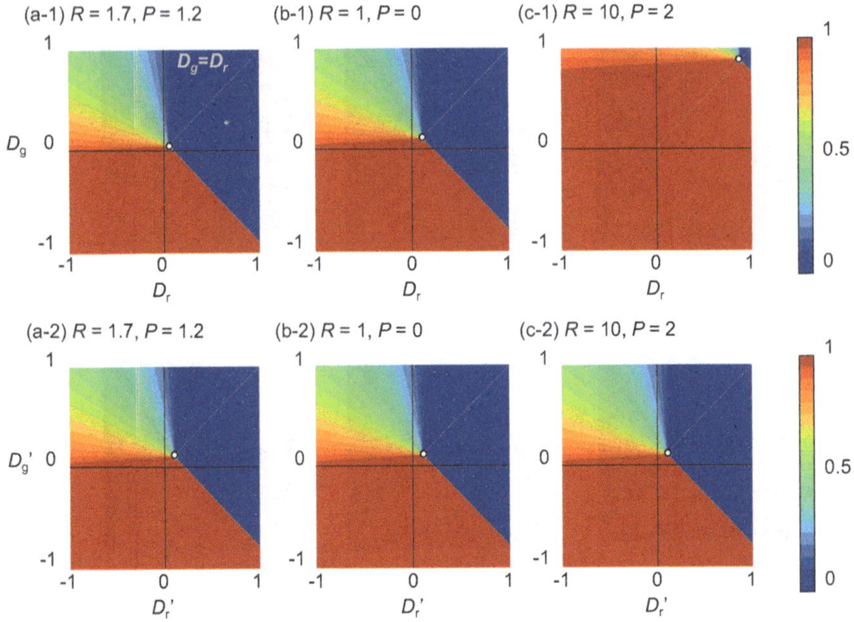

Fig. 2.14 Equilibrium cooperation fraction D_r–D_g (in the upper line) and D_r'–D_g' (in the lower line) diagrams of group selection for (**a**) $R = 1.7$, $P = 1.2$, (**b**) $R = 1$, $P = 0$, and (**c**) $R = 10$, $P = 2$ with the number of groups given by $m = 50$ and the maximum size of a group given by $n = 500$

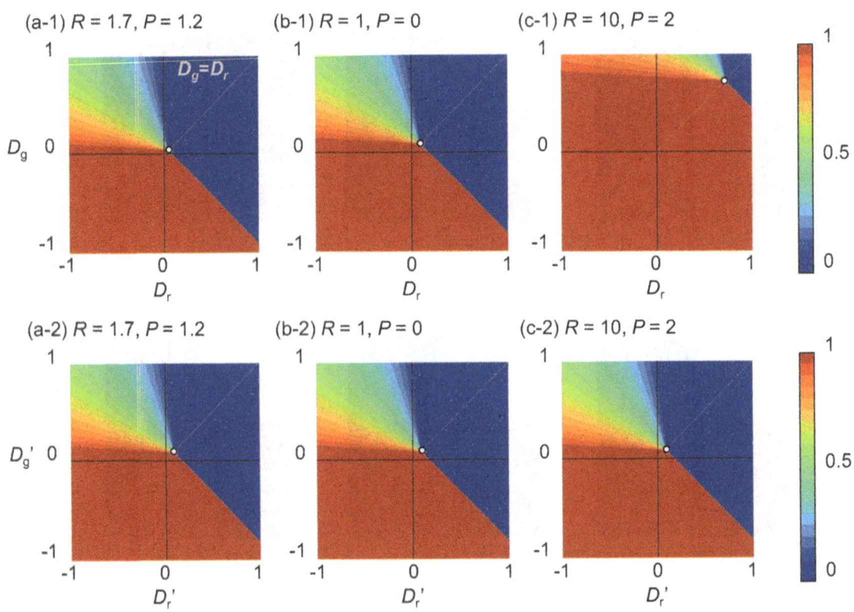

Fig. 2.15 Equilibrium cooperation fraction D_r–D_g (in the upper line) and D_r'–D_g' (in the lower line) diagrams of network reciprocity for (**a**) $R = 1.7$, $P = 1.2$, (**b**) $R = 1$, $P = 0$, and (**c**) $R = 10$, $P = 2$ with the number of neighbors given by $k = 12$

In the figure, each of the white circles indicates the boundary point of the four game classes: D-dominant, C-dominant, polymorphic, and bi-stable (hereafter "four-corners"). Each horizontal black broken line indicates the difference between the four-corners for $R = 1$ and $P = 0$, and for $D_g' = D_r' = 0$. Let us confirm that, through direct reciprocity, the original four game-classes (PD in the first quadrant, Chicken in the second quadrant, Trivial in the third quadrant, and SH in the fourth quadrant) shift to the positive side of the D_g-axis (D_g'-axis), depicted by the black dotted line. Shifting upward implies that the weaker region of PD becomes SH, which has a bi-stable equilibrium. In short, direct reciprocity can weaken a Chicken-type dilemma; GID ($D_g = T - R$ or $D_g' = (T - R)/(R - P)$). In the D_g–D_r diagrams (upper panels), the larger that $R - P$ becomes, the larger the upward shift that can be observed. In the D_g'–D_r' diagrams (lower panels), however, upward shifting does not depend on $R - P$. Surprisingly, the respective equilibria on the D_g'–D_r' diagrams are completely consistent with each other, despite the different $R - P$ values. Therefore, a set of parameters, D_g' and D_r', that considers D_g and D_r but also takes account of the influence from $R - P$, can be universally appropriate for evaluating the dilemma strength in a population with any reciprocity mechanisms. This observation is universally true for the four other reciprocity mechanisms discussed with reference to Figs. 2.12, 2.13, 2.14, and 2.15.

2.3.1.2 Indirect Reciprocity

Indirect reciprocity is based on reputation (Alexander 1987; Nowak and Sigmund 1998). Unlike direct reciprocity, whereby the focal player's decision is based on whether their opponent has offered cooperation or defection in the previous encounter, in indirect reciprocity, the focal agent's decision is determined on whether their opponent has offered cooperation or defection to another agent in the previous round. In fact, the focal agent chooses their strategy (offering C or D) based on their opponent's reputation, called their image score (IS). The parameter q denotes the probability of knowing the IS of another individual—in short, of knowing whether they are a cooperator or defector. Let us assume that a defector, D, always defects, whereas a cooperator, C, only defects when they know that their opponent is a defector. Thus, C cooperates with D with probability $1 - q$. The equivalent payoff matrix is:

$$\text{Indirect Reciprocity}\ \begin{matrix} C \\ D \end{matrix} \overset{\begin{matrix} C & \qquad\qquad D \end{matrix}}{\begin{pmatrix} R & (1-q)S+qP \\ (1-q)T+qP & P \end{pmatrix}}. \tag{2.32}$$

Hence, we can revise D_g' and D_r' to $D_{g\ \mathrm{DR}}'$ and $D_{r\ \mathrm{DR}}'$:

$$D'_{g\mathrm{IDR}} = \frac{\{(1-q)T + qP\} - R}{R - P}, \tag{2.33.1}$$

$$D'_{r\mathrm{IDR}} = \frac{\{(1-q)T + qP\} - \{(1-q)S + qP\}}{R - P}. \tag{2.33.2}$$

Assuming that $q = 0.1$, we obtain Figs. 2.12. In the case of indirect reciprocity, because of the structural similarity of the equivalent payoff matrix (Eq. (2.32)) to that of direct-reciprocity case (Eq. (2.30)), Fig. 2.12 is almost consistent with Fig. 2.11. In a nutshell, indirect reciprocity can weaken the Chicken-type dilemma.

2.3.1.3 Kin Selection

The concept of kin selection arose from the idea that evolutionary games are often played between individuals who are genetic relatives (Hamilton 1964). Consider a population in which the average relatedness between interacting individuals is given by r, which is a real number between 0 and 1. In such a population, the opponent's contribution that equals the r of the opponent's payoff can be joined to the focal agent's payoff. The equivalent payoff matrix is:

$$\text{Kin Selection} \quad \begin{matrix} & C & D \\ C & \multirow{2}{*}{} & \\ D & & \end{matrix} \begin{pmatrix} R & \frac{S + rT}{1 + r} \\ \frac{T + rS}{1 + r} & P \end{pmatrix}. \tag{2.34}$$

Hence, we can revise D'_g and D'_r to $D'_{g\ \mathrm{KS}}$ and $D'_{r\ \mathrm{KS}}$:

$$D'_{g\mathrm{KS}} = \frac{\left(\frac{T+rS}{1+r}\right) - R}{R - P}, \tag{2.35.1}$$

$$D'_{r\mathrm{KS}} = \frac{P - \left(\frac{S+rT}{1+r}\right)}{R - P}. \tag{2.35.2}$$

Assuming that $r = 0.1$, we obtain Figs. 2.13. Unlike direct and indirect reciprocities, each white broken line in Fig. 2.13 shows that the four-corners shift from the point $D_g' = D_r' = 0$ along the $D_g' = D_r'$ line (the 45-degree line). Shifting upward and to the right implies that the weaker PD region becomes Trivial, which has a C-dominated equilibrium, i.e., a dilemma-free state. In short, the kin-selection mechanism can weaken both the Chicken-type dilemma GID ($D_g = T - R$ or $D_g' = (T - R)/(R - P)$) and the SH-type dilemma, RAD ($D_r = P - S$ or $D_g' = (P - S)/(R - P)$).

2.3.1.4 Group Selection

Group selection is based on the idea that competition occurs not only between individuals, but also between groups.[9] Here, we use the approach described by Traulsen and Nowak (Traulsen and Nowak 2006). A population is subdivided into m groups, of which the maximum size is n. Individuals interact with others in the same group according to a 2×2 game. The fitness of an individual is $1 - \omega - \omega F$, where F is the payoff and ω is the intensity of selection. At each round, one individual from the entire population is chosen for reproduction proportional to fitness. That offspring is added to the same group. If the group reaches maximum size, it can split into two groups with a certain probability p. In this procedure, a randomly selected group dies to prevent the population from exploding. The maximum population size is defined as mn; with a probability of $1 - p$, however, the group does not divide, but a random individual from that group is chosen to die. We assume weak selection ($\omega < < 1$) and rare group splitting ($p << 1$) with large n and m. The equivalent payoff matrix is:

$$\text{Group Selection} \begin{matrix} \\ C \\ D \end{matrix} \begin{matrix} \begin{matrix} C & \quad\quad D \end{matrix} \\ \begin{pmatrix} (n+m)R & nS + mR \\ nT + mP & (n+m)P \end{pmatrix} \end{matrix}. \tag{2.36}$$

Hence, we can revise D'_g and D'_r to $D'_{g\ \mathrm{GS}}$ and $D'_{r\ \mathrm{GS}}$:

$$D'_{g\mathrm{GS}} = \frac{(nT + mP) - (m+n)R}{(m+n)R - (m+n)P}, \tag{2.37.1}$$

$$D'_{r\mathrm{GS}} = \frac{(m+n)P - (nS + mR)}{(m+n)R - (m+n)P}. \tag{2.37.2}$$

Assuming that $m = 50$ and $n = 500$, we obtain Figs. 2.14. We can confirm that group selection can weaken both the Chicken-type dilemma and the SH-type dilemma.

2.3.1.5 Network Reciprocity

Network reciprocity relies on two effects: (1) limiting the number of game opponents (diminishing anonymity), leading to increased mutual cooperation and (2) a local adaptation mechanism whereby a player copies a strategy from a neighbor linked to them through a network. These two effects explain how cooperators survive in a

[9]There have been many studies on this point. Because of space limitations, we only cite five; Wynne-Edwards (1962), Williams (1996), Wilson (1975), Maynard Smith (1976), Slatkin and Wade (1978).

network game of PD, even though players are required to use only the simplest strategy—either cooperation or defection (requiring only 1 bit of memory) (Nowak and May 1992). Therefore, hundreds of studies have reported on network reciprocity, primarily in the fields of theoretical biology and statistical physics.[10] The individuals of a population occupy the vertices of a graph; the edges denote who interacts with whom. Each individual interacts with all of their neighbors according to the standard payoff matrix, as in Eq. (2.2). The payoff for each agent is totaled over all games with their neighbors. An individual's fitness is given by $1 - \omega - \omega F$, where F is the payoff for the individual and ω ($\omega \in [0, 1]$) is the intensity of selection. Here, we consider evolutionary dynamics according to Death–Birth updating (DB) (Ohtsuki et al. 2006), whereby a random individual is chosen in each round to die; then, the neighbors compete for empty sites proportional to their fitness.

A calculation using pair approximation on regular graphs (with each vertex having k edges) leads to a deterministic differential equation that describes how the expected frequency of cooperation (defection) changes over time. This differential equation is actually a standard replicator equation with a modified payoff matrix (i.e., equivalent payoff matrix) (Ohtsuki and Nowak 2006). This payoff matrix is given by

$$\text{Network Reciprocity}\begin{matrix} \\ C \\ D \end{matrix}\begin{matrix} \quad C & \quad D \\ \left(\begin{matrix} R & S+H \\ T-H & P \end{matrix}\right) \end{matrix}, \tag{2.38}$$

where H in Eq. (2.38) is defined as follows:

$$H = \frac{(k+1)(R-P) - T + S}{(k+1)(k-2)}. \tag{2.39}$$

Hence, we can revise D'_g and D'_r to $D'_{g\ \mathrm{GS}}$ and $D'_{r\ \mathrm{GS}}$

$$D'_{g\mathrm{NR}} = \frac{(T-H) - R}{R-P}, \tag{2.40.1}$$

$$D'_{r\mathrm{NR}} = \frac{P - (S+H)}{R-P}. \tag{2.40.2}$$

Assuming that $k = 12$, we obtain Figs. 2.15. As in the cases of kin selection and group selection, we are able to confirm that network reciprocity can weaken both Chicken-type and SH-type dilemmas.

[10]Because of space limitations, we can cite here only five of the following: Hassell et al. (1994), Ebel and Bornholdt (2002), Santos and Pacheco (2005), Santos et al. (2006), Yamauchi et al. (2010).

The discussion above is based on a theoretical framework using pair approximation on regular graphs. The question of whether the concept of universal scaling for dilemma strength using D_g' and D_r' works well for other cases of various topologies was deliberately explored by Wang and his colleagues (2015), who partially relied upon the MAS approach varying not only in underlying topology, but also in strategy-updating rule. They found that the universal scaling concept given by D_g' and D_r' works reasonably well, even if various strategy-update rules are assumed. In assuming heterogeneous networks like scale-free networks and small-world graphs instead of homogeneous topologies like rings and lattices, the scaling concept malfunctions, because different number of playing games among agents that a heterogeneous network intrinsically allows significantly affects its evolutionary process. Thus, the underlying topology sometimes becomes more significant than the influence resulting from the game structure.

Here, let us visually confirm once again how the concept of a universal scaling for dilemma strength using D_g' and D_r' works for each of the five reciprocity mechanisms discussed above. Figure 2.16 shows how the dilemma on the default setting is assumed when no reciprocity mechanisms are implemented; $\left(D'_r, D'_g\right)$ is relaxed by each of the five mechanisms when presuming $w = 0.2$ for direct reciprocity, $q = 0.2$ for indirect reciprocity, $r = 0.2$ for kin selection, $m = 100$ & $n = 500$ for group selection, and $k = 4$ for network reciprocity. In panel (b) presuming $\left(D'_r, D'_g\right) = (0.1, 0.1)$ for the default that belongs to PD, we can confirm that either direct or indirect reciprocity changes PD to SH, while each of the other three mechanisms changes PD to Trivial. Panel (c) displays a different scenario starting from $\left(D'_r, D'_g\right) = (0.1, -0.1)$, which is SH. Here, kin selection, group selection, and network reciprocity convert that SH to Trivial. In panel (d), starting from a Chicken dilemma with $\left(D'_r, D'_g\right) = (0.1, -0.1)$, the five mechanisms successfully convert the scenario to Trivial.

Figure 2.17 visually explains how the original (default) D_g'–D_r' plane, classified by four colors and labels—PD, Chicken, SH, and Trivial—is distorted by each of the five mechanisms. Direct and indirect reciprocities suppress the Chicken-type dilemma and allow it to become either Trivial or SH. Meanwhile, kin selection, group selection, and network reciprocity relax both the Chicken-type and SH-type dilemmas.

2.3.2 *Concept of a Social Efficiency Deficit*

In the previous sub-section, we successfully introduced the idea of "dilemma strength " (hereafter DS) by means of a universal scaling by D_g' and D_r'. One thing to be noted is that DS can be defined only in the case of a 2×2 game. DS cannot be explicitly defined for other cases (i.e., more general social dilemma games such as multi-player games including the PGG), as well as more practical

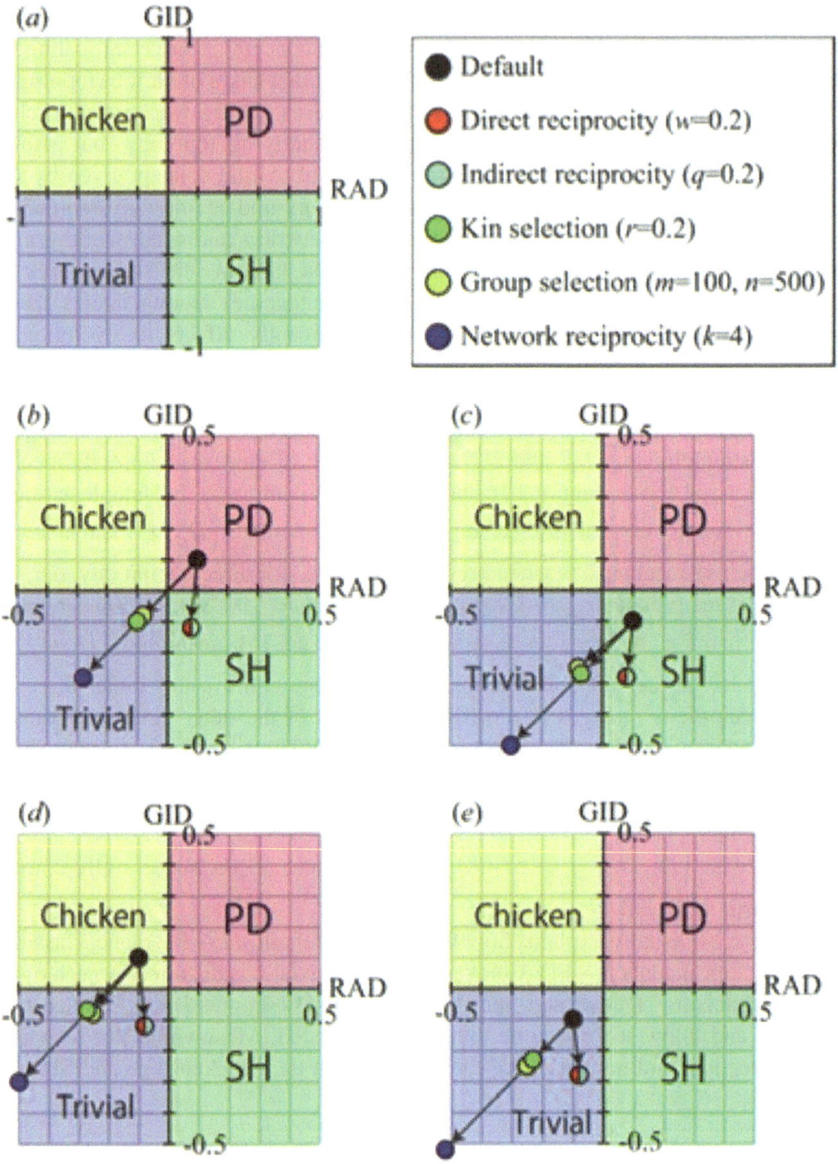

Fig. 2.16 Visual explanation of how each of the five reciprocity mechanisms relaxes the dilemma. Panel (**a**) shows the default phase diagram of four game classes. Other panels respectively show how a game of (**b**) PD, (**c**) Stag Hunt, (**d**) Chicken, and (**e**) Trivial being transferred by the five reciprocity mechanisms

application-oriented games like the vaccination game, which is one of the main topics of concern in the present book, where the time-dependent game structure cannot be given as a static 2×2 payoff matrix or multi-player game payoff function discussed in Sect. 2.2. Although DS is quite powerful, it has the crucial drawback of being inapplicable it to other general games.

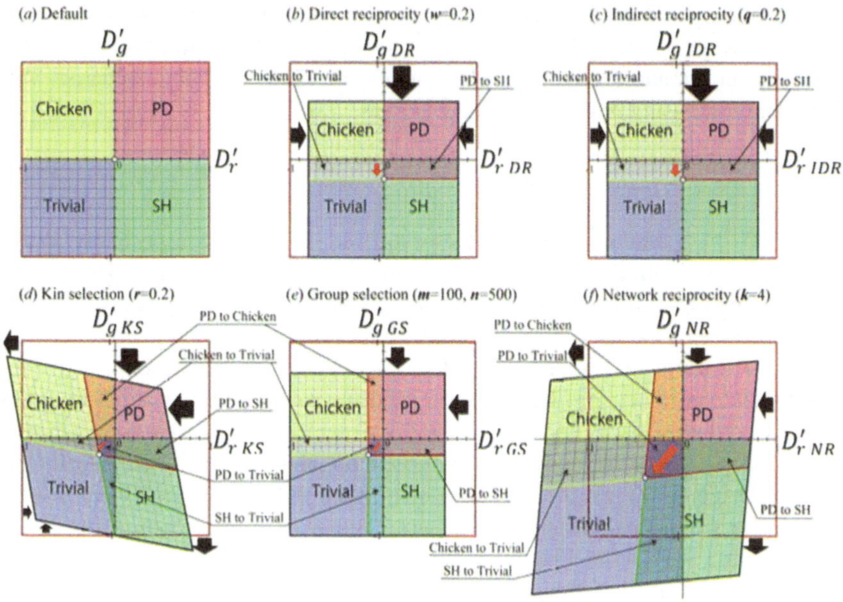

Fig. 2.17 Visual explanation of how each of the five reciprocity mechanisms relaxes the Chicken-type and SH-type dilemmas. Black arrows indicate how the original dilemma situation of the default setting that presumes a well-mixed and infinite population can be oppressed by the mechanism. The red arrow shows the shifting direction of the origin of the D_g'–D_r' plane. If it goes down and to the left, a smaller PD could be Trivial by the mechanism. Panel (**a**) shows the default setting. Other panels explain how each of the five mechanisms; (**b**) direct reciprocity, (**c**) indirect reciprocity, (**d**) kin selection, (**e**) group selection, and (**f**) network reciprocity, brings a skewed phase diagram to relax a social dilemma

Needless to say, it is important to examine whether or not a social dilemma works under a certain model-parameter setting. To remedy this accompanying pitfall, in this sub-section, we introduce a new index for explaining the presence of a social dilemma, termed the "social efficiency deficit" (SED)). In a word, SED indicates how much the payoff could be improved from NE toward a social optimum (SO); that is, SED is accounted for by the difference in payoffs between SO and NE. The term SO signifies the scenario with the maximum socially accumulated payoff of a game. If DS can be treated as a "prognostic" index for whether a social dilemma prevails,[11] SED would be thought as an "ex-post" index, since both the socially optimal payoff and the payoff at NE can always be estimated, regardless of model complexity.[12] Thus, one can predict SED and evaluate whether or not a social dilemma actually works by referring to whether SED is positive or zero.

[11]This is because for DS, D_g', and Dr' can be identified in advance of actual analysis. Whenever one knows the payoff matrix, DS can be evaluated.

[12]We can reproduce any game model as a form of MAS irrespective of game type, regardless of whether it is 2-payer or multi-player, 2-strategy or multi-strategy, complex or simple, or realistic or ideal. As long as a MAS model is established, we can explore Nash equilibrium as well as the socially optimal state, at least numerically.

The dilemma strength refers to a numerical measure of the extent of social viscosity needed to switch from NE to a socially optimal point, while the SED provides information regarding the payoff shortfall at NE compared with SO. Therefore, using δ, we can express SED as

$$\delta = \Pi^{\text{Social Opt.}} - \Pi^{\text{NE}}, \tag{2.41}$$

where $\Pi^{\text{Social Opt.}}$ and Π^{NE} indicate the payoffs at SO and NE, respectively. Referring to the definition of a NE, we should confirm that $\Pi^{\text{Social Opt.}} \geq \Pi^{\text{NE}}$. The absence or presence of a social dilemma corresponds to whether δ is zero or positive. Clearly, if NE coincides with SO, there is no incentive for the game players to try to improve their payoff, which corresponds to a dilemma-free situation, as we confirmed in the first paragraph of Sect. 2.3. This scenario yields no payoff-shortage at NE from SO, which accordingly insists upon the absence of a social dilemma. However, if there is a payoff gap between NE and SO, then it is still possible to improve the payoff from NE, which demonstrates the existence of a social dilemma. As previously mentioned, the scaling parameters for DS—D_g' and D_r'—were proposed for 2 × 2 games with a well-defined payoff matrix; however, games having more than two players, more than two strategies, or both are not tractable with the aid of DS parameters, even if the payoff matrix is well-defined. Hence, we can claim that SED is a general parameter for explaining the presence of a social dilemma, regardless of the game structure. Furthermore, if a concrete relationship between DS and SED can be established for a dilemma game, this would allow SED to conjecture not only the existence, but also the extent, of the social dilemma associated with that game. Nonetheless, in the situation where DS cannot be defined explicitly, SED still can illustrate the presence of a social dilemma.

Below, we discuss SED in detail and present its correct definition for several important games.

2.3.2.1 Donor and Recipient Game

Let us start with the most fundamental and common game: the Prisoner's Dilemma (PD). We have already recognized that a PD requires $T > R > P > S$; thus it meets with $D'_g = \frac{D_g}{R-P} = \frac{T-R}{R-P} > 0$ and $D'_r = \frac{D_r}{R-P} = \frac{P-S}{R-P} > 0$. Here, let us limit PD to refer to games that need R-reciprocity, not ST-reciprocity; this requires another condition: $S + T > 2R$.

In the following text, we are specifically concerned with the Donor and Recipient (D & R) game, of which the payoff matrix is given by

$$\begin{pmatrix} R(= b - c) & -c \\ b & 0 \end{pmatrix} \equiv \mathbf{M}_{\mathbf{D\&R}}. \tag{2.42}$$

Clearly, $\mathbf{M_{D\ \&\ R}}$ has the property $D_g = D_r$. Now, choosing $D_g = D_r \equiv \Delta$, we can easily find $b = R + \Delta$ and $c = \Delta$.

D'_g and D'_r can also be rewritten as

$$\begin{cases} D'_g = \dfrac{\Delta}{R} \\ D'_r = \dfrac{\Delta}{R} \end{cases}. \tag{2.43}$$

Hence, Eqs. (2.42) and (2.43) establish an inverse relationship between DS and SED—i.e., the higher DS corresponds to the lower SED and vice versa (see Fig. 2.18). In a game with stronger DS (see panels (a) and (c)), the transition from a defective state D to a cooperative state C requires abundant social viscosity to

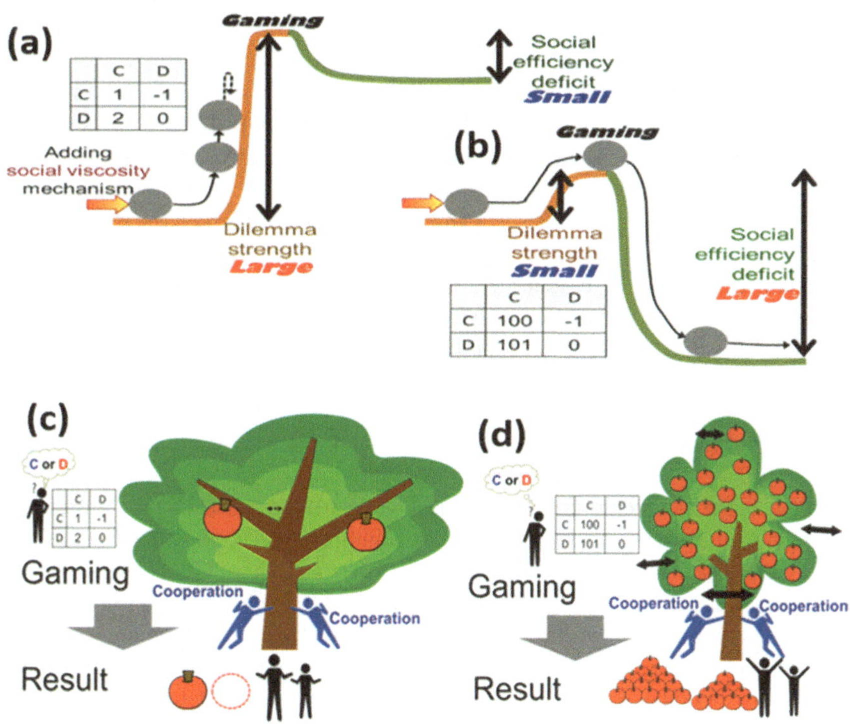

Fig. 2.18 Schematic explanation of (**a**) game with large DS and small SED, and (**b**) game with small DS and large SED observable in a PD game. The game (**a**) is having a stronger dilemma, consequently, requiring abundant social viscosity to overcome the dilemma but the payoff gain is not worth enough; on the other hand, game (**b**) is showing the opposite scenario. Game (**c**) and (**d**) are having the same payoff structures as that of (**a**) and (**b**), respectively, but portraying in a different fashion to understand the mechanism. In a PD game with large DS and small SED (**c**), even the dilemma being solved, the fruits brought by cooperation would not be much. Whereas in a game with small DS and large SED (**d**), the dilemma can be easily solved; thus, the fruits brought by cooperation would be much

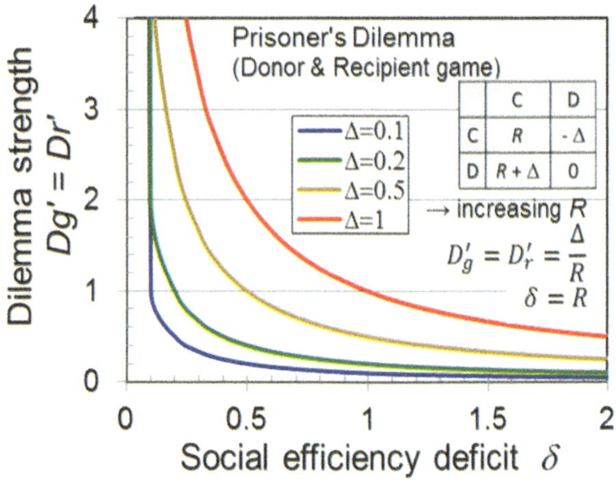

Fig. 2.19 SED vs. DS line graphs showing the inverse relationship between DS (D_g' and D_r') and SED for several Δ values in the case of a 2 × 2 D & R game. The DS shows a positive correlation with Δ values

overcome the dilemma; however, the payoff gain for this state transition is not so significant. By contrast, in a game with lower DS (see panels (b) and (d)), a smaller social viscosity can promote cooperation from defection and yields a higher payoff. This means that a huge payoff shortfall (i.e., higher SED) at NE is observed for the latter game; however, the former game has the opposite scenario. The above discussion can be justified graphically by Fig. 2.19, wherein a DS versus SED graph for a 2 × 2 D & R game exhibits an inverse relationship with several Δ values. That is, the DS decreases with the increase of δ and vice versa. Additionally, the higher Δ value corresponds to the higher DS, which is quite conceivable.

2.3.2.2 Public Goods Game

The PGG is the most representative multi-player game that basically[13] belongs to the n-PD class; thus, it features a D-dominant equilibrium. The payoff structure was already discussed in Fig. 2.6. It is worthwhile to note that a PGG can describe the multi-player case of a D & R game.

Let us suppose that, at every step of the game, G (>2) players are randomly selected to play the game from an infinite and well-mixed population. Assume that a cooperator (C) donates a cost, $c > 0$, for a public good and a defector (D) donates nothing. The total donation of the cooperators (among G players) is amplified by a factor r (the reward or amplifying factor). The amplified public good is then distributed equally to all players, regardless of cooperators and defectors. If n_C is the number of cooperators among G players, then the payoff for cooperators ($\pi_C(n_c)$) and defectors ($\pi_D(n_C)$) can be expressed by $\pi_C(j) = r.\frac{j}{G}.c - c$ and $\pi_D(j) = r.\frac{j}{G}.c$,

[13]Depending on whether $r < G$ or $r > G$, the game class becomes either n-PD or n-Trivial, as confirmed in Eq. (2.46).

respectively. Or, if we consider x to be the global fraction of cooperators (i.e., $(1-x)$ is the fraction of defectors), then we can have $\pi_C(x) = r.\, x.\, c - c$ and $\pi_D(x) = r.\, x.\, c$. Following Eqs. (2.22.1)–(2.23), we can write the average fitness or the average payoff for cooperators and defectors, respectively, as:

$$f_C = \sum\nolimits_{j=0}^{G-1} \binom{G-1}{j} x^j (1-x)^{G-1-j} \pi_C(j+1), \tag{2.44.1}$$

$$f_D = \sum\nolimits_{j=0}^{G-1} \binom{G-1}{j} x^j (1-x)^{G-1-j} \pi_D(j),$$
$$f_D = \sum_{j=0}^{G-1} \binom{G-1}{j} x^j (1-x)^{G-1-j} \pi_D(j). \tag{2.44.2}$$

With some algebraic simplifications, the payoff difference $f_C - f_D$ can be written as

$$f_C - f_D = c\left(\frac{r}{G} - 1\right). \tag{2.45}$$

From Eq. (2.45), it is easy to see that if $r < G$, then D dominates C and the average fitness is zero ($\pi_D(0) = 0$); therefore, the game becomes multi-player PD. Otherwise, C dominates D and it becomes a C-dominated Trivial game. The replicator equation for the game can be written as

$$\frac{dx}{dt} = x(1-x)(f_C - f_D) = cx(1-x)\left(\frac{r}{G} - 1\right). \tag{2.46}$$

This equation justifies only two types of equilibria for this linear PGG, namely $x = 0$ (D-dominant) and $x = 1$ (C-dominant). The sign of $\left(\frac{r}{G} - 1\right)$ determines the destiny of an evolution. In particular, if $r < G$, x always decreases; thus, NE is an all-defectors-state, $x = 0$. SO is evidently an all-cooperators state, $x = 1$. Eventually, we draw SED as

$$\delta = \pi_C(1) - \pi_D(0) = c(r-1) \text{for } r < G. \tag{2.47.1}$$

Contrarily, if $r > G$, x always increases; thus NE is an all-cooperators-state $x = 1$, which is consistent with SO. We draw SED as

$$\delta = \pi_C(1) - \pi_C(1) = 0 \text{for } r > G. \tag{2.47.2}$$

Eq. (2.47.2) is fully justified by the fact that none of the dilemmas ensures zero SED.

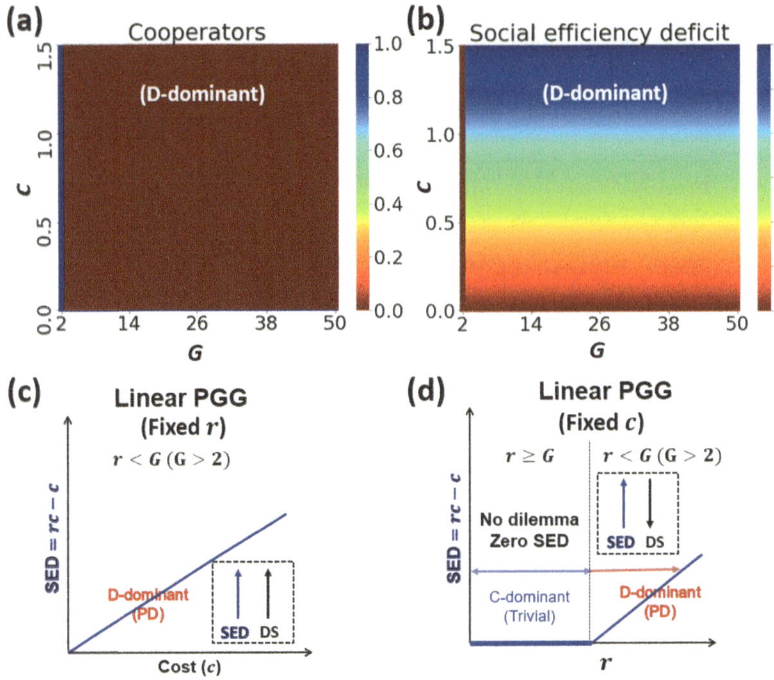

Fig. 2.20 (**a**) G (>2) versus c 2D heatmap for the fraction of cooperators in a PGG (with amplifying factor $r = 2.0$) with a linear-payoff function comprising the C-dominant (blue region) and D-dominant equilibria. (**b**) G versus c 2D heatmap for SED represents a monotonic increase of SED with cost in the case of D-dominant equilibrium; the C-dominant equilibrium has no SED. (**c**) A schematic depiction of SED and DS about the cost for a D-dominant equilibrium with a fixed $r < G$ $(G > 2)$. SED increases linearly with c. Also, SED and DS are positively correlated with the cost. Contrarily, (**d**) SED and DS change inversely with r when cost is fixed

Figure 2.20 shows a heat map of SED under varying G and c (panel (b)). As quantified above, the region of C-dominance, which appears dark blue in panel (a) and where $r > G$, is consistent with a SED of zero in panel (b).

Panel (c) displays how SED varies when c varies, while r remains constant. This shows that SED increases with an increase of c. Although we cannot define DS in PGG, we qualitatively understand that DS increases along with c. This is because the condition $\pi_D(n_c - 1) > \pi_C(n_c)$ becomes more likely, which is the condition incurring n-PD, as we discussed in Figs. 2.6 and 2.7. Note that, in this case, the direction in which SED increases is consistent with that of DS. This is unlike a 2×2 PD game, where DS is inversely proportional to SED, as discussed in Fig. 2.19.

Panel (d) displays how SED varies when r varies with c being held constant. SED increases with the increase of r in the region of $r < G$. We also know that DS increases with the decrease of r, also because the $\pi_D(n_c - 1) > \pi_C(n_c)$ condition becomes more likely. In contrast to panel (c), we should note that the direction of the increase of SED is inconsistent with that of DS.

This discussion of panels (c) and (d) teaches us that the relationship between SED and DS is not as simple as what we observed in 2×2 PD games.

2.3.2.3 PD with Social Viscosity

Let us explore how adding a mechanism to increase social viscosity to an original game changes the SED. We are concerned with the example of a PD with network reciprocity, based on what was discussed in sub-section 2.3.1.

Let us consider the Donor and Recipient game, defined by Eq. (2.42), and apply network reciprocity, quantified by Eqs. (2.38)–(2.39). Eqs. (2.40.1) and (2.40.2) can be replaced by

$$D'_{g\ \mathrm{NR}} = D'_{r\ \mathrm{NR}} = \frac{\Delta}{R} - \frac{kR - 2\Delta}{R(k+1)(k-2)}. \tag{2.48}$$

Because the D & R game belongs to PD, the SED is $\delta = R - 0 = R$. Here, we obtain the explicit relationship between SED (i.e., δ) and DS $\left(D'_g = D'_r\right)$, which is visually shown in Fig. 2.21. With the given values of Δ and k, solving Eq. (2.48) for R allows us to estimate the critical value of SED (let us say; R^*) at which the social dilemma disappears $\left(D'_{g\mathrm{NR}} = D'_{r\mathrm{NR}} = 0\right)$,

$$\delta^* = R^* = \frac{\Delta((k+1)(k-2)+2)}{k}. \tag{2.49}$$

Finally, by increasing δ ($=R$), we examine how DS varies with SED for different degree levels (k) when a social network is introduced in a 2×2 D & R game, as shown in Fig. 2.21. As k increases, the PD with network reciprocity converges toward the default case, i.e., PD without network reciprocity, which is quite likely. If $k = 4$, the dilemma diminishes at $\delta = 1.5$ (which can be justified using Eq. (2.49)), because the lower the degree level, the higher the possibility of cooperation; thus, PD with network reciprocity becomes a trivial game.

Again, let us confirm that SED works well as an indicator of whether or not a social dilemma exists behind the given game. Zero SED implies a Trivial game with a NE according to the SO.

2.3.2.4 Chicken Game

Let us start the discussion from scratch to draw a general result for the Chicken game, where $D_g > 0$ and $D_r < 0$ is imposed. Without loss of generality, the payoff matrix can be formulated as

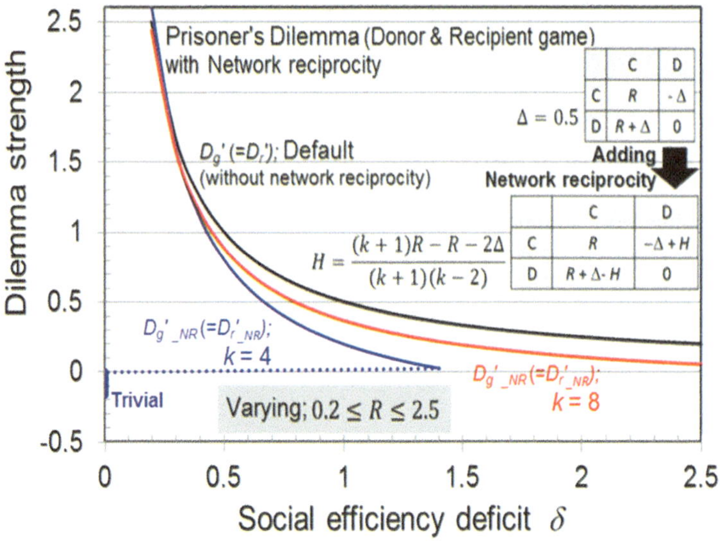

Fig. 2.21 How the relationship between DS (D_g' and D_r') and SED is skewed if network reciprocity is introduced in the case of a 2 × 2 D & R game. At degree level $k = 4$, the dilemma disappears with a little increase in SED and converts to a Trivial game. If we increase the degree level, the game converges to the default PD (without network reciprocity)

$$\mathbf{M} = \begin{pmatrix} R & P - D_r \\ R + D_g & P \end{pmatrix}. \tag{2.50}$$

Let x be the frequency of cooperators. From Eq. (2.21), the internal equilibrium for this game is defined as $x^* = D_r/(D_r - D_g)$. Note that under $D_g > 0$ and $D_r < 0$, the internal equilibrium becomes a NE. Meanwhile, the average payoff of a well-mixed and infinite population can be

$$\overline{\pi}(x) = (x \quad 1 - x)\mathbf{M}\begin{pmatrix} x \\ 1 - x \end{pmatrix} = x\pi_C + (1 - x)\pi_D, \tag{2.51}$$

where the average payoff of the cooperators is $\pi_C = xR + (1 - x)(P - D_r)$ and that for the defectors is $\pi_D = x(R + D_g) + (1 - x)P$. Therefore, $\overline{\pi}(x)$ is

$$\overline{\pi}(x) = x^2(D_r - D_g) + x(D_g - D_r + R - P) + P. \tag{2.52}$$

Now, $\overline{\pi}(x^*) = \frac{D_r^2}{(D_r - D_g)^2}(D_r - D_g) + \frac{D_r}{D_r - D_g}(D_g - D_r + R - P) + P$, which can be simplified as

$$\overline{\pi}(x^*) = x^*(D_g + R - P) + P. \tag{2.53}$$

Taking the derivative with respect to x in Eq. (2.51), we obtain,

$$\frac{d\bar{\pi}(x)}{dx} = 2x(D_r - D_g) + (D_g - D_r + R - P). \tag{2.54}$$

Now, $\left(\frac{d\bar{\pi}(x)}{dx}\right)_{x=x_m} = 2x_m(D_r - D_g) + (D_g - D_r + R - P) = 0$ yields

$$x_m = \frac{1}{2}\frac{(D_g - D_r + R - P)}{D_g - D_r} = \frac{1}{2}\left(1 + \frac{R - P}{D_g - D_r}\right). \tag{2.55}$$

If $R - P < D_g - D_r$, then $\frac{1}{2} < x_m < 1$. However, if $R - P \geq D_g - D_r$ then $x_m \geq 1$; but x_m cannot be more than one. Hence, the SO is achieved at $x_m = 1$ when $R - P \geq D_g - D_r$. This can be summarized as follows:

$$x_m = \begin{cases} \frac{1}{2}\left(1 + \frac{R - P}{D_g - D_r}\right), \text{if } R - P < D_g - D_r \\ 1, \qquad\qquad\qquad\quad \text{if } R - P \geq D_g - D_r \end{cases}. \tag{2.56}$$

Note that the condition $R - P < D_g - D_r$ is equivalent to $2R < S + T$, where $S = P - D_r$ and $T = R + D_g$. Similarly, $R - P \geq D_g - D_r$ is equivalent to $2R \geq S + T$. From Eq. (2.52), the socially optimal payoff is

$$\Pi^{\text{Social Opt.}} = \bar{\pi}(x_m) = \begin{cases} \frac{1}{4}\frac{(D_g - D_r + R - P)^2}{D_g - D_r} + P, & \textit{if } 2R < S + T \\ R, & \textit{if } 2R \geq S + T \end{cases}. \tag{2.57}$$

Presuming a Chicken game with the internal equilibrium as its NE, we already know $\Pi^{\text{NE}} = \bar{\pi}(x^*)$, as given by Eq. (2.52). Therefore, the SED can be written as

$$\begin{aligned} \delta &= \Pi^{\text{Social Opt.}} - \Pi^{\text{NE}} \\ &= \begin{cases} \frac{1}{4}\frac{(D_g - D_r + R - P)^2}{D_g - D_r} - x^*(D_g + R - P), & \text{if } 2R < S + T \\ R - P - x^*(D_g + R - P), & \text{if } 2R \geq S + T \end{cases}. \end{aligned} \tag{2.58}$$

Figure 2.22 shows SED along DS when presuming $\begin{pmatrix} R & \Delta_2 \\ R + \Delta_1 & 0 \end{pmatrix}$. Further imposing $\Delta_1 = \Delta_2 = \Delta > 0$, the game is Chicken. Note that $D_g = -D_r$. At a glance, as observed in the case of PD for Fig. 2.19, SED is inversely proportional to the DS D_g. Note that the curve of DS – SED (i.e., $D'_g - \delta$) is continuous, although at the point of meeting with the condition $2R = S + T$, the first equation of Eq. (2.58) becomes the second one, i.e., shifting from a game needing *ST*-reciprocity ($2R < S + T$) to a game needing *R*-reciprocity ($2R \geq S + T$).

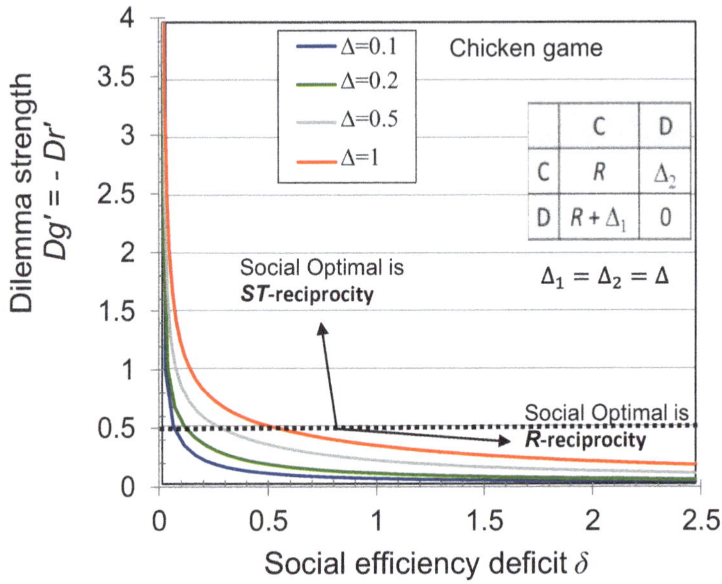

Fig. 2.22 SED vs. DS line graphs show the inverse relationship between DS (D'_g and $-D'_r$) and SED for several Δ (>0) values in the case of a 2×2 Chicken game

2.3.3 Application of SED

In the previous sub-section, we shed light on a 2×2 game and PGG that are both standard template models in the evolutionary game. Here, in this sub-section, we touch upon how SED can be applied to a real problem in which a realistic and complex game structure, unlike the static game represented by a simple payoff matrix, is presumed. The main theme of the present book is epidemiology and the vaccination game. Thus, it would be meaningful to discuss how the concept of SED can be used in the vaccination game. The vaccination game, which will be introduced in detail in Chap. 3, is one of the hottest areas of application of evolutionary game theory.

Here, let us briefly explain the structure of the vaccination game with the MFA, in which an individual is exposed to the risk of a seasonal influenza-like communicable disease spreading, and decides either to commit to a pre-emptive vaccination at the beginning of each season (implying cooperation) or not (defection). A vaccination works stochastically to bring perfect immunity (or nothing). An individual's decision whether or not to take a vaccine depends on the relative cost of vaccination $C_r(C_r = C_V/C_I;\ 0 \leq C_r \leq 1$; where C_V and C_I are the costs of vaccination and infection, respectively; without loss of generality, C_I can be set to 1) and vaccine effectiveness e ($0 \leq e \leq 1$). A cooperator (vaccinator) always obtains perfect immunity for $e = 1$. The spread of the infection obeys the well-known SIR dynamics (explained in detail in Chap. 3). Namely, in each season, a small initial fraction of

Table 2.3 Four fractions of individuals with their respective payoffs (values within brackets)

Strategy/state	Healthy	Infected
Vaccinated (V)	HV ($-Cr$)	IV($-C_r - 1$) ()
Non-vaccinated (NV)	SFR (0) (0)	FFR (-1)

infected individuals triggers the spread of the disease into the population with cooperators (vaccinators) and defectors co-existing. SIR dynamics, as formulated by a set of dynamical equations, control what fraction of individuals are consequently infected before a disease is eradicated. At the end of each epidemic season, the entire population is classified into four fractions of individuals: HV—vaccinated and healthy, IV—vaccinated but infected, SFR—successful free riders (non-vaccinated but healthy), and FFR—failed free riders (non-vaccinated and infected). Table 2.3 summarizes the four classes observed at the end of a season and their respective payoffs. These groups are prone to updating their strategy (committing to vaccinating or not) at the beginning of each season by evaluating their payoffs based on last season's experience following a so-called individual-based risk assessment (IB-RA) rule.[14]

Each season is steered by the local timescale (the days elapsed in a season) in the extent of a disease spreading, final epidemic size (FES), at each season, while repeated season handled by the global time-series lets fraction of vaccinators, called vaccination coverage, time evolve. We perform a simulation episode until we reach a social equilibrium (Nash equilibrium; NE), at which point we measure the FES, vaccination coverage (VC), and average social payoff (ASP) in the final season. ASP results from the infected cost and the vaccination cost over the entire society. We perform such simulation by varying C_r and e.

2.3.3.1 Derivation of SED

We calculate the ASP at a social equilibrium (NE) for every combination of C_r and e. We also estimate the ASP at a SO without considering the game approach by taking the maximum ASP for each pair-setting of C_r and e for vaccination coverage (x) ranging from 0 to 1 and then calculate the difference between ASP at SO and NE (as defined in Eq. (2.59)) to derive SED (i.e., δ). Figure 2.23 schematically explains the derivation of SED for a vaccination game. Suppose that $\mathrm{ASP}^{\mathrm{NE}}_{C_{r_i}, e_j}$ and $\mathrm{ASP}^{\mathrm{Social\ Opt.}}_{C_{r_i}, e_j}$ stand for the ASP at NE (panel (a)) and ASP (without game approach)

[14]Although the details are also explained in Chap. 3, it can be seen as a sort of social-learning process. An individual references their own payoff and that of their neighbor as quantified by MFA, and determines whether to copy their neighbor's strategy or keep their own based on the Fermi function, of which the argument is defined as the payoff difference.

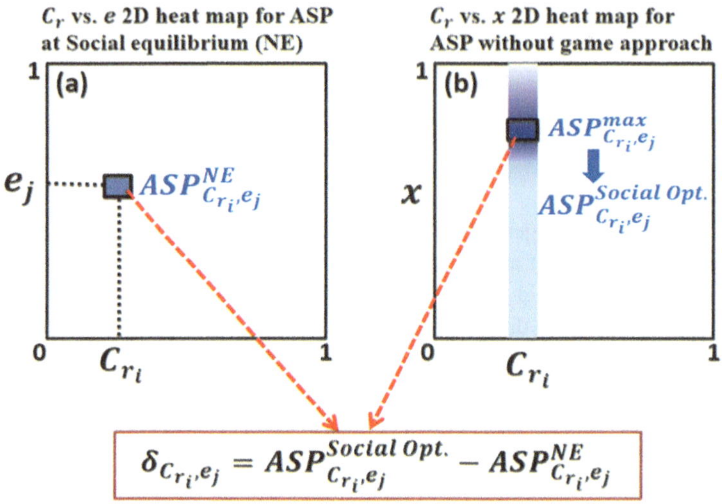

Fig. 2.23 Schematic of the derivation of SED. Panel (**a**) represents the average social payoff (ASP) at NE for a (C_{ri}, e_j) pair, whereas panel (**b**) depicts the ASP (without game approach) for that particular C_{ri} with vaccination coverage x $(0 \leq x \leq 1)$ for each e_j; at this C_{r_i}, the blue region surrounded by the rectangle is assumed to have the maximum ASP, which is called the ASP at social optimum (SO) for the pair (C_{ri}, e_j). Finally, the difference between ASP at SO and ASP at NE yields the SED at (C_{ri}, e_j)

at SO (panel (b)), respectively, for each pair (C_{ri}, e_j). Then, we can define SED at (C_{ri}, e_j) as

$$\delta_{C_{r_i},e_j} = \mathrm{ASP}^{\text{Social Opt.}}_{C_{r_i},e_j} - \mathrm{ASP}^{\text{NE}}_{C_{r_i},e_j}. \tag{2.59}$$

2.3.3.2 Discussion

By estimating the SEDs for each pair of (C_{r_i}, e_j), we generate a 2D heat map for SED to visualize how SED varies as a function of C_r and e (Fig. 2.24). The triangular region enclosed by blue dashed lines has no SED (consequently no dilemma) because in this case, a lower vaccine effectiveness does not inspire people to vaccinate at all, subsequently leading to a D-dominant trivial state as SO (panel (c)). With the game approach, the counterpart of this triangular region in the VC heat map (panel (b)) also has a D-dominant NE that, in accordance, yields an identical ASP (panels (c)–(d)) to that of observed at SO; that is, the payoff at NE cannot be further improved, and consequently, possesses no social dilemma at all. However, another region enclosed by red dashed lines (panel (a)) (especially for low cost) appears to have no SED. The equivalent region in the VC-phase diagram (panel (b)) displays a C-dominant NE; that is, all people vaccinate, albeit the effectiveness is not

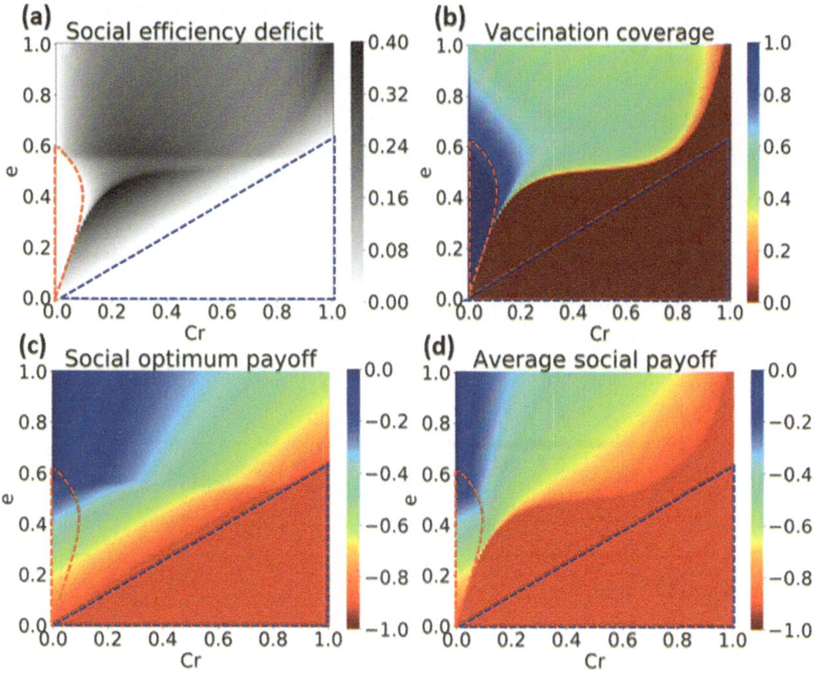

Fig. 2.24 Panel (**a**) C_r versus e 2D heat-map of SED for the vaccination game. The regions enclosed by blue (triangular region) and red dashed lines have no SED, and consequently no social dilemma; (**b**) C_r versus e 2D heat-map of vaccination coverage (VC) at social equilibrium (NE) displaying regions with blue and red dashed lines as D-dominant and C-dominant in NE, respectively; (**c**) C_r versus e 2D heat-map of ASP at social optimum (SO); and (**d**) C_r versus e 2D heat-map of ASP at social equilibrium (NE)

very high. Moreover, the ASPs (panels (c)–(d)) associated with this region are almost identical at NE as well as SO, which therefore confers no SED—i.e., no social dilemma is occurring. This implies that the region featuring no SED is C-dominant trivial, which differs from the previous blue-triangle region, in which no SED resulted from the D-dominant trivial-game structure.

Meanwhile, the remaining region in panel (a) possesses certain levels of SED that indicate the presence of a social dilemma whereby we can perceive non-monotonic changes in SED if the vaccination cost is relatively low. These non-monotonic phenomena can be explained using the line graphs in Fig. 2.25 that reveal how SED, VC, and ASP at SO and NE (with different cost levels) vary as functions of e. Clearly, VC is correlated with C_r and e. With a lower vaccination cost, VC becomes maximal, even with a medium level of effectiveness; however, afterward, it monotonically decreases because the situation with a more effective vaccine might inspire

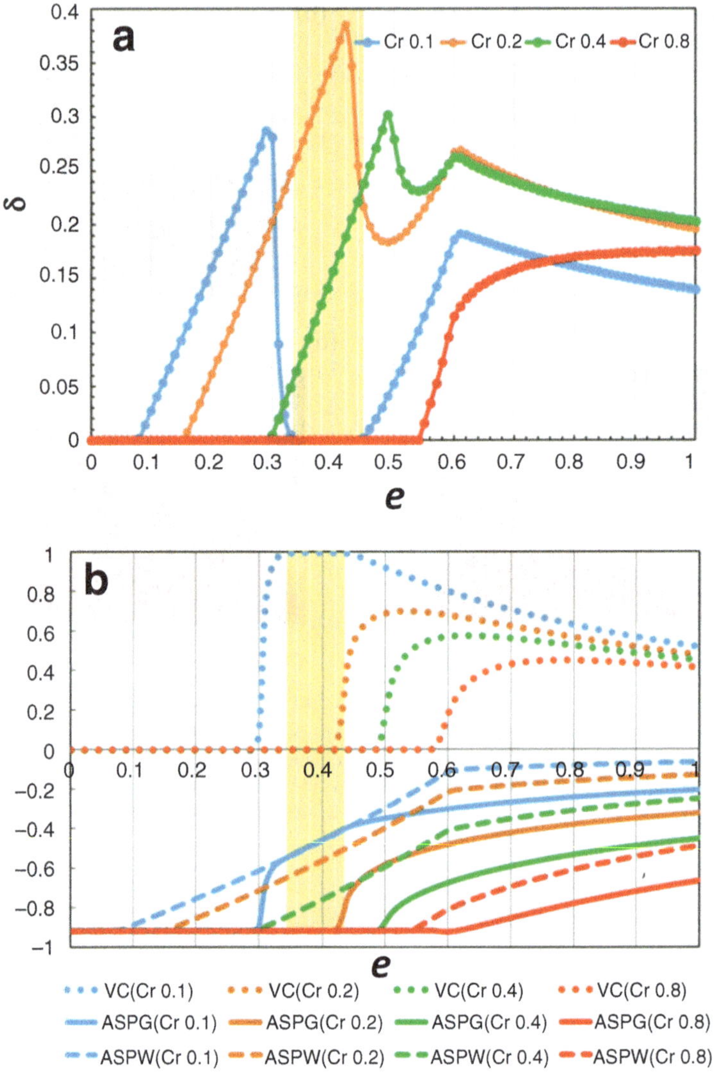

Fig. 2.25 Panel (**a**) e versus δ line graphs showing how SED changes with different cost levels C_r; (**b**) line graphs above the x-axis represent the vaccination coverage (VC) at NE corresponding to several costs, C_r, and line graphs below the x-axis display the corresponding ASPs at NE (solid lines) and SO (dashed lines) when e varies from 0 to 1. ASPG and ASPW stand for ASPs with and without game approaches, respectively. The region filled in light yellow depicts the C-dominant NE (everyone commits to a vaccine) wherein both payoffs merge; this corresponds to the cost $C_r = 0.1$, and consequently there is no SED (no dilemma)

some people to freeride on so-called herd immunity.[15] These phenomena yield an increasing tendency in ASP with *e*. By examining the difference between the two payoffs in Fig. 2.25b (below the x-axis), one can easily conceive of the mechanism of the non-monotonic tendency in SED about *e* (panel (a)). More specifically, let us consider the case $C_r = 0.1$ (blue-colored line graphs). Here, the highest payoff difference can be observed when *e* is nearly 0.3. This, in turn, exhibits the maximum gap between two ASPs (panel (b)); in other words, the maximum SED (panel (a)); afterwards, both payoffs merge (blue solid and dashed lines coincide within the light-yellow region in Fig. 2.25b) for a certain range of *e* that accordingly leads to zero SED (no social dilemma). However, another SED peak appears at $e = 0.6$, which can also be justified by observing the payoff difference (blue solid and dashed lines) at $e = 0.6$ from Fig. 2.25b. Conversely, the situation with an expensive vaccine exhibits no SED (consequently no dilemma) until the vaccine effectiveness reaches a certain level, because no one here intends to take an expensive vaccine that without a satisfactory level of effectiveness.

All discussions above shed clear light on how SED contributes to evolutionary game theory. This enables complex games to be analyzed in terms of whether a social dilemma works behind the model's surface from the mathematical viewpoint, and to elucidate the nature of this dilemma. The concept of DS is certainly a powerful theoretical framework for quantifying a social dilemma. However, it is only applicable to 2-player and 2-strategy games. SED lacks this drawback because it applies universally to any games, regardless of the number of players and strategies or the complexity. Thus, it is very powerful.

References

Alexander R (1987) The biology of moral systems. Aldine De Gruyter, New York

Ebel H, Bornholdt S (2002) Coevolutionary games on networks. Phys Rev E 66:056118

Hamilton WD (1963) The evolution of altruistic behavior. Am Nat 97:354–356

Hamilton WD (1964) The genetical evolution of social behavior. J Theor Biol 7:1–16

Hassell MP, Comins HN, May RM (1994) Species coexistence and self-organizing spatial dynamics. Nature 313:10–11

Kabir KMA, Tanimoto J, Wang Z (2018) Influence of bolstering network reciprocity in the evolutionary spatial prisoner's dilemma game: a perspective. Eur Phys J B 91:312

Maynard Smith J (1976) Group selection. Q Rev Biol 51:277–283

Nowak MA (2006) Five rules for the evolution of cooperation. Science 314:1560–1563

Nowak MA, May RM (1992) Evolutionary games and spatial chaos. Nature 359:826–829

Nowak MA, Sigmund K (1998) Evolution of indirect reciprocity by image scoring. Nature 393:573–577

Ohtsuki H, Nowak MA (2006) The replicator equation on graphs. J Theor Biol 243:86–97

[15] With a vaccination coverage above a threshold level, an individual with no vaccination can hardly be infected. Commitment to vaccination from the social majority enables those non-vaccinating individuals to be protected from infection. This is called "herd immunity" and discussed in Chap. 3.

Ohtsuki H, Hauert C, Lieberman E, Nowak MA (2006) A simple rule for the evolution of cooperation on graphs and social networks. Nature 441:502–505

Santos FC, Pacheco JM (2005) Scale-free networks provide a unifying framework for the emergence of cooperation. Phys Rev Lett 95:098104

Santos FC, Pacheco JM, Lenaerts T (2006) Cooperation prevails when individuals adjust their social ties. PLoS Comput Biol 2:1284–1291

Slatkin M, Wade MJ (1978) Group selection on a quantitative character. Proc Natl Acad Sci U S A 75:3531–3534

Tanimoto J (2014) Mathematical analysis of environmental system. Springer, Tokyo

Tanimoto J (2015) Fundamentals of evolutionary game theory and its applications. Springer, Tokyo

Tanimoto J, Sagara H (2007) Relationship between dilemma occurrence and the existence of a weakly dominant strategy in a two-player symmetric game. Biosystems 90(1):105–114

Taylor M, Nowak MA (2007) Transforming the dilemma. Evolution 61(10):2281–2292

Traulsen A, Nowak MA (2006) Evolution of cooperation by multilevel selection. Proc Natl Acad Sci U S A 103:10952–10955

Trivers R (1971) The evolution of reciprocal altruism. Q Rev Biol 46:35–37

Trivers R (1985) Social evolution. Benjamin/Cummings, Menlo Park, CA

Wakiyama M, Tanimoto J (2011) Reciprocity phase in various 2×2 games by agents equipped with 2-memory length strategy encouraged by grouping for interaction and adaptation. Biosystems 103(1):93–104

Wang Z, Kokubo S, Tanimoto J, Fukuda E, Shigaki K (2013) Insight on the so-called spatial reciprocity. Phys Rev E 88:042145

Wang Z, Kokubo S, Jusup M, Tanimoto J (2015) Universal scaling for the dilemma strength in evolutionary games. Phys Life Rev 14:1–30

Williams GC (1996) Adaption and natural selection: a critique of some current evolutionary thought. Princeton Univ. Press, Princeton, NJ

Wilson DS (1975) A theory of group selection. Proc Natl Acad Sci U S A 72:143–146

Wynne-Edwards VC (1962) Animal dispersion in relation to social behavior. Oliver and Boyd, Edinburg

Yamauchi A, Tanimoto J, Hagishima A (2010) What controls network reciprocity in the prisoner's dilemma game? Biosystems 102(2-3):82–87

Chapter 3
Fundamentals of Mathematical Epidemiology and the Vaccination Game

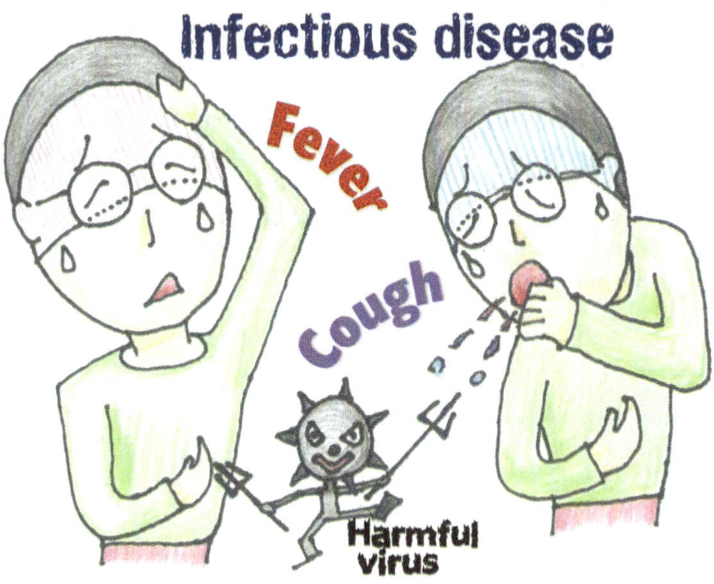

This chapter presents the fundamentals of mathematical epidemiology and vaccination games. Mathematical epidemiology is based on a set of ordinary differential equations (ODEs) that are used to depict the global time evolution of a disease. This field has roots back to the SIR model of Kermack and McKendrick (1927), which was established in 1927. The vaccination game is a unified framework that combines mathematical epidemiology with evolutionary game theory in order to describe how human behavior influences the dynamics of disease spread and how these dynamics affect behavior in turn. Fu and his colleagues (2011) introduced the "vaccination game" framework based on multi-agent simulation (MAS).

J. Tanimoto, *Sociophysics Approach to Epidemics*, Evolutionary Economics and Social Complexity Science 23, https://doi.org/10.1007/978-981-33-6481-3_3

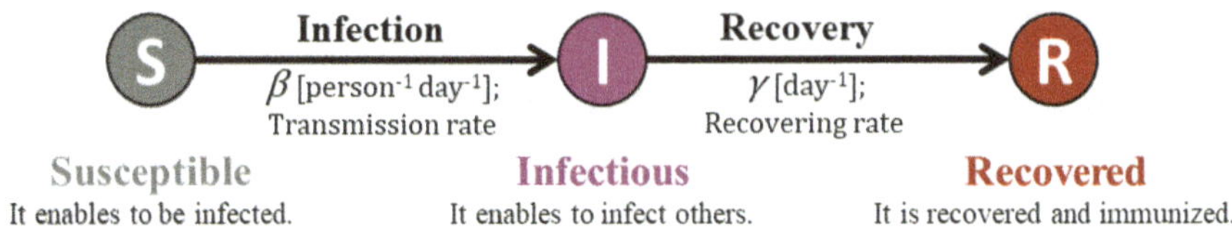

Fig. 3.1 Block-chart of the SIR model

3.1 Basic Model: SIR, SIS, and SEIR

3.1.1 Formulation of the SIR Model

Let us presume a well-mixed (i.e., non-spatially structured) and infinite population. The assumption of such a hypothetical mother population has been called the "mean-field approximation" (MFA)) in the field of physics. Because we are concerned with a sufficiently short time-scale (i.e., short duration) as compared with a human life, we can presume that the total population is constant, which implies that the effects of both birth rate and mortality can be negligible. We consider the time evolution of subpopulations of susceptible, infectious, and recovered (sometimes called "removed") people, denoted by $S(t)$, $I(t)$, and $R(t)$, respectively. These categories refer to people who have not yet been infected or immunized (i.e., are capable of infection), literally infectious people, and people who have recovered from the disease and are now immune, respectively. We can establish each dynamical equations as:

$$\begin{aligned}\frac{dS(t)}{dt} &= -\beta S(t)I(t);\\ \frac{dI(t)}{dt} &= \beta S(t)I(t) - \gamma I(t);\\ \frac{dR(t)}{dt} &= \gamma I(t);\end{aligned} \tag{3.1}$$

where β is the transmission rate [day^{-1} person^{-1}] and γ is the recovery rate [day^{-1}] (meaning the inverse of the number of days needed to recover from the disease state). The total population, N, should be normalized to unity. Thus, the following relation is imposed as a constraint:

$$N = S(t) + I(t) + R(t) = 1. \tag{3.2}$$

In this theoretical framework, the states are called "compartments," and so the framework is called a compartment model or an ODE model The flow between the compartments of the SIR model is drawn in Fig. 3.1.

Supposing a transient situation in which a small number of initial infectious people $I(0)(\cong 0)$ emerge (i.e., there is an initially large susceptible population, $S(0)$

($\cong$1), and initially few recovered people, $R(0)(\cong 0)$), we can establish the transient dynamics of $I(t)$ by

$$\frac{dI(t)}{dt} = (\beta S(0) - \gamma)I(t). \tag{3.3}$$

This equation is easily solved as

$$I(t) = I(0) \cdot \exp\left[(\beta S(0) - \gamma)t\right], \tag{3.4}$$

which implies that the transient infectious population increases in an exponential manner in obeyance to Malthus's law. Obviously, for

$$\frac{\beta S(0)}{\gamma} \cong \frac{\beta}{\gamma} \equiv R_0 > 1, \tag{3.5}$$

the disease cannot be suppressed, i.e., it becomes endemic (nevertheless $I(\infty) = 0$, because we presume no new individuals entering S). By contrast, $R_0 < 1$ ensures that the disease goes to extinction. R_0 is called basic reproduction number, which physically describes the number of secondarily infected people triggered by a single infected person.

In the following sections and chapters, we introduce various ODE models, including SEIR, ones containing other compartments such as Q (quarantine), and much more complex models with many compartments. In such cases, the focal dynamics for defining R_0 cannot be represented by a single like Eq. (3.5). Rather, we need to analyze the "next-generation matrix," $\mathbf{FV}^{-1}$. $\mathbf{F}$ comes from that Jacobi matrix at disease-free equilibrium (DFE)) for the "newly infected generation vector," while $\mathbf{V}$ comes from that of the "transferring vector." The focal matrix and vector can be established from the original set of dynamical equations like Eq. (3.1). The counterpart threshold to $\frac{\beta S(0)}{\gamma}$ in Eq. (3.5) (i.e., the basic reproduction number, R_0), can be drawn from the maximal eigenvalue of $\mathbf{FV}^{-1}$, which is called the spectral diameter and denoted by the output of function $\rho[]$. The product $\mathbf{FV}^{-1}$ physically indicated "newly generated infected people up until the eradication of the disease," since $\mathbf{F}$ roughly represents the infection-generation rate from a physics point of view, whereas $\mathbf{V}^{-1}$ roughly implies the expected time at which the infection is eradicated. Noting this point, it would be understandable for $\rho[\mathbf{FV}^{-1}]$ to mean the basic reproduction number.[1]

Incidentally, the first and the third equations of Eq. (3.1) can eliminate $I(t)$ through the rendering

[1]A reader can consult any standard textbooks concerned with mathematical epidemiology. One such example is: Brauer et al. (2019).

$$\frac{1}{S(t)}\frac{dS(t)}{dt} = -R_0\frac{dR(t)}{dt}. \tag{3.6}$$

By taking the integral with respect to t for both sides of Eq. (3.6) with a time range of $[0, T]$ and presuming that $T \to \infty$, we finally obtain

$$S(\infty) = S(0) \cdot \exp\left[-R_0(R(\infty) - R(0))\right] = \exp\left[-R_0 R(\infty)\right]. \tag{3.7}$$

To this end, we can identify the final epidemic size (FES)) as below:

$$R(\infty) = 1 - \exp\left[-R_0 R(\infty)\right]. \tag{3.8}$$

Despite being a transcendental equation, Eq. (3.8) can be solved by numerically. In particular, we can predict the FES using only R_0.

3.1.2 *Herd Immunity*

Let us presume a situation in which a fraction x of the population is immunized at the beginning of the epidemic. Those immunized people may have been brought to immunity by a pre-emptive vaccination (artificial immunity) or by being infected once before (immunity obtained by infection; in this book, we call this naturally obtained immunity). Subsequently by referring to Eq. (3.8), we can identify FES by

$$R(x, \infty) = (1 - x)\left(1 - \exp\left[-R_0 R(x, \infty)\right]\right). \tag{3.9}$$

Hence, the critical immunity fraction meets with $R_0 < 1$ in order to eradicate the epidemic; x_H, is quantified by

$$x_H = 1 - \frac{1}{R_0}. \tag{3.10}$$

For instance, if the pre-emptive-vaccination coverage in a certain population is more than x_H, an individual who does not commit to a vaccination can avoid infection in a season, which is called an effect of herd immunity. Figure 3.2 visually shows the time evolution of S, I, and R, as well as the herd-immunity level.

3.1.3 *Formulation of the SIS Model*

Unlike the SIR process, the SIS model is premised on the idea that an infected individual does not acquire a lasting immunity, but simply returns to the susceptible state (see Fig. 3.3). Thus, its dynamics can be depicted as:

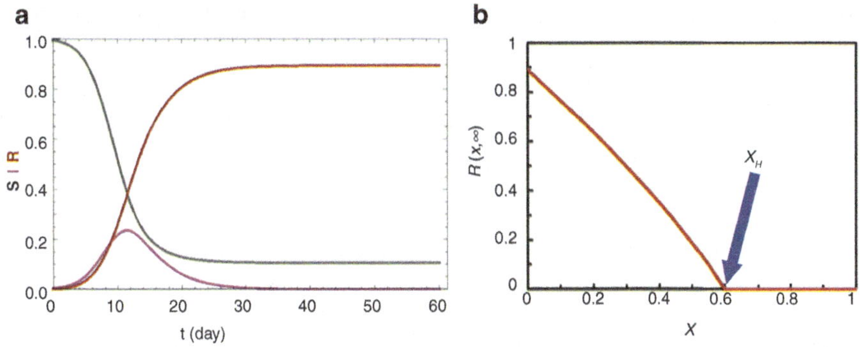

Fig. 3.2 Numerical examples of the SIR model, presuming $R_0 = 2.5$; (**a**) time evolutions of respective compartments $S(t)$, $I(t)$, and $R(t)$; (**b**) $R(x, \infty)$ varying with the initial fraction of immunized people, x. The intersection with the X-axis indicates the level of herd immunity, x_H

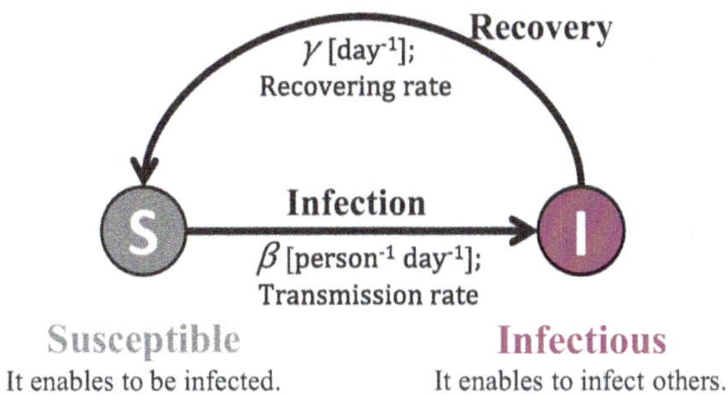

Fig. 3.3 Block-chart of the SIS model

$$\begin{aligned}\frac{dS(t)}{dt} &= -\beta S(t)I(t) + \gamma I(t),\\ \frac{dI(t)}{dt} &= \beta S(t)I(t) - \gamma I(t).\end{aligned} \tag{3.11}$$

We presume that the total population is preserved at $N = 1$:

$$N = S(t) + I(t) = 1. \tag{3.12}$$

Substituting Eq. (3.12) into Eq. (3.11) to eliminate S, we find the following logistic equation:

$$\frac{dI(t)}{dt} = (\beta - \gamma)I(t) - \gamma I(t)^2. \tag{3.13}$$

This equation is easily solved. Taking $t \to \infty$, we can conclude that if $R_0 = \frac{\beta}{\gamma} < 1$, the disease goes to extinction, i.e., $I(\infty) = 0$, which is called the DFE; if $R_0 = \frac{\beta}{\gamma} > 1$, the disease does not die out, i.e., $I(\infty) = 1 - \frac{1}{R_0}$; this is called endemic equilibrium (EE)).

3.1.4 Formulation of the SEIR Model

We started with the SIR model, which is the most fundamental compartmental ODE model. The SIS model should be regarded as comparably important to the SIR model, since it has a recursive loop from I to S. This is substantial despite being a subtle difference. The SIR model only allows an individual to flow along one direction from S to R. Hence, at equilibrium, i.e., $t \to \infty$, there are no infectious individuals, $I(\infty) = 0$, even though $R_0 = \frac{\beta}{\gamma} > 1$ is valid. Conversely, the equilibrium of the SIS model presuming $R_0 = \frac{\beta}{\gamma} > 1$ allows a non-zero $I(\infty)$ due to the existence of the backward loop to S.

In this sub-section, let us introduce another standpoint. Some infectious diseases have a latent period between the states of S and I, in which an individual is exposed to the pathogen and ingests it, but shows no symptoms (under the incubation period) and capacity to infect others (under the latent period); this is called the exposed state (E). In a nutshell, an individual in state E is infected but latently infectious (not infectious). The dynamical equations of such a process, called the SEIR model (see Fig. 3.4), can be represented as

$$\begin{aligned} \frac{dS(t)}{dt} &= -\beta S(t)I(t), \\ \frac{dE(t)}{dt} &= \beta S(t)I(t) - \epsilon E(t), \\ \frac{dI(t)}{dt} &= \epsilon E(t) - \gamma I(t), \end{aligned} \tag{3.14}$$

where ϵ is the probability of an exposed individual becoming infectious, with the inverse, $1/\epsilon$, indicating the latent period [day]. One worthwhile thing to confirm is that the latent period differs from what is called the incubation period. As Fig. 3.5 explains, the latent period relates to whether a focal person has the capability of infecting others, while the incubation period relates to whether their situation is symptomatic. In general, there are many diseases for which the infectious state starts in the incubation period, and/or whose infectious period ends during the

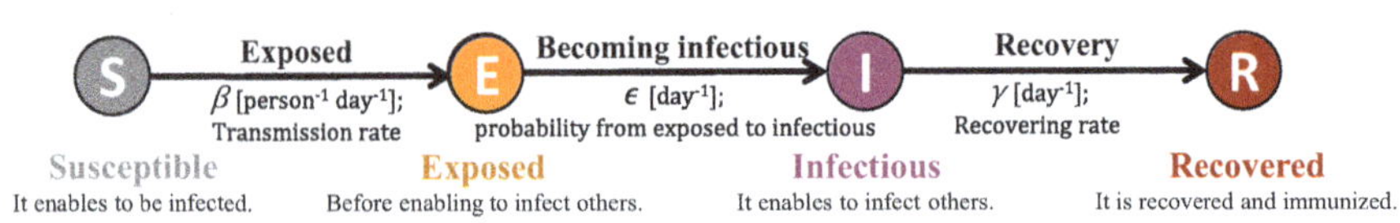

Fig. 3.4 Block-chart of the SEIR model

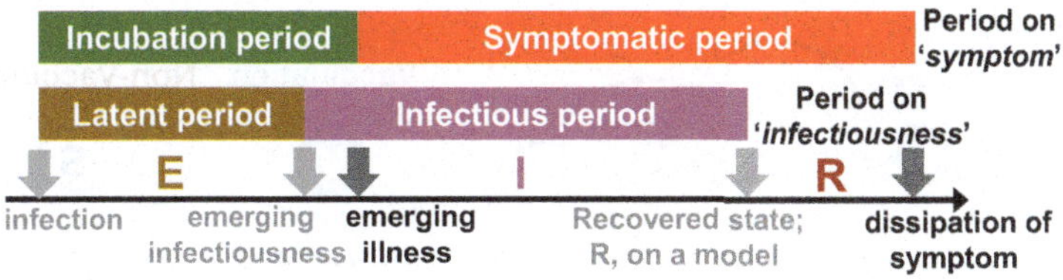

Fig. 3.5 Meanings of various symptom- and infectiousness-related periods in the SEIR model

symptomatic period; however, there are several diseases for which the incubation and latent periods can be consistent.

Obeying Eq. (3.14), the basic reproduction number, R_0, can be given as

$$R_0 = \frac{\epsilon\beta}{\epsilon\gamma} = \frac{\beta}{\gamma}. \tag{3.15}$$

To briefly sum up, we have learned only three fundamental ODE models: SIS, SIR, and SEIR. It can be said, though, that we have learned every aspect of the ODE model. Because of its inherent flexibility, the concept of ODE modeling gives you a powerful tool to depict any epidemic processes by adding other compartments.

3.2 Theoretical Framework of a Vaccination Game

This section introduces the so-called vaccination game from a theoretical standpoint. Although there had been previous seminal works, Fu and his colleagues are generally credited with the inception of the "vaccination game" and with solidly and clearly establishing its framework in 2011 (Fu et al. 2011). For this, they utilized a MAS approach. Yet the concept of a vaccination game does not rely on whether a MAS or ODE-based approach is assumed. Below, we present the theoretical framework of the vaccination game at first prior to the MAS approach, which is extended a bit from what Fu et al. presumed. As an important premise for the ODE version of the vaccination game, we assume that a population is well-mixed and infinite, and thus does not have any social network as its underlying topology. Later, Sect. 3.4 relaxes this assumption.

In the baseline vaccination game, we assume two timescales: the local timescale and the global timescale. At the local timescale, the spread of an epidemic is of concern, whereas the global timescale concerns an individual's behavior toward the epidemic. More precisely, the global timescale administrates repeating seasons in view of the alteration of human behavior, while the local timescale depicts the daily evolution of the epidemic. Between consecutive seasons (i.e., at the end of a one season), an individual alters their strategy (meaning their attitude or behavior, e.g., whether to commit to a vaccine or to do nothing in the next season) based on their experience in the current season. This implies the vaccination introduced here is a

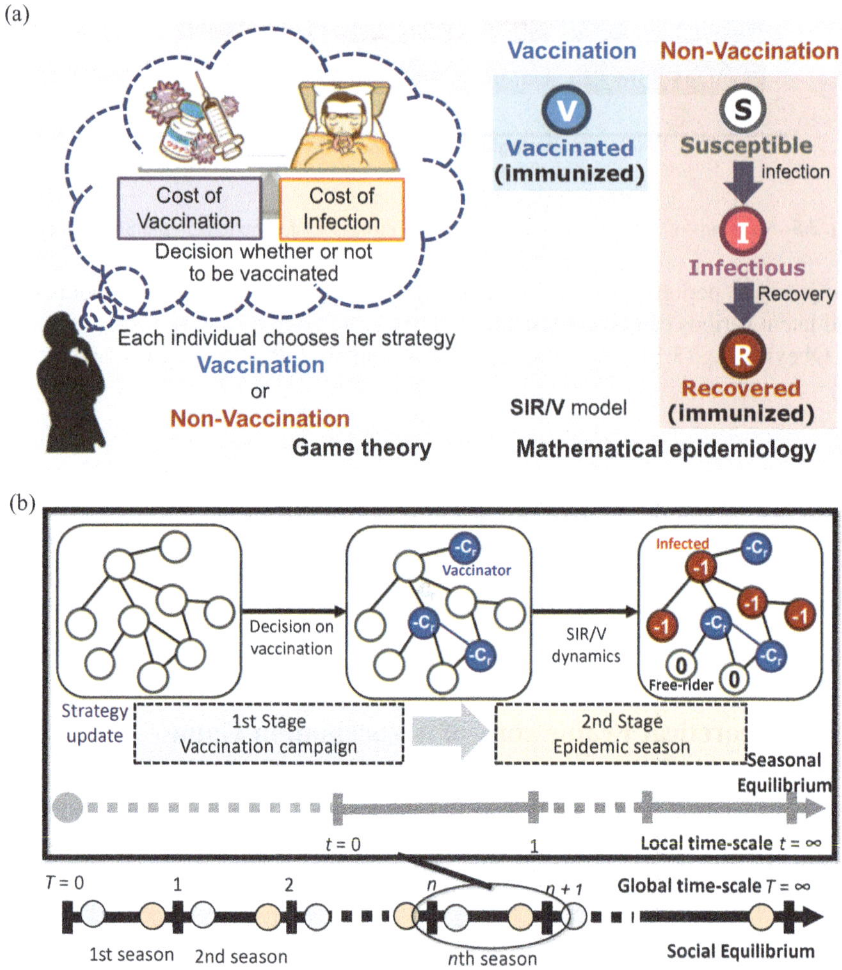

Fig. 3.6 Holistic structure of the vaccination game; (**a**) the vaccination game dovetails a mathematical-epidemiological model (ODE model) with evolutionary game theory; (**b**) the former frame is handled at the local time-scale, while the latter is handled at the global time-scale

pre-emptive vaccination. This framework may be appropriate for analyzing a seasonal epidemic like influenza. For the sake of the explanation below, we rely on the SIR dynamics to depict what happens in a single season. However, depending on the features of the target epidemic, any ODE model, whether SIR, SIS, SEIR, or other, can be applied to the framework. To describe the time evolution of individuals' attitudes on a global timescale, we rely on evolutionary game theory. Figure 3.6 schematically presents the holistic structure of the vaccination game.

As is well recognized, the protective efficacy of a flu vaccine persists for less than a year because of waning antibodies and year-to-year changes in the circulating

Table 3.1 Payoff structure determined at the end of an epidemic season

Strategy/state	Healthy	Infected
Vaccinated	$-C_r$	$-C_r - 1$
Non-vaccinated	0	-1

virus. Therefore, under a voluntary vaccination program where the vaccine cost is not covered by the public-h ealth insurance but by individual expense, each individual must decide every year whether to be vaccinated. Thus, the dynamics of our model consist of two stages: the first stage is a vaccination campaign and the second is an epidemic season.

The First Stage: The Vaccination Campaign
Here, in this stage, each individual decides whether to get vaccinated before the beginning of the seasonal epidemic, i.e., before any individuals are exposed to the epidemic strain. Vaccination imposes a cost, C_{v}, upon each individual who decides to be vaccinated. The cost of vaccination includes the monetary cost and other perceived risks, such as adverse side effects. If an individual is infected, they incur the cost of infection, C_{i}.

The Second Stage: The Epidemic Season
Here, at the beginning of this stage, the epidemic strain enters the population, and a number I_0 of randomly selected susceptible individuals are identified as the initially infected ones. Then, the epidemic spreads according to SIR dynamics. As discussed in the following sub-section, vaccination does not deliver a perfect state of immunization to a vaccine because of the stochastic feature of the vaccination. Although an individual who does not commit to a vaccination is exposed to a relatively higher risk of infection, even a vaccinated individual may be infected during a season. If this is the case, the vaccinated individual pays not only the vaccination cost, C_{v}, but also the infection cost, C_{i}. By contrast, an individual can avoid infection despite not committing a vaccination, paying nothing at all. Such an individual should be called a successful free rider who relies upon herd immunity without any cost burden. To simplify the evaluation of each individual's payoff, without loss of generality, we rescale the cost by defining a relative cost of vaccination, $C_r = C_v/C_i$ $(0 \leq C_r \leq 1;$ $C_i = 1)$. Table 3.1 delivers the payoff structure of the current vaccination game, wherein there are four states: vaccinated and healthy, vaccinated but infected, non-vaccinated but healthy, and non-vaccinated and infected.

The most important target of the vaccination game is the social dilemma called the "paradox of epidemiology" or "vaccination dilemma," which is schematically shown in Fig. 3.7.

Any rational individual has a strong incentive to exploit the public good by free-riding on herd immunity. However, this incentive, wherein the individual pays nothing but still obtains a benefit, only works as long as the majority in the community spontaneously commit to the vaccination. By contrast, if the majority disregards vaccination, then doing nothing is no longer a better option, because infection is likely. Thus, spontaneous vaccination becomes the rational option. This

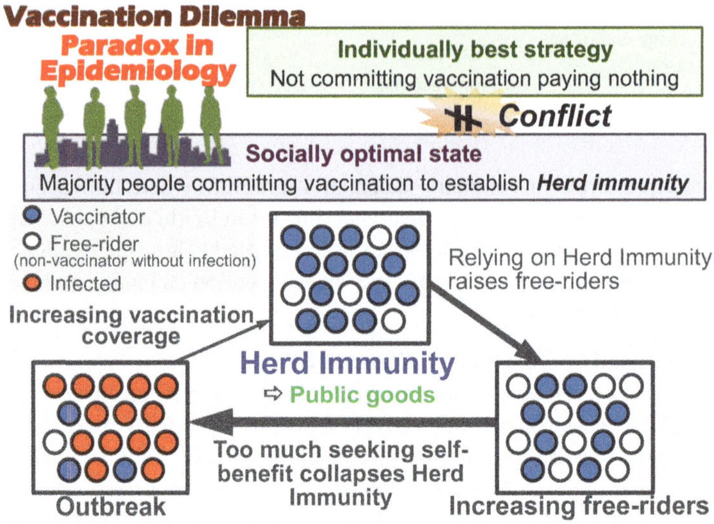

Fig. 3.7 Visual explanation of the 'paradox of epidemiology' of the 'vaccination dilemma', where herd immunity is likened to a public good

difference implies that the best choice for an individual is to always adopt the strategy of the social minority: either to free ride when herd immunity is well established or to get vaccinated when most people neglect to do so. This situation obviously contains the structure of a minority game, as Vardavas pointed out (Vardavas et al. 2007). A minority game,[2] originally defined as the El Farol Bar problem (Brian Arthur 1994), is a typical social dilemma that can be observed in many real situations; the most heavily concentrated applications are in financial markets, including currency and stock markets. In a minority game, any individual has the incentive to adopt the strategy of the minority under any circumstance.

3.2.1 Two Models to Represent Stochastic Vaccination: Effectiveness and Efficiency

An unvaccinated individual faces the risk of being exposed to infection during a season, whereas the risk to a vaccinated individual depends on the vaccination. If a vaccine provides an individual with perfect immunity to the disease during a season, a vaccinated individual has no risk of infection. Although some studies have presumed this situation for simplicity, it would not be realistic. In general, vaccination works in a stochastic way. A model of this might use one of the two concepts.

[2]A reader can consult with, for example; Challet et al. (2005).

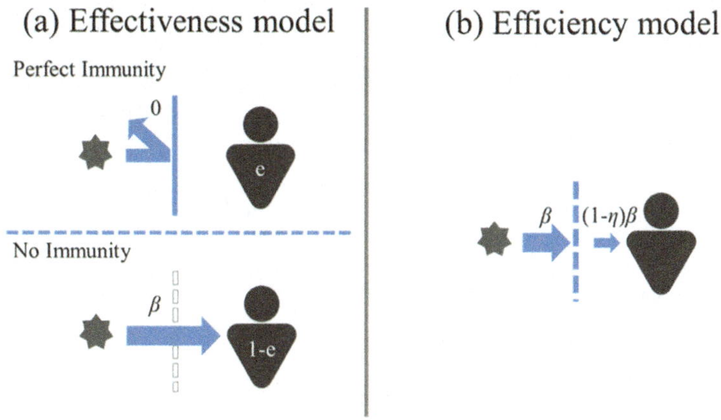

Fig. 3.8 Schematic explanation of (**a**) an effectiveness model and (**b**) an efficiency model

One is stochasticity obeying a lottery-like process, meaning that whether an injection will bring perfect immunity or not is stochastic. Let us call this concept the "effectiveness model." Let us introduce effectiveness, e, as the probability of brining perfect immunity through vaccination, meaning, so-to-speak, the "winning rate" of an injection. Another concept is that a vaccination stochastically works to protect an individual from being infected. We call this concept an "efficiency model." Separate from effectiveness, efficiency (η) indicates how a vaccination can reduce transmission rate, β, in a season. Figure 3.8 explains both concepts.

In practice, an efficiency model will likely present an intermediate measure like wearing a mask, gargling, or alteration of other lifestyle habits to protect against infection, rather than emulating a vaccination. Those intermediate protecting measures might be thought to be less effective than a vaccine injection but may be less costly. Moreover, some vaccinations' stochastic behavior may be represented by an efficiency model rather than an effectiveness model. Thus, in the following discussion, we imagine a case where the efficiency model is applied to vaccination and not just to intermediate protecting measures.

3.2.1.1 Effectiveness Model

A vaccinated population is separated into two classes: immune individuals obtaining perfect immunity and non-immune ones failing to get immunity. Let the effectiveness of the vaccine and the vaccination coverage be e ($0 \leq e \leq 1$) and x, respectively. Note that the dynamics of this case obey the SIR model, Eq. (3.1). The fraction of vaccinated individuals with immunity must be ex, while that of non-immune individuals is $(1 - ex)$. Recalling Eq. (3.9), we can express the FES, R, at the end of a season in relation to both x and time t at equilibrium ($t = \infty$) as.

Table 3.2 Fractions of four types of individual using the effectiveness model

Strategy/state	Healthy	Infected
Vaccinated	$x(e+(1-e)\exp)[-R_0R(x,\infty)]$	$x(1-e)(1-\exp)[-R_0R(x,\infty)]$
Non-vaccinated	$(1-x)\exp[-R_0R(x,\infty)]$	$(1-x)\exp[-R_0R(x,\infty)]$

$$R(x,\infty)=(1-ex)(1-\exp[-R_0R(x,\infty)]). \tag{3.16}$$

However, if we intend to formulate precise dynamics, we must modify the SIR dynamics to SVIR dynamics as below:

$$\begin{cases} \dfrac{dS(x,t)}{dt}=-\beta S(x,t)I(x,t), \\ \dfrac{dV(x,t)}{dt}=-\beta(V(x,t)-eV(x,0))I(x,t), \\ \dfrac{dI(x,t)}{dt}=\beta S(x,t)I(x,t)+\beta(V(x,t)-eV(x,0))I(x,t)-\gamma I(x,t), \\ \dfrac{dR(x,t)}{dt}=\gamma I(x,t). \end{cases} \tag{3.17}$$

Because $V(x,0)=x$, the transcendental equation for $R(x,\infty)$, which is drawn from Eq. (3.17) and gives its analytical solution, is entirely consistent with what can be drawn from Eq. (3.9), which is Eq. (3.16).

Let us confirm that the time variable, t, obeys the local timescale. At the end of every season, any individual is classified into one of the four categories: vaccinated and healthy, vaccinated but infected, non-vaccinated but healthy, and non-vaccinated and infected. $R(x,\infty)$ gives the respective fractions of the four different types of individual depending on whether they are vaccinated or non-vaccinated and whether they are healthy or infected, as summarized in an explicit way in Table 3.2.

3.2.1.2 Efficiency Model

The efficiency model is somewhat more complex than the effectiveness model. Let the efficiency for avoiding infection be η ($0\le\eta\le1$), indicating how defense measures can decrease the probability of being infected. A non-vaccinated susceptible individual may become infected if they are exposed to infectious individuals at a disease-transmission rate of β [day^{-1} person^{-1}]. An individual committing the provision (vaccination) who is in S may also become infectious at a rate of $(1-\eta)\beta$. An infected individual recovers at the recovery rate γ [day^{-1}].

The modified SIR model should be called the SVIR model. With the introduction of a provision (vaccination) to the efficiency model, we have.

Table 3.3 Fractions of four types of individual using the efficiency model

Strategy/state	Healthy	Infected
Vaccinated	$x \exp[-(1-\eta)R_0R(x,\infty)]$	$x(1-\exp)[-(1-\eta)R_0R(x,\infty)]$
Non-vaccinate	$(1-x)\exp[-R_0R(x,\infty)]$	$(-x)(1-\ \exp[R_0R(x,)\infty])$

$$\begin{cases} \dfrac{dS(x,t)}{dt} = -\beta S(x,t)I(x,t), \\ \dfrac{dV(x,t)}{dt} = -(1-\eta)\beta V(x,t)I(x,t), \\ \dfrac{dI(x,t)}{dt} = \beta S(x,t)I(x,t) + (1-\eta)\beta V(x,t)I(x,t) - \gamma I(x,t), \\ \dfrac{dR(x,t)}{dt} = \gamma I(x,t), \end{cases} \tag{3.18}$$

with the presumed initial values $S(x,0) = 1 - x$, $V(x,0) = x$, and $I(x,0) = 0$. Variable x indicates the fraction of committing to a provision, i.e., vaccination coverage. We impose the following constraint:

$$S(x,t) + V(x,t) + I(x,t) + R(x,t) = 1. \tag{3.19}$$

Because the population is not completely susceptible, it is accurate to use a control reproduction number, R_c, instead of the basic reproduction number, R_0. In this case, R_c can be estimated as

$$R_c = \frac{\beta}{\gamma}[S(x,0) + (1-\eta)V(x,0)] = R_0[S(x,0) + (1-\eta)V(x,0)]. \tag{3.20}$$

By solving Eqs. (3.18) and (3.19) with the above initial conditions, we can obtain the FES and other variables observed at equilibrium in a season:

$$S(x,\infty) = (1-x)\exp[-R_0R(x,\infty)], \tag{3.21}$$

$$V(x,\infty) = x\exp[-(1-\eta)R_0R(x,\infty)], \tag{3.22}$$

$$R(x,\infty) = 1 - (1-x)\exp[-R_0R(x,\infty)] - x\exp[-(1-\eta)R_0R(x,\infty)]. \tag{3.23}$$

In the limit of this process, the respective fractions of the four different types of individuals at equilibrium are summarized in Table 3.3.

Comparing Tables 3.2 and 3.3, it is worth noting that the "success probability of free riding" is always given by $(1 - x)\exp[-R_0R(x,\infty)]$, regardless of whether we presume a perfect vaccination, an imperfect one, or even an intermediate defense measure.

3.2.2 *Strategy-Updating Rule*

Following upon what happened in the second stage of the previous season, which obeys the SIR or SVIR process, an individual brushes up his/her strategy either committing vaccination or not in the first stage of the current season (see Fig. 3.6b).

According to psychology, the relationship between stimulus (input) to a human body and response (output) commonly features "convexity," which can alternatively be called "non-linearity"; this can be described as a human responding with less sensitivity when exposed to larger stimuli. The Weber–Fechner law (van der Helm 2010), the most well-known psychophysics relation of the human response to a physical stimulus, reproduces this principle.

Here, the stimuli are the expected costs resulting from vaccination and infection. More precisely, we presume that the focal agent recognizes how much their payoff (cost) is in relation to a "comparison target". This assumption implies that the range of stimuli (or inputs) can be in both negative or positive. Moreover, the response (output) is the probability of whether the focal agent will commit to a vaccination. Noting those points, although the Weber–Fechner law suggests that a logarithmic-function can represent the stimulus—response relation, it would be acceptable for a sigmoid-function to represent this dynamic, as it has both convexity and positive and negative regions of stimulus. Here, the aforementioned "comparison target" refers to the reference payoff to one's own payoff. Although there might be several possibilities, let us presume two concepts in the following discussion.

The first is the payoff of one of the focal player's acquaintances—meaning their neighbors, if we assume a social network connecting individuals. But, so far in this section, we have presumed the MFA, which presumes no spatial structure at all. Thus, we have to seek an alternative to "a neighbor" and "neighbors," which we will discuss later.

Another concept is that of direct comparison between alternatives. In practice, a player favors vaccination if the average payoff for vaccinators is better than that of non-vaccinators. More realistically, an individual likely observes those average payoffs, for which information can be found through the media.

The first concept resulting from the comparison of one's own payoff with that others can be said to be some sort of "social learning," or "social imitation," while the second concept results from direct comparison and can be called "self-estimation"; it does not yield a strategy-copying probability like the first concept, but rather directly defines the probability of committing to a vaccination.

All of the above-discussed background prompts us to introduce the three following strategy-update rules.

3.2.2.1 Individual-Based Risk Assessment (IB-RA)

This concept is based on social learning and was originally introduced by Fu et al. (2011) in line with a MAS approach, whereby agents, spatially distributed on an underlying network and exposed to infectious risk, learn whether to commit to a vaccination from one of their neighbors. This idea exactly reflects the assumption of the vast majority of studies dealing with spatial prisoner's dilemma (SPD) games and is called pairwise comparison based on a Fermi function (as described by PW-Fermi). Agent i randomly selects agent j from his neighbors. Let us assume that their payoffs are π_i and π_j, respectively. The probability of agent i copying agent j's strategy, s_j, whether vaccination or non-vaccination, is represented by $P\,(s_i \leftarrow s_j)$ and is defined as

$$P\left(s_i \leftarrow s_j\right) = \frac{1}{1 + \exp\left[\frac{\pi - r_j}{\kappa}\right]}, \tag{3.24}$$

where κ indicates the sensitivity to the gain difference. For $\kappa \rightarrow \infty$ (weak selection pressure), an individual i is insensitive to the payoff difference $\pi_i - \pi_j$ against another individual j and the probability $P(s_i \leftarrow s_j)$ approaches 1/2 asymptotically, regardless of the payoff difference. For $\kappa \rightarrow 0$ (strong selection pressure), individuals are sensitive to the payoff difference, and they certainly copy the successful strategy that earns the higher payoff, even if the difference between payoffs is very small. We assume that $\kappa = 0.1$ throughout this book.

If agent i does not copy the strategy of agent j, he maintains his own strategy.

The idea above can be applied to the MAS model, but not to ODE model we are now discussing, because there is no spatial structure; rather, MFA is presumed. Therefore, we should modify the above concept to the following MFA-based framework.

In the present framework, there are four classes of individual in relation to cost burden: (1) a successful free rider (SFR) who pays nothing, (2) a failed free-rider (FFR) who pays -1, (3) an infected vaccinator (IV) who pays $-C_r-1$, and (4) a healthy vaccinator (HV) who pays $-C_r$ (see payoff matrix provided in Table 3.1). A player may assume one of the two strategies: vaccination (hereinafter V) or non-vaccination (hereinafter NV). Thus, the transition probability that affects the time transition of vaccination coverage, x, which should be considered in the IB-RA rule, is covered by one of the following eight cases:

$$P(HV \leftarrow SFR) = \frac{1}{1 + \exp\left[-(0 - (-C_r))/\kappa\right]}, \tag{3.25.1}$$

$$P(HV \leftarrow FFR) = \frac{1}{1 + \exp\left[-(-1 - (-C_r))/\kappa\right]}, \tag{3.25.2}$$

$$P(IV \leftarrow SFR) = \frac{1}{1 + \exp\left[-(0 - (-C_r - 1))/\kappa\right]}, \tag{3.25.3}$$

$$P(IV \leftarrow FFR) = \frac{1}{1 + \exp\left[-(-1 - (-C_r - 1))/\kappa\right]}, \tag{3.25.4}$$

$$P(SFR \leftarrow HV) = \frac{1}{1 + \exp\left[-(-C_r - 0)/\kappa\right]}, \tag{3.25.5}$$

$$P(SFR \leftarrow IV) = \frac{1}{1 + \exp\left[-(-C_r - 1 - 0)/\kappa\right]}. \tag{3.25.6}$$

$$P(FFR \leftarrow HV) = \frac{1}{1 + \exp\left[-(-C_r - (-1))/\kappa\right]}, \tag{3.25.7}$$

$$P(FFR \leftarrow IV) = \frac{1}{1 + \exp\left[-(-C_r - 1 - (-1))/\kappa\right]}. \tag{3.25.8}$$

Note that ($St2$) in $P(St1 \leftarrow St2)$ means the state of a focal individual amid {SFR, FFR, IV, HV}, while ($St1$) is the state of the targeted individual.

3.2.2.2 Strategy-Based Risk Assessment (SB-RA)

This concept was also introduced in line with the MAS approach by Fukuda and her colleagues (Fukuda et al. 2014). Let us imagine a situation in which information regarding the consequences of adopting a certain strategy is disclosed to society by the media, and everyone in the population has access to these consequences. Then, individuals no longer rely upon the payoff of any one specific neighbor. Instead, in adapting their strategy, they tend to assess the risk based on a socially averaged payoff that results from adopting a certain strategy.

To reflect the above situation, rather than Eq. (3.24), we use

$$P(s_i \leftarrow s_j) = \frac{1}{1 + \exp\left[\frac{\pi - <\pi S_j>}{\kappa}\right]}, \tag{3.26}$$

where $< \pi_{S_j} >$ is the average payoff obtained by averaging a collective payoff over individuals who adopt the same strategy as that of a randomly selected neighbor j of individual i. Although the sampling number can be a control parameter that ranges from only one individual (i.e., only one of i's neighbors, j) to all individuals among the whole population who adopt the same strategy as j, throughout this book, we presume that sampling is taken for the entire population.

As with the case of IB-RS, we should modify the concept above to the following framework based on MFA for ODE models. Unlike IB-RA, we must know the average payoffs of vaccinators and non-vaccinators besides the cases of SFR, FFR, IV, HV, which have fixed payoffs of 0, 1, $-C_r - 1$, and $-C_r$, respectively. As mentioned above, the states of those additional two are represented by the abbreviations V and NV, respectively. Vaccinators can be called cooperators, while non-vaccinators should be called defectors in view of their dedication to the herd immunity. Hence, we introduce the notation $\langle \pi_C \rangle$ for the average payoff of vaccinators V (cooperators, C) and $\langle \pi_D \rangle$ for that of non-vaccinators NV (defectors; D).

Because of the imperfectness of the vaccine, $\langle \pi_C \rangle$ can range from $-C_r - 1$ to $-C_r$. By contrast, $\langle \pi_C \rangle$ ranges from -1 to 0. Consequently, the transition probability that affects the time transition of vaccination coverage, x, which should be considered in the SB-RA rule, is instead covered by one of the four following cases:

$$P(\mathrm{HV} \leftarrow \mathrm{NV}) = \frac{1}{1 + \exp\left[-(\langle \pi_D \rangle - (-C_r))/\kappa\right]}, \tag{3.27.1}$$

$$P(\mathrm{IV} \leftarrow \mathrm{NV}) = \frac{1}{1 + \exp\left[-(\langle \pi_D \rangle - (-C_r - 1))/\kappa\right]}, \tag{3.27.2}$$

$$P(\mathrm{SFR} \leftarrow \mathrm{V}) = \frac{1}{1 + \exp\left[-(\langle \pi_C \rangle - 0)/\kappa\right]}, \tag{3.27.3}$$

$$P(\mathrm{FFR} \leftarrow \mathrm{V}) = \frac{1}{1 + \exp\left[-(\langle \pi_C \rangle - (-1))/\kappa\right]}. \tag{3.27.4}$$

Note that ($St2$) in $P(St1 \leftarrow St2)$ denotes the state of a focal individual amid {V, NV}, while ($St1$) denotes the state of the target individual {SFR, FFR, IV, HV}.

3.2.2.3 Direct Commitment (DC)

Direct commitment is the representative framework for "self-estimation," which differs from "social learning (imitation)" such as IB-RA and SB-RA. Applying MFA, the transition probability that we now must consider is one of the following:

$$P(\mathrm{V} \leftarrow \mathrm{NV}) = \frac{1}{1 + \exp\left[-(\langle \pi_D \rangle - \langle \pi_C \rangle)/\kappa\right]}, \tag{3.28.1}$$

$$P(\mathrm{NV} \leftarrow \mathrm{V}) = \frac{1}{1 + \exp\left[-(\langle \pi_C \rangle - \langle \pi_D \rangle)/\kappa\right]}. \tag{3.28.2}$$

The three update rules for mutual comparison are presented in Fig. 3.9 for mutual comparison.

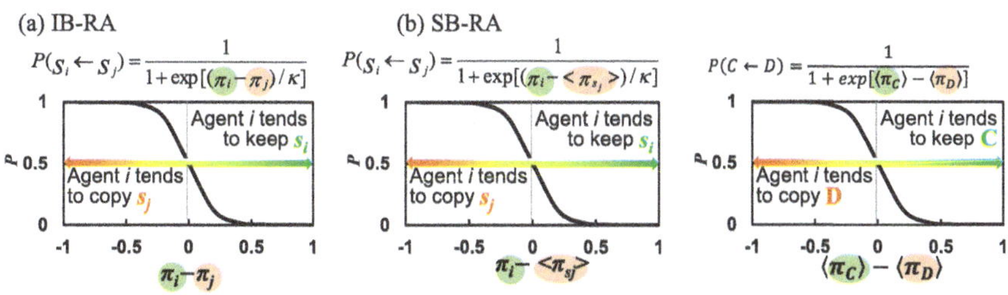

Fig. 3.9 Strategy update based on the concept of social imitation; (**a**) IB-RA, (**b**) SB-RA and (**c**) DC when presuming the MAS approach

3.2.3 Global Dynamics for Strategy Updating

Returning to Fig. 3.6, we should now turn our attention to the global dynamics of varying x in repeating seasons based on what happened in the previous season. Strategy updating takes place after each epidemic season as defined above, thereby inevitably increasing or decreasing x.

Above, we introduced two different epidemic models, namely the effectiveness and efficiency models, as well as three different updating rules, i.e., IB-RA, SB-RA, and DC. Hence, we establish the following six different types of dynamics:

Effectiveness model + IB-RA:

$$\begin{aligned} \frac{dx}{dt} &= x(1-x)(e+(1-e)\exp[-R_0R(x,\infty)])\exp[-R_0R(x,\infty)](P(\mathrm{SFR}\leftarrow\mathrm{HV})-P(\mathrm{HV}\leftarrow\mathrm{SFR})) \\ &+x(1-x)(e+(1-e)\exp[-R_0R(x,\infty)])(1-\exp[-R_0R(x,\infty)])(P(\mathrm{FFR}\leftarrow\mathrm{HV})-P(\mathrm{HV}\leftarrow\mathrm{FFR})) \\ &+x(1-x)(1-e)(1-\exp[-R_0R(x,\infty)])\exp[-R_0R(x,\infty)](P(\mathrm{SFR}\leftarrow\mathrm{IV})-P(\mathrm{IV}\leftarrow\mathrm{SFR})) \\ &+x(1-x)(1-e)(1-\exp[-R_0R(x,\infty)])^2(P(\mathrm{FFR}\leftarrow\mathrm{IV})-P(\mathrm{IV}\leftarrow\mathrm{FFR})) \end{aligned}, \tag{3.29}$$

Efficiency model + IB-RA:

$$\begin{aligned} \frac{dx}{dt} &= x(1-x)\exp[-(1-\eta)R_0R(x,\infty)]\exp[-R_0R(x,\infty)](P(\mathrm{SFR}\leftarrow\mathrm{HV})-P(\mathrm{HV}\leftarrow\mathrm{SFR})) \\ &+x(1-x)\exp[-(1-\eta)R_0R(x,\infty)](1-\exp[-R_0R(x,\infty)])(P(\mathrm{FFR}\leftarrow\mathrm{HV})-P(\mathrm{HV}\leftarrow\mathrm{FFR})) \\ &+x(1-x)(1-\exp[-(1-\eta)R_0R(x,\infty)])\exp[-R_0R(x,\infty)](P(\mathrm{SFR}\leftarrow\mathrm{IV})-P(\mathrm{IV}\leftarrow\mathrm{SFR})) \\ &+x(1-x)(1-\exp[-(1-\eta)R_0R(x,\infty)])(1-\exp[-R_0R(x,\infty)])(P(\mathrm{FFR}\leftarrow\mathrm{IV})-P(\mathrm{IV}\leftarrow\mathrm{FFR})) \end{aligned}, \tag{3.30}$$

Effectiveness model + SB-RA:

$$\begin{aligned} \frac{dx}{dt} &= -x(1-x)(e+(1-e)\exp[-R_0R(x,\infty)])P(\mathrm{HV}\leftarrow\mathrm{NV}) \\ &-x(1-x)(1-e)(1-\exp[-R_0R(x,\infty)])P(\mathrm{IV}\leftarrow\mathrm{NV}) \\ &+x(1-x)\exp[-R_0R(x,\infty)]P(\mathrm{SFR}\leftarrow\mathrm{V}) \\ &+x(1-x)(1-\exp[-R_0R(x,\infty)])P(\mathrm{FFR}\leftarrow\mathrm{V}) \end{aligned}, \tag{3.31}$$

Efficiency model + SB-RA:

$$\begin{aligned} \frac{dx}{dt} &= -x(1-x)\exp[-(1-\eta)R_0R(x,\infty)]P(\mathrm{HV}\leftarrow\mathrm{NV}) \\ &-x(1-x)(1-\exp[-(1-\eta)R_0R(x,\infty)])P(\mathrm{IV}\leftarrow\mathrm{NV}) \\ &+x(1-x)\exp[-R_0R(x,\infty)]P(\mathrm{SFR}\leftarrow\mathrm{V}) \\ &+x(1-x)(1-\exp[-R_0R(x,\infty)])P(\mathrm{FFR}\leftarrow\mathrm{NV}) \end{aligned}, \tag{3.32}$$

Effectiveness or efficiency model + DC:

$$\frac{dx}{dt} = -xP(\mathrm{V} \leftarrow \mathrm{NV}) + (1 - x)P(\mathrm{NV} \leftarrow \mathrm{V}). \quad (3.33)$$

Note that Eq. (3.33) is common to both the effectiveness and efficiency models.

Something worth noting is that Eq. (3.33) is qualitatively consistent with what are called replicator dynamics, one of the most common concepts in evolutionary game theory for expressing a system's dynamics as we discussed in Chap. 2.

All the above dynamical equations, which together constitute a set of ODEs, can be solved numerically. We introduce a so-called explicit scheme[3] for the time-varying terms to obtain a numerical solution for the vaccination coverage at social equilibrium after infinite seasons have elapsed.

One technical point worth noting is that as small of a discretized time-step Δt as possible should be introduced in an actual numerical process, despite requiring a long computational time. If an inappropriately large Δt is presumed for either local (dealing with SIR or SVIR) or global dynamics (dealing with strategy updating), unrealistic fluctuating time evolution may occur, leading to an erroneous prediction at social equilibrium, even though numerical divergence does not happen.

The result is shown in Sect. 3.3 and compared with the result coming from the MAS approach.

3.3 MAS Approach to the Vaccination Game

As mentioned above, the vaccination game originally assumed a MAS approach. MAS models can implement more realistic scenarios than ODE models in various respects. The most important advantage vis-à-vis an ODE model is that a MAS model can take account of spatial structure, i.e., the underlying network structure connecting individuals. The basic concept of the MAS approach is that each individual is generated as an agent in a computational world *on-silica*, which is called an artificial society. Thus, a MAS model is so flexible that it can adjust to any specific situations that must be emulated for a real-world problem. This fact is very advantageous, but simultaneously imposes a limitation. One constraint is that a sufficient number of agents should be generated to avoid the unwilling bias resulting from the finiteness of a population. For instance, a MAS considering all people living on Earth (generating 8 billion agents for an artificial society) does not seem practically feasible. But we should generate at least more than several 10^4 agents to avoid the abovementioned bias.

[3]A reader can consult with any standard textbooks concerned with applied mathematics for the numerical approach; for example: Tanimoto (2014).

3.3.1 Spatial Structure When Taking the MAS Approach

Despite some limitations, MAS is a powerful tool for considering the spatial structure of a finite population size.

Let us imagine a spatial structure for the whole population. This structure is represented by a network consisting of nodes (vertices) and links (edges). The dynamics of SIR or SVIR (hereafter called SIR/V) on a spatially structured population are not captured by a system of ODEs; thus, we numerically simulate the spread of an epidemic on a network using the Gillespie algorithm (Gillespie 1977) customized to the extended SIR/V model. Implementation of the Gillespie algorithm is crucially important in the current dynamic system in a MAS world, where a finite population is presumed and the probabilities of a single infection or of a certain agent recovering depend upon the product of the population size and the stochastic parameters—the effective transmission rate, β_e, and the effective recovery rate; γ_e. We can easily suppose that, in such a stochastic system, the time for the next event to happen after a single event has happened becomes smaller with the increase of the total population. The Gillespie algorithm can consider such a stochastic situation, which stipulates a local-scale time-step (see the bottom panel of Fig. 3.6) for the next event. Therefore, Δt for the time-evolution process in a season does not have a fixed value; rather, it is flexible. β_e and γ_e are compared with β and γ in Eq. (3.1) in the next sub-section.

Meanwhile, concerning the underlying social network, we can account for any topology in the MAS model. Many practically and theoretically important topologies have been discussed, including rings (a representative 1D regular graph)), lattices (a representative 2D regular graph), random regular networks (RRG), Barabási–Albert scale-free (BA-SF) networks (Barabási and Albert 1999), Erdős–Rényi random (E-R random) graphs (Bollobás 1985), and small-world (SW) networks (Watts and Strogatz 1998). Figure 3.10 illustrates some of them.

Incidentally, various properties have been defined to evaluate the topology and complexity of networks. These include average degree $<k>$, degree distribution $P(k)$, average path length, average cluster coefficient, assortative coefficient, degree centrality, betweenness centrality, and so forth. In rings, lattices, or RRGs, each vertex has the same degree k, i.e., the same number of links, and so is called a homogenous degree-distributed network. The degree distribution of a scale-free graph obeys a scale-free distribution, and that of the E-R random graph obeys a Poisson distribution. These are classified as heterogeneous degree-distributed networks. A scale-free network has a small number of agents (called "hub" agents) with a huge number of links, while the vast majority of agents have a small number of links. This is why its degree distribution is scale-free. And it is the substance of scale-free network. A small-word network can be constructed from a regular graph such as a ring or a lattice. Starting from such a regular graph and serving default links with a "short-cut" probability (usually presumed to be a small value) and re-connecting them to a randomly selected location, we obtain a small-world graph (this is done by the Watts–Strogatz SW algorithm; however, there are several other

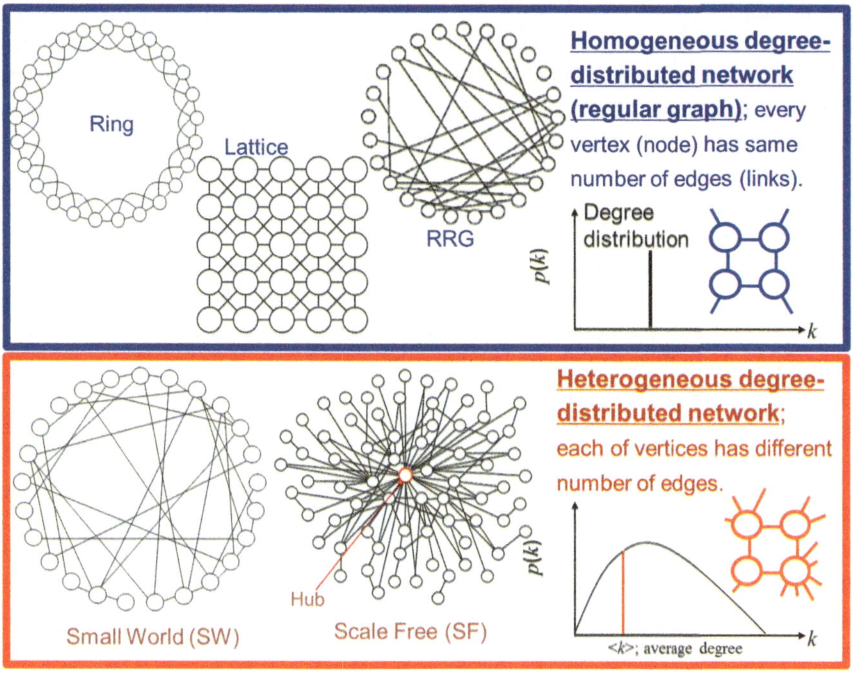

Fig. 3.10 Representative topologies for the MAS model

algorithms for generating SW graphs, such as the Newman–Watts SW algorithm). One of the most important characteristics of a SW network is that it has a quite small average path length, as compared with the original regular graph. It is understood that some complex networks observed in real human social systems can be described as either scale-free or small-world. Some recent work (Matsuzawa et al. 2017) proposed an alternative concept for a conventional SW network, whereby a rebuilding link obeys another stochastic principle instead of a fully random short-cut, for which a less-distant short-cut is rather more likely than a far-remote one.

3.3.2 *Effective Transmission Rate, β_e, and Effective Recovery Rate, γ_e*

Returning to the definition of basic reproduction number $R_0 = \frac{\beta}{\gamma}$, and maintaining the recovery rate, γ, at a certain value (for example; $\gamma = 1/3$ in the case of seasonal influenza, which implies that an infected individual recovers within 3 days on average), the extent to which a disease is transmittable is controlled by R_0. If we presume a seasonal influenza, $R_0 = 2.5$ is common as a general assumption, which leads to $\beta = 0.833$ as long as we assume that $\gamma = 1/3$. But this specific value, β, should not be applied for a MAS presuming a certain underlying network and a certain population size N, because, in MAS, the transmission rate is determined by accounting for whether a focal agent is infected by a neighbor according to the Gillespie algorithm. This is obviously not a daily event, and this neighbor may or may not be uniquely infected; there might be other infected neighbors of the focal agent. Returning to the original definition, which has physical units of [day^{-1} person^{-1}], β describes the probability with which an infected person can infect another person within a single day. In a MAS process obeying the Gillespie algorithm, we should consider the "effective transmition rate," β_e, rather than β. Therefore, we must go through the following process for each different toplogy, for each different average degree, and for each different population size N in order to identify β_e. Note that we presume that $\gamma_e = \gamma$.

By assuming R_0 and solving the transcendental Eq. (3.8), we can fix the final epidemic size $R(\infty)$ with no vaccinated (immunized) individuals if the dynamics obey the SIR model in view of MFA. For example, presuming $R_0 = 2.5$, we obtain $R(\infty) \cong 0.89$. In a MAS setting, we randomly place I_0 agents as initial infected individuals on the network, with all agents being non-vaccinators. The final epidemic size (hereafter, FES) is observed in this specific simulation episode. We repeat this simulation with a sufficient number of trials (more than 100) to obtain a robust ensemble average of FES. By varying β_e, we are consequently able to obtain the relation β_e-FES, from which we can read that the crossing point with an FES of 0.89 comes from $R(\infty) \cong 0.89$, which is exactly what we need (i.e., β_e). Figure 3.11 presents a couple of examples.

3.3.3 Result of the Vaccination Game; Comparison Between the MAS and ODE Models

Let us begin by discussing the result from the ODE model. Figure 3.12 shows the FES for different levels of vaccination coverage using the effectiveness and efficiency models. From Fig. 3.12, the so-called critical vaccination coverage, x_c, which eradicates the epidemic spread, can be read from the border of the extinct phase, at which FES = 0. This border suggests the critical vaccination coverage for suppressing the spread of an infection, which can be determined analytically as $x_c = (1 - 1/R_0)/(1 - \eta)$ for the efficiency model and as $x_c = (1 - 1/R_0)/e$ for the effectiveness model. Clearly, as long as a less-reliable defense measure is provided, say $\eta < 0.6$ or $e < 0.4$ for example, we cannot avoid the breakout of an epidemic, even if all individuals use a particular defense measure.

One notable point is that if the same fraction for either η or e is imposed and the same vaccination cost is presumed, the effectiveness model works slightly better to suppress FES than does the efficiency model. Figures 3.13 and 3.14, which relate to the effectiveness and efficiency models, respectively, give the final epidemic size (FES); left-hand panels), vaccination coverage (VC); central panels), and average

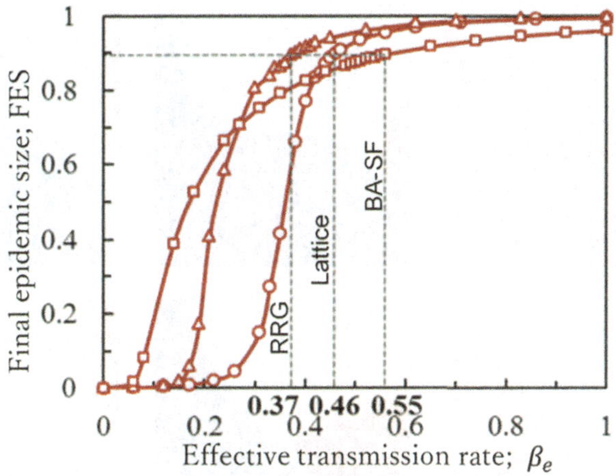

Fig. 3.11 The final proportions of infected individuals as a function of the effective transmission rate β when no individuals are vaccinated on each network: square lattice (circles), random regular graph (RRG) (triangles), Barabási–Albert scale-free (BA-SF) network (squares). For the lattice (circles), the population size is $N = 70 \times 70$ with a von Neumann neighborhood, the recovery rate is $\gamma = 1/3$ [day^{-1}], and the initial infected population is $I_0 = 5$. For RRG (triangles): population size $N = 4900$, degree $k = 4$, recovery rate $\gamma = 1/3$ [day^{-1}], and initial infected population is $I_0 = 5$. For a BA-SF network (squares), population size $N = 4900$, average degree $< k > = 4$, recovery rate $\gamma = 1/3$ [day^{-1}], and initial infected population $I_0 = 5$. Each plotted point represents an average over 100 runs

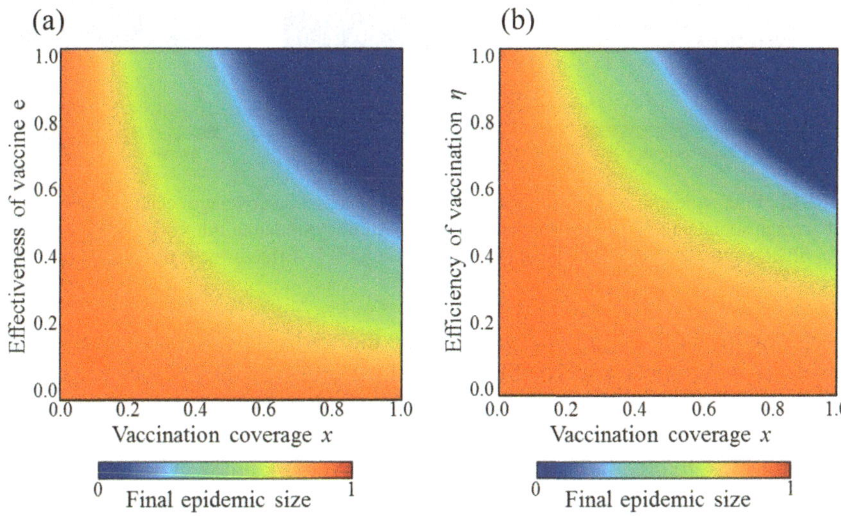

Fig. 3.12 Final epidemic size according to vaccination coverage and (**a**) effectiveness e (effectiveness of a vaccination) or (**b**) efficiency η (efficiency of an intermediate-defense measure). We presume $R_0 = 2.5$, which is applied consistently in this study

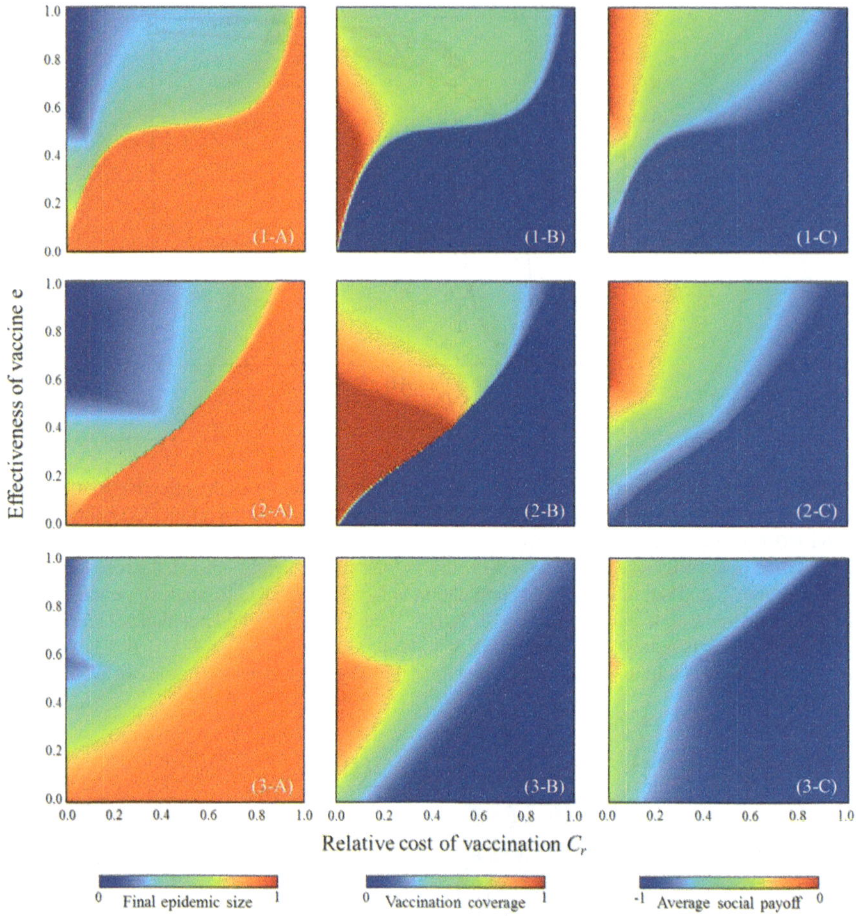

Fig. 3.13 Final epidemic size (left-hand panels; *-A), vaccination coverage (central panels; *-B), and average social payoff (right-hand panels; *-C) for strategy-updating rule IB-RA (upper panels; 1-*), SB-RA (middle panels; 2-*), and DC (lower panels; 3-*). Effectiveness model is applied

social payoff (ASP; right-hand panels) for the various strategy-updating rules, namely IB-RA (upper panels), SB-RA (middle panels), and DC (lower panels). The social-average payoff consists of the vaccination and infection costs, which theoretically range from $-C_r-1$ to 0.

The regions colored uniformly in light red (FES), dark blue (VC), and light blue (ASP) indicate that a pandemic is taking place, with most individuals not committing to vaccination; thus, an almost full-scale spread of the infection occurs. Roughly speaking, these regions emerge when a smaller effectiveness (efficiency) is presumed or a larger cost is imposed. This seems quite natural because most individuals tend to shy away from vaccination if it is less reliable and/or too expensive. The border between each of these monotone regions and the remaining region implies a

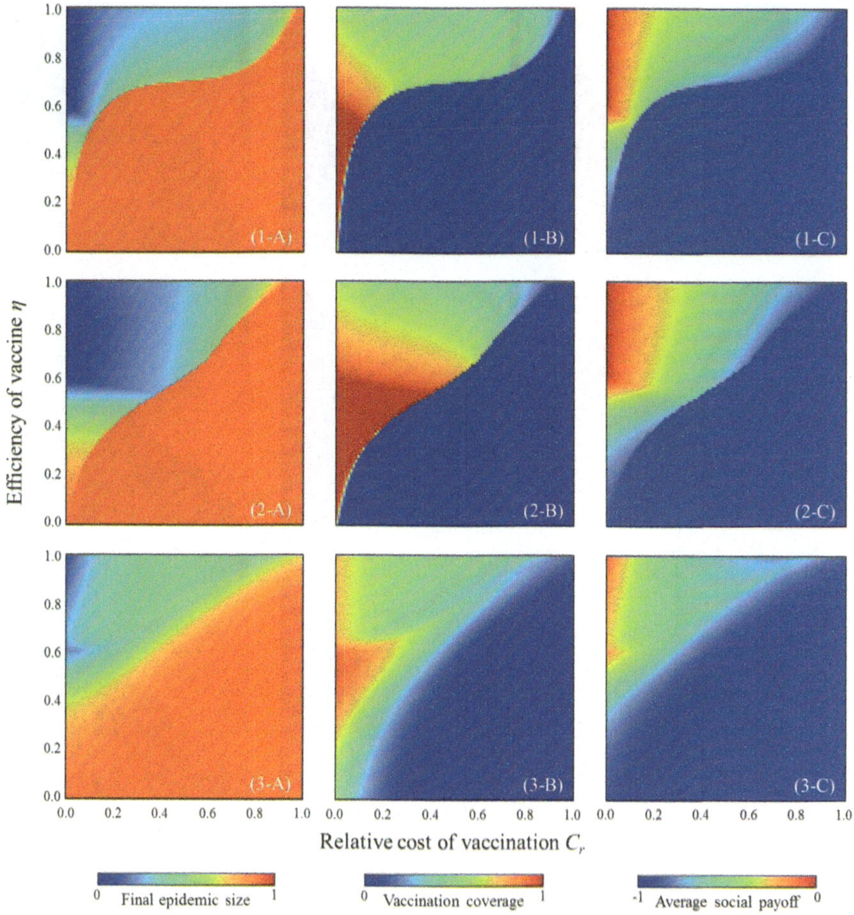

Fig. 3.14 Final epidemic size (left-hand panels; *-A), vaccination coverage (central panels; *-B), and average social payoff (right-hand panels; *-C) for the strategy-updating rules IB-RA (upper panels; 1-*), SB-RA (middle panels; 2-*), and DC (bottom panels; 3-*). The efficiency model is applied

combination of critical effectiveness (efficiency) and critical vaccination cost to control the spread of an epidemic, causing an obvious phase change (between the pandemic phase and the controlled phase). Interestingly, in the controlled phase, a lower effectiveness (efficiency) can realize higher VC, which is also helped by the effect of lower cost. Even if a large fraction of individuals relies on vaccination, the epidemic cannot be eradicated because of the lower vaccine reliability.

The detailed tendencies of the three update rules differ, although the overall tendency remains the same to some extent. Comparing the effectiveness model and the efficiency model, the latter has a wider pandemic phase at first glance. This implies that a vaccination with a certain η is less effective at suppressing the

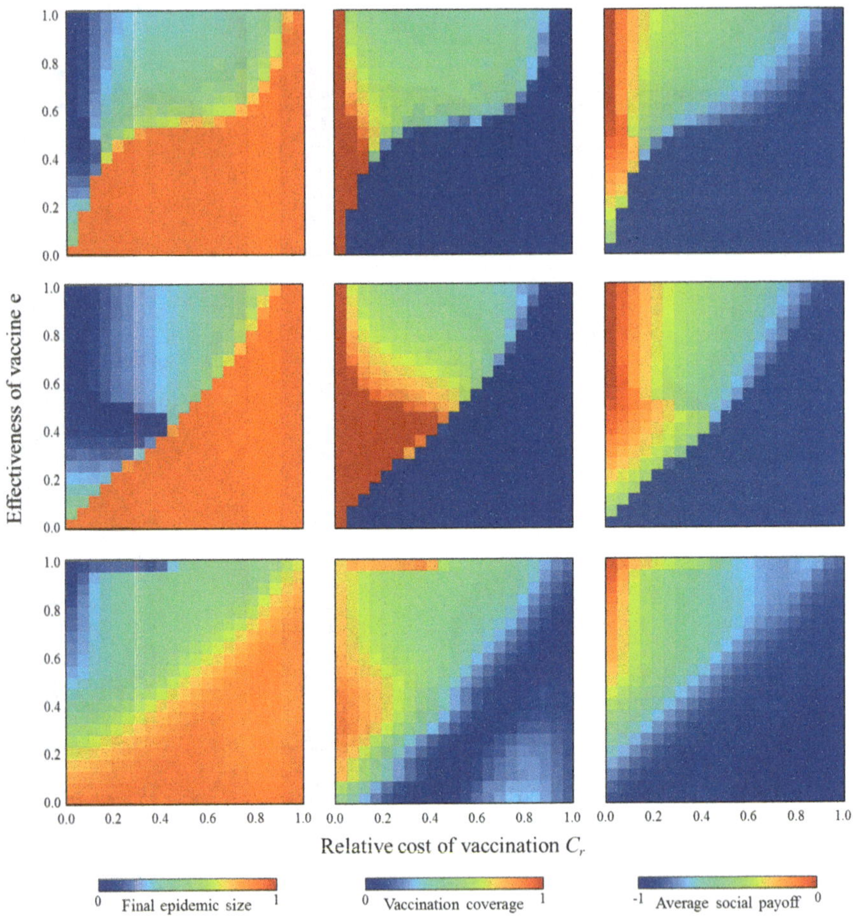

Fig. 3.15 MAS result on final epidemic size (left-hand panels; *-A), vaccination coverage (central panels; *-B), and average social payoff (right-hand panels; *-C) for strategy-updating rule IB-RA (upper panels; 1-*), SB-RA (middle panels; 2-*), and DC (lower panels; 3-*). Effectiveness model is applied. We presumed a perfect graph with $N = 1000$, $\beta = 0.00088$, and $\gamma = 1/3$

spread of an epidemic than imperfect vaccination, with e being defined as having the same numerical value as η, as presented in Fig. 3.12.

To validate this theoretical framework using ODE modeling, we conducted a series of numerical simulations based on the MAS approach previously discussed. Because we assumed a well-mixed population, we presumed a perfect graph as the underlying network connecting the agents. Following what we discussed in Sect. 3.3.2, we set $\beta_e = 0.00088$, which was determined to be the minimum effective transmission rate that exceeds the pre-defined threshold FES of 0.9 without vaccination. The results are shown in Figs. 3.15 and 3.16. In general, all of these results are, respectively, consistent with Figs. 3.13 and 3.14, although subtle discrepancies arise from the fact that the simulation presumes a finite population size of $N = 1000$.

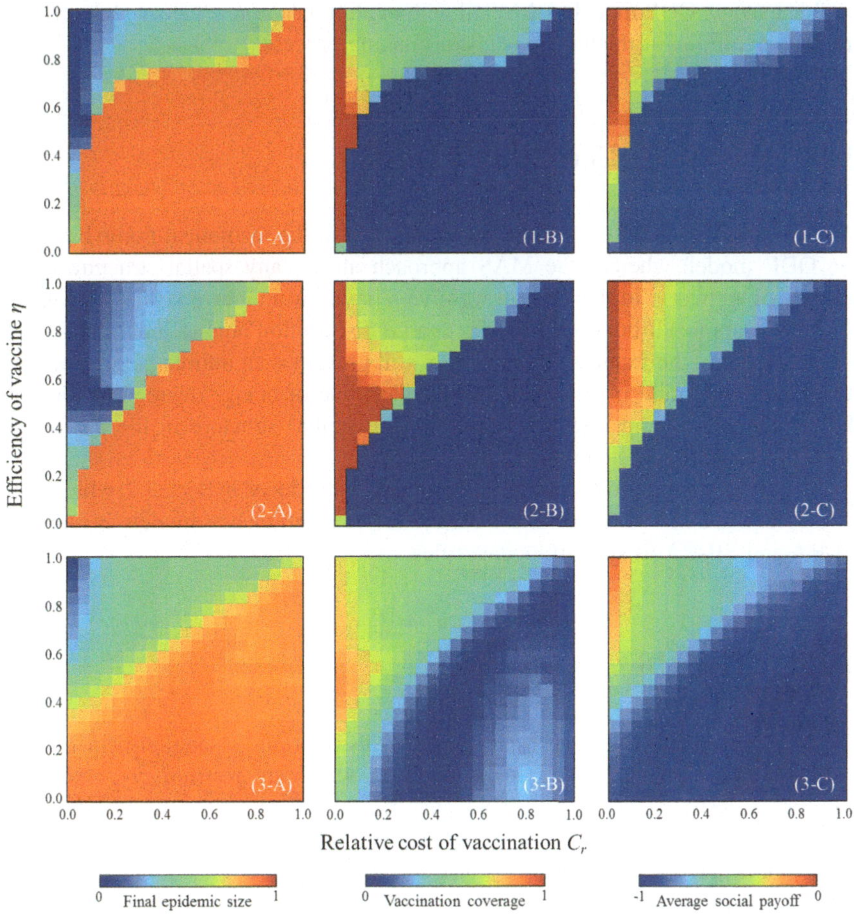

Fig. 3.16 MAS result for the final epidemic size (left-hand panels; *-A), vaccination coverage (central panels; *-B), and average social payoff (right-hand panels; *-C) for the strategy-updating rules IB-RA (upper panels; 1-*), SB-RA (middle panels; 2-*), and DC (lower panels; 3-*). The efficiency model is applied. We presumed a perfect graph with $N = 1000$, $\beta = 0.00088$, and $\gamma = 1/3$

The validation proves that the MAS approach based on the Gillespie algorithm well represents the result from the theoretical approach based on the ODE model and evolutionary game theory. Inversely speaking, the theoretical approach dovetailing the ODE model with evolutionary game theory well emulates the predictions for the vaccination game that were firmly traced by the MAS model. The latter statement might be thought more important than the former one, since the theoretical framework for the vaccination game is more convenient, more flexible, and less cumbersome than that for the MAS model. In fact, when dealing with an epidemic other than the one that we are implicitly presuming here (influenza), one should seek to establish a new ODE model to dovetail with evolutionary game theory in describing the time evolution of the human decision-making process.

3.4 Effect of the Underlying Topology

We presume a well-mixed and infinite population in the theoretical frame based on the ODE model, whereas the MAS approach allows any spatial structure as an underlying network. In this section we give a powerful framework that still takes a theoretical approach but considers the degree distribution among individuals. Precisely speaking, the framework given here still presumes an infinite population, but relaxes the constraint of being "well-mixed" to one of being "degree-distributed." This allows us to implement the effect of the spatial structure in the theoretical framework for the vaccination game.

In the following discussion, we apply what we introduced in Sect. 3.2, which is an SIR/V model combined with the effectiveness and efficiency models in view of IB-RA and SB-RA as strategy-update rules.

3.4.1 Degree Distribution

We consider two of the most typical degree distributions (spatial structures); the Poisson and the power-law degree distributions, which are, respectively, associated with Erdős–Rényi random graph (E-R random)[12] and the Barabási-Albert scale-free (BA-SF)[11] networks. In Poisson's degree distribution $P(k) = \exp(-\langle k \rangle)\langle k \rangle^k / k!$, most modes have a connectivity k close to the mean value, $\langle k \rangle = \sum_k kP(k)$. Conversely, the power-law degree distribution $P(k) \sim k^{-3}$ is often used to reproduce the real-world networks observed in complex social networks. Our focus on these two-degree distributions aims to reveal how different, non-homogeneous degree-distribution patterns quantitatively affect the spread of disease. As below, in our actual numerical analysis, we presume that the average connectivity of the degree distributions is assimilated to $\langle k = 8 \rangle$. To bring the power-law degree distribution near BA-SF with $\langle k \rangle = 8$, we consider the approximation $P(k) = A/k(k + 1)(k + 2)$ derived from the master equation by Albert and Albert-László (2002) and also assume that the minimum and maximum degrees are $k_{\min} = 3$ and 4 ($P(3) = P(4)$) and $k_{\max} = 100$, respectively. Here, A is a normalization constant.

3.4.2 Networked SIR Model

First, through a brief review of the SIR network model, we determine the dynamics of epidemic spreading. At each time-step, each node adopts one of the three possible states, and during a unit time-step, susceptible nodes connected to an infected one are infected with a rate of β. Meanwhile, the infected nodes are recovered with a rate γ. Defining the effective spreading rate $\lambda = \beta/\gamma$ (which does not represent the basic reproduction number, despite being defined in the same way, Eq. (3.5)), without loss of generality, we set $\gamma = 1$. To take the heterogeneity induced by the presence of nodes with different degrees into account, we consider the time evolution of the density of the susceptible $S_k(t)$, infected $I_k(t)$, and recovered $R_k(t)$ nodes of degree k. These variables are connected through normalization as follows:

$$S_k(t) + I_k(t) + R_k(t) = 1. \tag{3.34}$$

Using the MFA, we form the following set of coupled differential equations:

$$\begin{cases} \dfrac{dS_k(t)}{dt} = -\lambda k S_k(t)\Theta(t), \\ \dfrac{dI_k(t)}{dt} = -\lambda k S_k(t)\Theta(t) \quad - I_k(t), \\ \dfrac{dR_k(t)}{dt} = I_k(t), \end{cases} \tag{3.35}$$

where $\Theta(t)$ represents the probability that any given link points to an infected site. This quantity can be computed in a self-consistent way. The probability that a link points to a node with s links is proportional to $sP(s)$; thus, the probability that a randomly chosen link points to an infected node is given by (Moreno et al. 2002; Pastor-Satorras et al. 2003)

$$\Theta(t) = \frac{\sum_k (k-1)P(k)I_k(t)}{\sum_s sP(s)} = \frac{\sum_k (k-1)P(k)I_k(t)}{\langle k \rangle}. \tag{3.36}$$

In this approximation, we neglect the connectivity correlations in the network (i.e., the probability that a link points to an infected node is considered independent of the connectivity of the node from which the link is emanating). Combined with the initial conditions $S_k(0) = 1$, $I_k(0) \sim 0$, and $R_k(0) = 0$, we obtain

$$S_k(t) = S_k(0)\exp\left[-\lambda k \phi(t)\right] = \exp\left[-\lambda k \phi(t)\right],$$

where the auxiliary function $\phi(t)$ is defined as,

$$\begin{aligned} \phi(t) &= \int_0^t \Theta(I(t'))dt' = \frac{1}{\langle k \rangle}\sum_{k=1}^{\infty}(k-1)P(k)\int_0^t I_k(t')dt' \\ &= \frac{1}{\langle k \rangle}\sum_{k=1}^{\infty}(k-1)P(k)R_k(t). \end{aligned} \tag{3.37}$$

We differentiate $\phi(t)$ as follows:

$$\begin{aligned}\frac{d\phi(t)}{dt} &= \frac{1}{\langle k\rangle}\sum_{k=1}^{\infty}(k-1)P(k)\frac{dR_k(t)}{dt}\\ &= \frac{1}{\langle k\rangle}\sum_{k=1}^{\infty}(k-1)P(k)I_k(t)\\ &= \frac{1}{\langle k\rangle}\sum_{k=1}^{\infty}(k-1)P(k)(1-S_k(t)-R_k(t))\\ &= \frac{1}{\langle k\rangle}\sum_{k=1}^{\infty}(k-1)P(k)(1-S_k(t))-\phi(t)\\ &= \frac{1}{\langle k\rangle}\sum_{k=1}^{\infty}(k-1)P(k)(1-\exp[-\lambda k\phi(t)])-\phi(t).\end{aligned} \tag{3.38}$$

Since $I_k(\infty) = 0$ and $d\phi(t)/dt = 0$ are intuitively approved, we obtain the following self-consistent equation for $\phi(\infty)$ from Eq. (3.38):

$$\begin{aligned}\phi(\infty) &= \frac{1}{\langle k\rangle}\sum_{k=1}^{\infty}(k-1)P(k)(1-\exp[-\lambda k\phi\infty)])\\ &= 1-\frac{1}{\langle k\rangle}-\frac{1}{\langle k\rangle}\sum_{k=1}^{\infty}(k-1)P(k)\exp[-\lambda k\phi\infty].\end{aligned}$$

The final fractions are respectively expressed as

$$\begin{aligned}S_k(\infty) &= S_k(0)\exp[-\lambda k\phi(\infty)] = \exp[-\lambda k\phi(\infty)];\\ R_k(\infty) &= 1-S_k(\infty) = 1-\exp[-\lambda k\phi(\infty)].\end{aligned} \tag{3.39}$$

With respect to Eq. (3.38), there is a trivial solution $\phi(\infty) = 0$. To obtain a non-zero solution, the condition;

$$\frac{d}{d\phi(\infty)}\left(1-\frac{1}{\langle k\rangle}-\frac{1}{\langle k\rangle}\sum_{k=1}^{\infty}(k-1)P(k)\exp[-\lambda k\phi(\infty)]\right)\Bigg|_{\phi(\infty)=0} > 1, \tag{3.40}$$

must be satisfied, leading to

$$\lambda\frac{\langle k^2\rangle-\langle k\rangle}{\langle k\rangle} > 1. \tag{3.41}$$

This inequality defines the so-called epidemic threshold

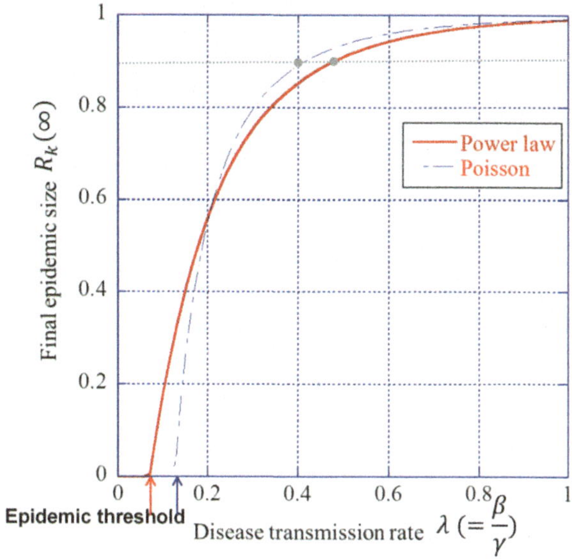

Fig. 3.17 Final epidemic size as a function of the transmission rate λ, when no individual is vaccinated on each network, following the power-law and Poisson degree distributions, which have an average number of degrees $<k> = 8$. The arrows point to the critical points of the effective spreading rate that can lead to epidemic spreading, known as the epidemic threshold

$$\lambda_c = \frac{\langle k \rangle}{\langle k^2 \rangle - \langle k \rangle}. \tag{3.42}$$

Figure 3.17 shows the FES $R(\infty)$, which results from summing up all degrees of $R_k(\infty)$, Eq. (3.39), as a function of the transmission rate λ when no individual is vaccinated on each network, following the power-law and Poisson degree distributions. From this figure, we confirm that the epidemic threshold described by Eq. (3.42) is consistent with the critical point of the transmission rate, which can lead to the spread of the epidemic. Moreover, we calibrate the transmission rate λ such that the FES across a network becomes 0.89, which is 0.40 for the Poisson distribution and 0.48 for the power-law distribution. These are required for comparison with the MAS result, as discussed in Sub-section 3.3.2. A comparison between the MFA and MAS results will be discussed in Sub-section 3.4.6.

3.4.3 Networked SIR/V Process with an Effectiveness Model

A vaccinated population is separated into two classes: perfectly immune individuals and non-immune individuals. Let the effectiveness of the vaccine, the density of a healthy vaccinated individual of degree k, and the VC of that individual with degree k be e $(0 \le e \le 1)$, V_k, and x_k, respectively. The fraction of vaccinated individuals with perfect immunity must be ex_k, while that of non-immune individuals is $(1 - ex_k)$. Thus, Eq. (3.17) based on an MFA that presumes no spatial structure should be modified as below:

$$\begin{cases} \dfrac{dS_k(t)}{dt} = -\lambda k S_k(t)\Theta(t). \\ \dfrac{dV_k(t)}{dt} = -\lambda k(V_k(t) - eV_k(0))\Theta(t). \\ \dfrac{dI_k(t)}{dt} = -\lambda k S_k(t)\Theta(t) - eV_k(0))\Theta(t) - I_k(t), \\ \dfrac{dR_k(t)}{dt} = I_k(t), \end{cases} \tag{3.43}$$

with assumed initial conditions $S_k(0) = 1 - x_k$, $V_k(10) = x_k$, $I_k(0)$~0, and $R_k(0) = 0$. The following constraint is required:

$$S_k(t) + V_k(t) + I_k(t) + R_k(t) = 1. \tag{3.44}$$

Combined with these initial conditions and following normalization, we obtain

$$S_k(t) = S_k(0)\exp[-\lambda k\phi(t)] = (1 - x_k)\exp[-\lambda k\phi(t)], \tag{3.45}$$

$$\begin{aligned} V_k(t) &= V_k(0)(e + (1 - e)\exp[-\lambda k\phi(t)]) \\ &= x_k(e + (1 - e)\exp[-\lambda k\phi(t)]). \end{aligned} \tag{3.46}$$

We modify the time evolution of $\phi(t)$ from Eq. (3.38) as:

$$\begin{aligned} \frac{d\phi(t)}{dt} &= \frac{1}{\langle k\rangle}\sum_{k=1}^{\infty}(k-1)\frac{dR_k(t)}{dt} \\ &= \frac{1}{\langle k\rangle}\sum_{k=1}^{\infty}(k-1)P(k)I_k(t) \\ &= \frac{1}{\langle k\rangle}\sum_{k=1}^{\infty}(k-1)P(k)(1 - S_k(t) - V_k(t) - R_k(t)) \\ &= \frac{1}{\langle k\rangle}\sum_{k=1}^{\infty}(k-1)P(k)(1 - S_k(t) - V_k(t)) - \phi(t) \\ &= \frac{1}{\langle k\rangle}\sum_{k=1}^{\infty}(k-1)P(k)(1 - S_k(0)\exp[-\lambda k\phi(t)] - V_k(0)(e + (1-e)\exp[-\lambda k\phi(t)])) - \phi(t). \end{aligned} \tag{3.47}$$

Because $I_k(\infty) = 0$ and $\lim_{t\to\infty} d\phi(t)/dt = 0$, we obtain from Eq. (3.47) the following self-consistent equation for $\phi(\infty)$:

$$
\begin{aligned}
\phi(\infty) &= \frac{1}{\langle k\rangle}\sum_{k=1}^{\infty}(k-1)P(k)(1-S_k(0)\exp[-\lambda k\phi(\infty)]-V_k(0)(e+(1-e)\exp[-\lambda k\phi(\infty)]))\\
&= 1-\frac{1}{\langle k\rangle}-\frac{1}{\langle k\rangle}\sum_{k=1}^{\infty}(k-1)P(k)(S_k(0)\exp[-\lambda k\phi(\infty)]+V_k(0)(e+(1-e)\exp[-\lambda k\phi(\infty)]))\\
&= 1-\frac{1}{\langle k\rangle}-\frac{1}{\langle k\rangle}\sum_{k=1}^{\infty}(k-1)P(k)((1-x_k)\exp[-\lambda k\phi(\infty)]+x_k(e+(1-e)\exp[-\lambda k\phi(\infty)])).
\end{aligned}
\tag{3.48}
$$

The FES and the other fractions can be expressed as

$$S_k(\infty) = S_k(0)\exp[-\lambda k\phi(\infty)] = (1-x_k)\exp[-\lambda k\phi(\infty)], \tag{3.49}$$

$$
\begin{aligned}
V_k(\infty) &= V_k(0)\\
&\times(e+(1-e)\exp[-\lambda k\phi(\infty)] = x_k(e+(1-e)\exp[-\lambda k\phi(\infty)]),
\end{aligned}
\tag{3.50}
$$

$$R_k(\infty) = (1-ex_k)(1-\exp[-\lambda k\phi(\infty)]). \tag{3.51}$$

As explained in the previous sub-section, the condition for having a non-zero solution imposes the following inequality:

$$
\begin{aligned}
&\frac{d}{d\phi(\infty)}\left(1-\frac{1}{\langle k\rangle}-\frac{1}{\langle k\rangle}\sum_{k=1}^{\infty}(k-1)P(k)((1-x_k)\exp[-\lambda k\phi(\infty)]\right.\\
&\left.+x_k(e+(1-e)\exp[-\lambda k\phi(t)]))\right)\Big|_{\phi(\infty)=0} > 1.
\end{aligned}
\tag{3.52}
$$

Assuming no dependence upon the degree k of the vaccinated individuals,

$$\lambda(1-ex)\frac{\langle k^2\rangle-\langle k\rangle}{\langle k\rangle} > 1. \tag{3.53}$$

This inequality defines the critical VC, which can eradicate the spread of an epidemic:

$$x_c = \left(1-\frac{\langle k\rangle}{\lambda(\langle k^2\rangle-\langle k\rangle)}\right)\Big/ e. \tag{3.54}$$

Note that the assumption that critical VC does not depend on degree is only presumed so as to draw Fig. 3.17, as discussed later. In a nutshell, Eqs. (3.52)–(3.54) are only required to obtain Fig. 3.17, which is irrelevant to the following discussion of the vaccination game.

Table 3.4 Fractions of four types of individuals in the effectiveness model

Strategy/state	Healthy	Infected
Vaccinated	$x_k(e + (1 - e)\exp[-\lambda k\phi(\infty)])$	$x_k(1 - e)(1 - \exp[-\lambda k\phi(\infty)])$
Non-vaccinated	$(1 - x_k)\exp[-\lambda k\phi(\infty)])$	$(1 - x_k)(1 - \exp[-\lambda k\phi(\infty)])$

The respective fractions of the four different types of individuals, depending on whether they are vaccinated or non-vaccinated and whether they are healthy or infected, are summarized in Table 3.4, which is the counterpart to Table 3.2 based on an MFA with no spatial structure.

3.4.4 Networked SIR/V Process with an Efficiency Model

Let the efficiency of a defense against contagion be η $(0 \leq \eta \leq 1)$, describing the extent to which the defensive measure can decrease the probability of infection. A non-vaccinated susceptible individual may become infected if they are exposed to infectious individuals with a disease-transmission rate λk [day^{-1} person^{-1}]. Vaccinated (that is, prepared) individuals in S who use the intermediate defense measure may also become infectious with probability $(1 - \eta)\lambda k$.

The SVIR process to describe such a condition is expressed as

$$\begin{cases} \frac{dS_k(t)}{dt} = -\lambda k S_k(t)\Theta(t), \\ \frac{dV_k(t)}{dt} = -(1-\eta)\lambda k V_k(t)\Theta(t), \\ \frac{dI_k(t)}{dt} = \lambda k S_k(t)\Theta(t) + (1-\eta)\lambda k V_k(t)\Theta(t) - I_k(t), \\ \frac{dR_k(t)}{dt} = I_k(t), \end{cases} \tag{3.55}$$

with the assumed set of initial conditions $S_k(0) = 1 - x_k$, $V_k(0) = x_k$, $I_k(0) \sim 0$, and $R_k(0) = 0$. The following constraint is requisite:

$$S_k(t) + V_k(t) + I_k(t) + R_k(t) = 1. \tag{3.56}$$

Combined with these initial conditions and through normalization, we obtain

$$S_k(t) = S_k(0)\exp[-\lambda k\phi(t)] = (1 - x_k)\exp[-\lambda k\phi(t)], \tag{3.57}$$

$$V_k(t) = V_k(0)\exp[-(1-\eta)\lambda k\phi(t)] = x_k \exp[-(1-\eta)\lambda k\phi(t)]. \tag{3.58}$$

We modify the time evolution of $\phi(t)$, as described by Eq. (3.38), as

$$\begin{aligned}
\frac{d\phi(t)}{dt} &= \frac{1}{\langle k \rangle}\sum_{k=1}^{\infty}(k-1)\frac{dR_k(t)}{dt} = \frac{1}{\langle k \rangle}\sum_{k=1}^{\infty}(k-1)P(k)I_k(t) \\
&= \frac{1}{\langle k \rangle}\sum_{k=1}^{\infty}(k-1)P(k)(1-S_k(t)-V_k(t)-R_k(t)) \\
&= \frac{1}{\langle k \rangle}\sum_{k=1}^{\infty}(k-1)P(k)(1-S_k(t)-V_k(t))-\phi(t) \\
&= \frac{1}{\langle k \rangle}\sum_{k=1}^{\infty}(k-1)P(k) \\
&\quad \times (1-S_k(0)\exp[-\lambda k\phi(t)]-V_k(0)\exp[-(1-\eta)\lambda k\phi(t)]) \\
&\quad -\phi(t).
\end{aligned} \tag{3.59}$$

Because $I_k(\infty) = 0$ and $\lim_{t \to \infty} d\phi(t)/dt = 0$, we obtain from Eq. (3.59) the following self-consistent equation for $\phi(\infty)$:

$$\begin{aligned}
\phi(\infty) &= \frac{1}{\langle k \rangle}\sum_{k=1}^{\infty}(k-1)P(k)(1-S_k(0)\exp[-\lambda k\phi(\infty)]-V_k(0)\exp[-(1-\eta)\lambda k\phi(\infty)]) \\
&= 1-\frac{1}{\langle k \rangle}-\frac{1}{\langle k \rangle}\sum_{k=1}^{\infty}(k-1)P(k)(S_k(0)\exp[-\lambda k\phi(\infty)]+V_k(0)\exp[-(1-\eta)\lambda k\phi(\infty)]) \\
&= 1-\frac{1}{\langle k \rangle}-\frac{1}{\langle k \rangle}\sum_{k=1}^{\infty}(k-1)P(k)((1-x_k)\exp[-\lambda k\phi(\infty)]+x_k\exp[-(1-\eta)\lambda k\phi(\infty)]).
\end{aligned} \tag{3.60}$$

The FES and other fractions can be expressed as

$$S_k(\infty) = S_k(0)\exp[-\lambda k\phi(\infty)] = (1-x_k)\exp[-\lambda k\phi(\infty)], \tag{3.61}$$

$$V_k(\infty) = V_k(0)\exp[-(1-\eta)\lambda k\phi(\infty)] = x_k\exp[-(1-\eta)\lambda k\phi(\infty)], \tag{3.62}$$

$$R_k(\infty) = 1-(1-x_k)\exp[-\lambda k\phi(\infty)])-x_k\exp[-(1-\eta)\lambda k\phi(\infty)]). \tag{3.63}$$

As in the previous sub-section, the condition with a non-zero solution imposes the following inequality:

$$\begin{aligned}
&\frac{d}{d\phi(\infty)}\left(1-\frac{1}{\langle k \rangle}-\frac{1}{\langle k \rangle}\sum_{k=1}^{\infty}(k-1)P(k)((1-x_k)\exp[-\lambda k\phi(\infty)]\right. \\
&\qquad\qquad \left. +x_k\exp[-(1-\eta)\lambda k\phi(\infty)])\right)\Big|_{\phi(\infty)=0} > 1.
\end{aligned} \tag{3.64}$$

Assuming no dependence on the degree k of vaccinated individuals,

$$\lambda(1-\eta x)\frac{\langle k^2\rangle - \langle k\rangle}{\langle k\rangle} > 1. \tag{3.65}$$

This inequality defines the critical VC, which can eradicate the spread of the epidemic:

$$x_c = \left(1 - \frac{\langle k\rangle}{\lambda(\langle k^2\rangle\langle k\rangle)}\right)/\eta. \tag{3.66}$$

Again, note that the above assumption that the critical VC has no dependence on degree is only presumed so as to draw Fig. 3.17, discussed later. In a nutshell, Eqs. (3.64)–(3.66) are only required to obtain Fig. 3.17, which is irrelevant to the following discussion of the vaccination game.

The respective fractions of the four individuals, depending on whether they are vaccinated or non-vaccinated and whether they are healthy or infected, are summarized in Table 3.5, which is the counterpart to Table 3.3 based on the MFA with no spatial structure.

Figure 3.18 shows the FES depending on the vaccination levels in both the effectiveness and efficiency models for three different population structures: the power-law degree distribution, the Poisson distribution, and the well-mixed population (Fig. 3.12). Although the vaccination coverage (VC) x_k depends on degree k in a real decision-making process for whether to take the vaccination, as we discuss later, we hypothetically assume uniform VC here. From Fig. 3.18, the so-called critical VC that eradicates the spread of an epidemic can be read from the border of the extinct phase, where FES $= 0$. These borders suggest the critical VC to suppress the spread of the infection, which are analytically given by Eqs. (3.54) and (3.66), respectively.

Notably, for the power-law degree distribution (left panels; (*-A)), infectious diseases can easily spread vis-à-vis the other two spatial structures, which is consistent with previous studies (Cardillo et al. 2013; Li 2017). Thus, the protective effect of pre-emptive vaccination against contagion is relatively low if individuals have a spatial structure featuring a power-law distribution rather than a Poisson distribution or a well-mixed population.

3.4.5 *Payoff Structure and Global Dynamics for Strategy Updating*

As a vaccination game, Table 3.1 is also applicable even considering the spatial structure amid individuals. However, we should modify the expected payoffs in the form of the ASP, e.g., the average cooperative (vaccinated) payoff and the average defective (non-vaccinated) payoff for the respective provisions, namely imperfect

Table 3.5 Fractions of four individuals in the case of the efficiency model

Strategy/state	Healthy	Infected
Vaccinated	$x_k \exp[-(1-\eta)\lambda k\phi](\infty)]$	$x_k(1 - \exp[-(1-\eta)\lambda k\phi](\infty)])$
Non-vaccinated	$(1-x_k)\exp[-\lambda k\phi(\infty)]$	$(1-x_k)(1 - \exp[-\lambda k\phi(\infty)])$

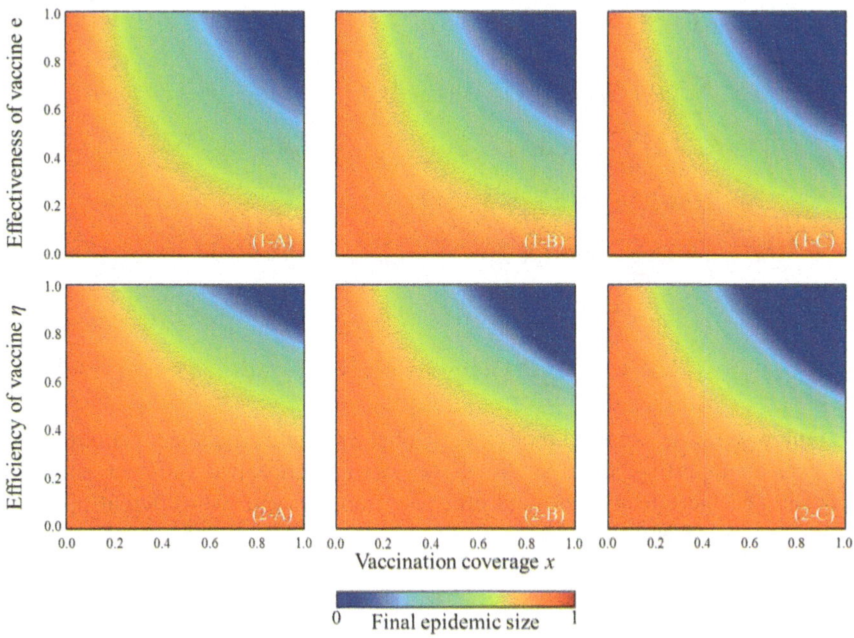

Fig. 3.18 Final epidemic size according to vaccination coverage and (1-*) effectiveness e (effectiveness of vaccination) or (2-*) efficiency η (efficiency of a defense against the contagion) in three different population structures: (*-A) power-law degree distribution; (*-B) Poisson distribution; and (*-C) well-mixed population. We assume that $\lambda = 4.8$ and 3.9 for power-law and Poisson distributions, respectively. For a well-mixed population, the reproduction number $R_0 = 2.5$ is applied, as in Fig. 3.11

vaccination and defense against contagion. We have to modify both the case with the effectiveness model and that with the efficiency model.

(Effectiveness model)

$$\langle \pi \rangle = -C_r \sum_k P(k) x_k \left(e + (1-e)\exp\left[-(\lambda k\phi(\infty)\right]\right)$$
$$-(C_r+1)\sum_k P(k) x_k (1-e)\left(1 - \exp\left[-\lambda k\phi(\infty)\right]\right)$$
$$-\sum_k P(k)(1-x_k)\left(1 - \exp\left[-\lambda k\phi(\infty)\right]\right), \quad (3.67)$$

$$\langle \pi_C \rangle = -C_r \sum_k P(k) x_k \big(e + (1-e) \exp\big[-(\lambda k \phi(\infty)\big]\big)$$
$$- (C_r + 1) \sum_k P(k) x_k (1-e)(1 - \exp[-\lambda k \phi(\infty)]), \tag{3.68}$$

$$\langle \pi_D \rangle = -\sum_k P(k)(1 - x_k)(1 - \exp[-\lambda k \phi(\infty)]). \tag{3.69}$$

(Efficiency model)

$$\langle \pi \rangle = -C_r \sum_k P(k) x_k \exp[-(1-\eta)\lambda k \phi(\infty)]$$
$$- (C_r + 1) \sum_k P(k) x_k (1 - \exp[-(1-\eta)\lambda k \phi(\infty)]) - \sum_k P(k)$$
$$\times (1 - x_k)(1 - \exp[-\lambda k \phi(\infty)]), \tag{3.70}$$

$$\langle \pi_C \rangle = -C_r \sum_k P(k) x_k \exp[-(1-\eta)\lambda k \phi(\infty)]$$
$$- (C_r + 1) \sum_k P(k) x_k (1 - \exp[-(1-\eta)\lambda k \phi(\infty)]), \tag{3.71}$$

$$\langle \pi_D \rangle = -\sum_k P(k)(1 - x_k)(1 - \exp[-\lambda k \phi(\infty)]). \tag{3.72}$$

Following these, we must modify the global dynamical equation to control what happens in repeating seasons. As above, we introduce two different epidemic models, i.e., the effectiveness and efficiency models, and two different updating rules, i.e., IB-RA and SB-RA:

Effectiveness model + IB-RA:

$$\frac{dx_k}{dt} = -x_k(e + (1-e)\exp[-\lambda k \phi(\infty)]) \frac{1}{\langle k \rangle} \sum_k kp(k)(1 - x_k) \exp[-\lambda k \phi(\infty)] P(\mathrm{HV} \leftarrow \mathrm{SFR})$$
$$-x_k(e + (1-e)\exp[-\lambda k \phi(\infty)]) \frac{1}{\langle k \rangle} \sum_k kp(k)(1 - x_k)(1 - \exp[-\lambda k \phi(\infty)]) P(\mathrm{HV} \leftarrow \mathrm{FFR})$$
$$-x_k(1-e)(1 - \exp[-\lambda k \phi(\infty)]) \frac{1}{\langle k \rangle} \sum_k kp(k)(1 - x_k) \exp[-\lambda k \phi(\infty)] P(\mathrm{IV} \leftarrow \mathrm{SFR})$$
$$-x_k(1-e)(1 - \exp[-\lambda k \phi(\infty)]) \frac{1}{\langle k \rangle} \sum_k kp(k)(1 - x_k)(1 - \exp[-\lambda k \phi(\infty)]) P(\mathrm{IV} \leftarrow \mathrm{FFR})$$
$$+(1 - x_k) \exp[-\lambda k \phi(\infty)] \frac{1}{\langle k \rangle} \sum_k kp(k) x_k (e + (1-e)\exp[-\lambda k \phi(\infty)]) P(\mathrm{SFR} \leftarrow \mathrm{HV})$$
$$+(1 - x_k) \exp[-\lambda k \phi(\infty)] \frac{1}{\langle k \rangle} \sum_k kp(k) x_k (1-e)(1 - \exp[-\lambda k \phi(\infty)]) P(\mathrm{SFR} \leftarrow \mathrm{IV})$$
$$+(1 - x_k)(1 - \exp[-\lambda k \phi(\infty)]) \frac{1}{\langle k \rangle} \sum_k kp(k) x_k (e + (1-e)\exp[-\lambda k \phi(\infty)]) P(\mathrm{FFR} \leftarrow \mathrm{HV})$$
$$+(1 - x_k)(1 - \exp[-\lambda k \phi(\infty)]) \frac{1}{\langle k \rangle} \sum_k kp(k) x_k (1-e)(1 - \exp[-\lambda k \phi(\infty)]) P(\mathrm{FFR} \leftarrow \mathrm{IV})$$
$$\tag{3.73}$$

Efficiency model + IB-RA:

$$\begin{aligned}\frac{dx_k}{dt} &= -x_k \exp[-(1-\eta)\lambda k\phi(\infty)]\frac{1}{\langle k\rangle}\sum_k kp(k)(1-x_k)\exp[-\lambda k\phi(\infty)]P(\mathrm{HV}\leftarrow\mathrm{SFR})\\ &-x_k \exp[-(1-\eta)\lambda k\phi(\infty)]\frac{1}{\langle k\rangle}\sum_k kp(k)(1-x_k)(1-\exp[-\lambda k\phi(\infty)])P(\mathrm{HV}\leftarrow\mathrm{FFR})\\ &-x_k(1-\exp[-(1-\eta)\lambda k\phi(\infty)])\frac{1}{\langle k\rangle}\sum_k kp(k)(1-x_k)\exp[-\lambda k\phi(\infty)]P(\mathrm{IV}\leftarrow\mathrm{SFR})\\ &-x_k(1-\exp[-(1-\eta)\lambda k\phi(\infty)])\frac{1}{\langle k\rangle}\sum_k kp(k)(1-x_k)(1-\exp[-\lambda k\phi(\infty)])P(\mathrm{IV}\leftarrow\mathrm{FFR})\\ &+(1-x_k)\exp[-\lambda k\phi(\infty)]\frac{1}{\langle k\rangle}\sum_k kp(k)x_k\exp[-(1-\eta)\lambda k\phi(\infty)]P(\mathrm{SFR}\leftarrow\mathrm{HV})\\ &+(1-x_k)\exp[-\lambda k\phi(\infty)]\frac{1}{\langle k\rangle}\sum_k kp(k)x_k(1-\exp[-(1-\eta)\lambda k\phi(\infty)])P(\mathrm{SFR}\leftarrow\mathrm{IV})\\ &+(1-x_k)(1-\exp[-\lambda k\phi(\infty)])\frac{1}{\langle k\rangle}\sum_k kp(k)x_k\exp[-(1-\eta)\lambda k\phi(\infty)]P(\mathrm{FFR}\leftarrow\mathrm{HV})\\ &+(1-x_k)(1-\exp[-\lambda k\phi(\infty)])\frac{1}{\langle k\rangle}\sum_k kp(k)x_k(1-\exp[-(1-\eta)\lambda k\phi(\infty)])P(\mathrm{FFR}\leftarrow\mathrm{IV})\end{aligned}, \tag{3.74}$$

Effectiveness model + SB-RA:

$$\begin{aligned}\frac{dx_k}{dt} &= -x_k(e+(1-e)\exp[-\lambda k\phi(\infty)])\frac{1}{\langle k\rangle}\sum_k kp(k)(1-x_k)P(\mathrm{HV}\leftarrow\mathrm{NV})\\ &-x_k(1-e)(1-\exp[-\lambda k\phi(\infty)])\frac{1}{\langle k\rangle}\sum_k kp(k)(1-x_k)P(\mathrm{IV}\leftarrow\mathrm{NV})\\ &+(1-x_k)\exp[-\lambda k\phi(\infty)]\frac{1}{\langle k\rangle}\sum_k kp(k)x_kP(\mathrm{SFR}\leftarrow\mathrm{V})\\ &+(1-x_k)(1-\exp[-\lambda k\phi(\infty)])\frac{1}{\langle k\rangle}\sum_k kp(k)x_kP(\mathrm{FFR}\leftarrow\mathrm{V})\end{aligned}, \tag{3.75}$$

Efficiency model + SB-RA:

$$\begin{aligned}\frac{dx_k}{dt} &= -x_k\exp[-(1-\eta)\lambda k\phi(\infty)]\frac{1}{\langle k\rangle}\sum_k kp(k)(1-x_k)P(\mathrm{HV}\leftarrow\mathrm{NV})\\ &-x_k(1-\exp[-(1-\eta)\lambda k\phi(\infty)])\frac{1}{\langle k\rangle}\sum_k kp(k)(1-x_k)P(\mathrm{IV}\leftarrow\mathrm{NV})\\ &+(1-x_k)\exp[-\lambda k\phi(\infty)]\frac{1}{\langle k\rangle}\sum_k kp(k)x_kP(\mathrm{SFR}\leftarrow\mathrm{V})\\ &+(1-x_k)(1-\exp[-\lambda k\phi(\infty)])\frac{1}{\langle k\rangle}\sum_k kp(k)x_kP(\mathrm{FFR}\leftarrow\mathrm{V})\end{aligned}, \tag{3.76}$$

3.4.6 Result of the Networked Vaccination Game; Comparison of Different Degree Distributions

Figures 3.19, 3.20, 3.21, and 3.22 show the FES (left-hand panels), VC (central panels), and ASP (right-hand panels) drawn on the 2D plane of either the relative vaccination cost vs. effectiveness (Figs. 3.19 and 3.21) or relative vaccination cost vs. efficiency (Figs. 3.20 and 3.22) graph for three different degree distributions: the power-law degree distribution (upper panels), the Poisson distribution (middle panels), and a well-mixed population (lower panels). Figures 3.19 and 3.20 show results considering IB-RA, while Figs. 3.21 and 3.22 show those considering SB-RA

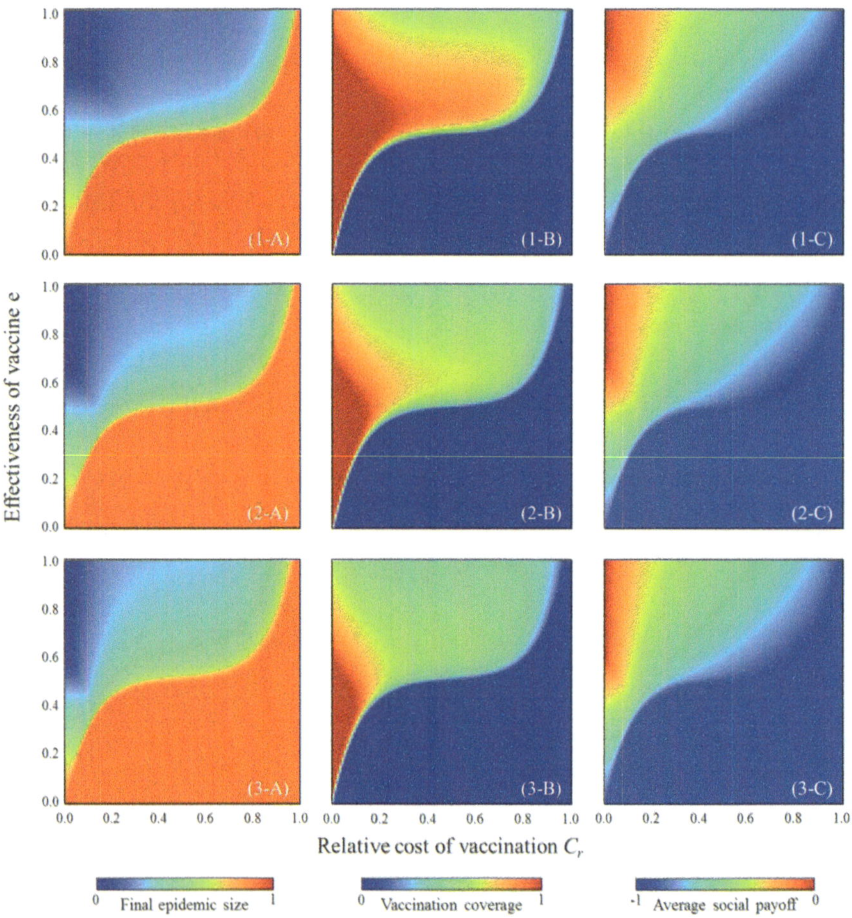

Fig. 3.19 Final epidemic size (left-hand panels; *-A), VC (central panels; *-B), and ASP (right-hand panels; *-C) in population structure: power-law-degree distribution (upper panels; 1-*), Poisson distribution (middle panels; 2-*), and well-mixed population (lower panels; 3-*). An effectiveness model is applied. IB-RA is applied for a strategy-updating rule

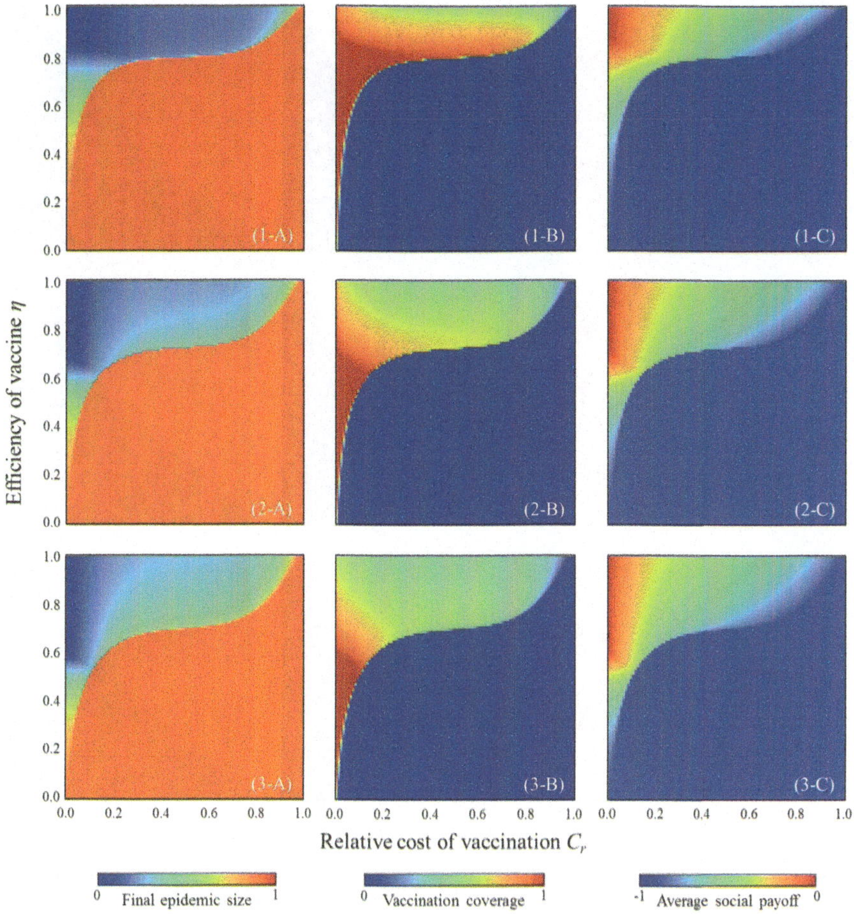

Fig. 3.20 Final epidemic size (left-hand panels; *-A), VC (central panels; *-B), and ASP (right-hand panels; *-C) in population structure: power-law degree distribution (upper panels; 1-*), Poisson distribution (middle panels; 2-*), and well-mixed population (lower panels; 3-*). The efficiency model is applied. IB-RA is applied for the strategy-updating rule

as the strategy-updating rule. Note that Fig. 3.19 panels (3-*) [Effectiveness + IB-RA] are consistent with Fig. 3.13 (1-*). Likewise, note that Fig. 3.20 panels (3-*) [Efficiency + IB-RA] are consistent with Fig. 3.14 (1-*); Fig. 3.21 panels (3-*) [Effectiveness + SB-RA] are consistent with Fig. 3.13 (2-*); and Fig. 3.22 panels (3-*) [Efficiency + SB-RA] is consistent with Fig. 3.14 (2-*). We presume an average degree of 8 for both the power-law (say, the BA-SF graph) and the Poisson (E-R random graph) distributions.

In all figures, light red in FES, dark blue in VC, and light red in ASP indicate that a pandemic is taking place, with most individuals not relying on vaccination. Thus, almost-full-scale spread of the infection is inevitable. Roughly speaking, these regions emerge when a smaller effectiveness (or efficiency) is assumed or a larger

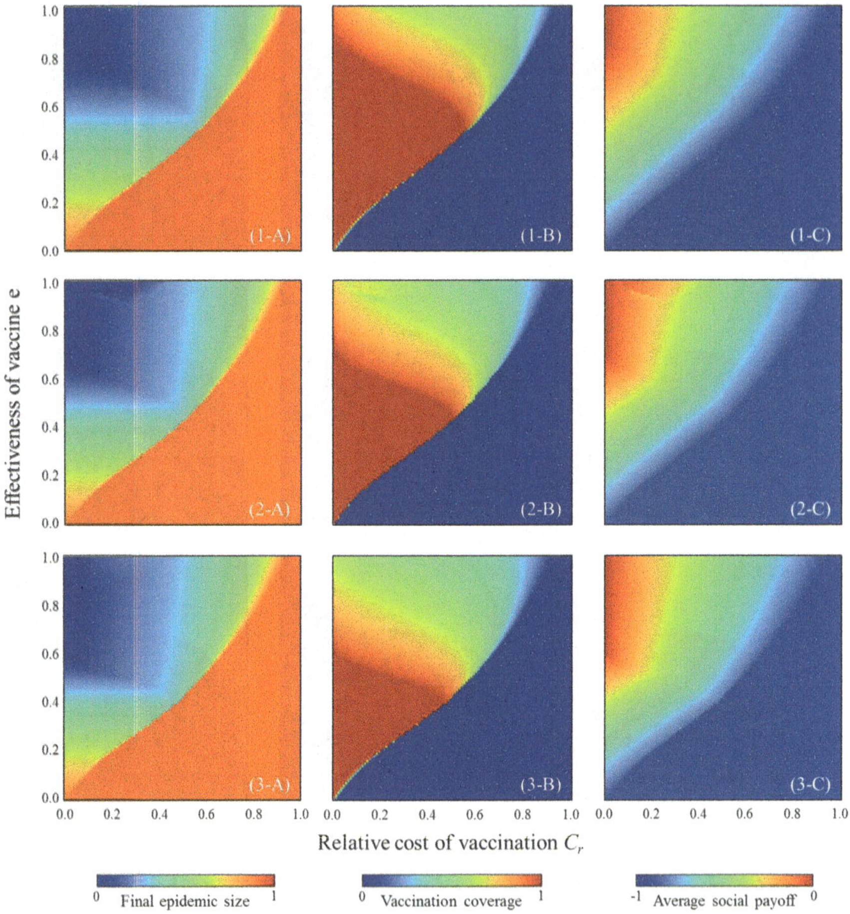

Fig. 3.21 Final epidemic size (left-hand panels; *-A), VC (central panels; *-B), and ASP (right-hand panels; *-C) in population structure: power-law degree distribution (upper panels; 1-*), Poisson distribution (middle panels; 2-*), and well-mixed population (lower panels; 3-*). The effectiveness model is applied. SB-RA is applied for the strategy-updating rule

cost is imposed. This seems quite natural, because most individuals tend to avoid committing vaccination.

The border between each of these monotone regions and the remaining area implies a critical line suggesting a combination of critical effectiveness (efficiency) and critical vaccination cost to appropriately control the spread of an epidemic; this implies a phase change between the pandemic and controlled phases. As far as the controlled phase is concerned, interestingly, a lower effectiveness (or efficiency) can realize a higher VC. Such an ironic situation can be justified as follows: the lower reliability of the protecting measure makes more individuals commit to it due to its uncertainty. From the opposite viewpoint, even if a large fraction of agents take a

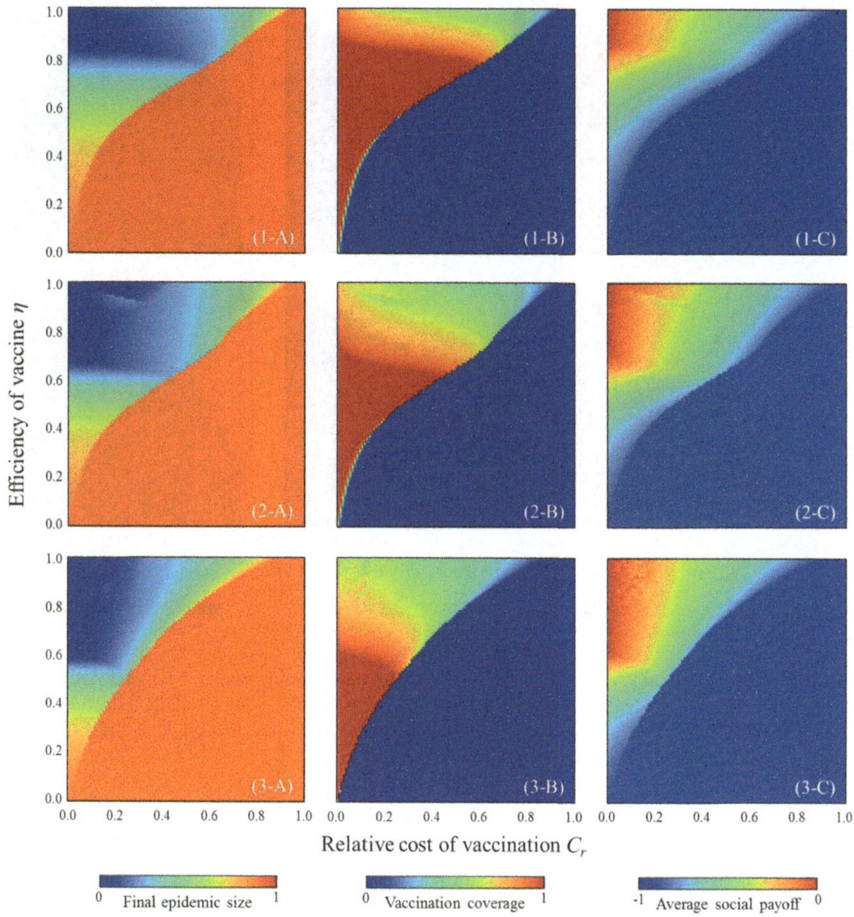

Fig. 3.22 Final epidemic size (left-hand panels; *-A), VC (central panels; *-B), and ASP (right-hand panels; *-C) in population structure: power-law degree distribution (upper panels; 1-*), Poisson distribution (middle panels; 2-*), and well-mixed population (lower panels; 3-*). The efficiency model is applied. SB-RA is applied for the strategy-updating rule

measure, an epidemic cannot be eradicated because of the lower reliability of the vaccination itself.

Comparing the different network topologies, the power-law distribution shows the largest VC, followed by the Poisson distribution. The well-mixed population has the smallest VC in the region of the controlled phase. This tendency, more precisely the order of these three topologies, seems acceptable. This is because the more heterogeneous the degree distribution becomes, the more degree agents exist. Specifically, the power-law distribution lets a hub agent work as a so-called super spreader. This tendency can be observed to some extent in all cases, regardless of the employed strategy-update rule or whether the efficiency or effectiveness model is assumed, although the case presuming SB-RA with the effectiveness model

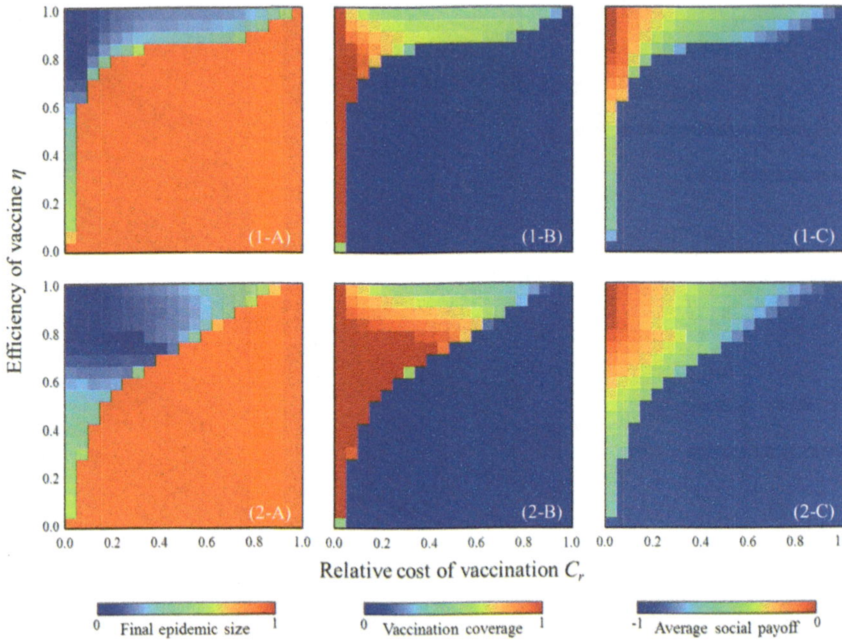

Fig. 3.23 Final epidemic size (left-hand panels; *-A), VC (central panels; *-B), and ASP (right-hand panels; *-C) for strategy-updating rules IB-RA (upper panels; 1-*) and SB-RA (lower panels; 2-*). Effectiveness model is applied. We assumed a BA-SF network with $N = 10^4$, $\langle k \rangle = 8$, $\beta_e = 0.19$, and $\gamma = 1/3$

(Fig. 3.21 (*-B)) seems subtle. The abovementioned tendency for VC implies that the power-law distribution brings less ASP than does the Poisson distribution, which is less than the well-mixed population shows.

The efficiency models have a wider pandemic phase at first glance. This implies that the defense against contagion with a given η is less effective for suppressing epidemic diffusion than is imperfect vaccination, which is consistent with what we discussed in Sub-section 3.3.3.

To validate our theoretical result, we conducted a series of numerical simulations based on the MAS approach. To compare the power-law degree distribution, we assumed a BA-SF network with an average degree $\langle k \rangle = 8$ as the underlying network connection among agents. Following previous studies, we set $\beta_e = 0.19$, which was determined as the minimum effective transition rate that exceeded a pre-defined threshold FES of 0.89 without vaccinated individuals. This result is obtained by an ensemble average of ten independent realizations starting from different initial conditions. Figures 3.23 and 3.24 show the results for the effectiveness and efficiency models, respectively. Generally, all four results are consistent with Figs. 3.19 (1-*), 3.21 (1-*), 3.20 (1-*), and 3.22 (1-*), although subtle discrepancies arise from the fact that the simulation assumed a finite population size of $N = 10^4$.

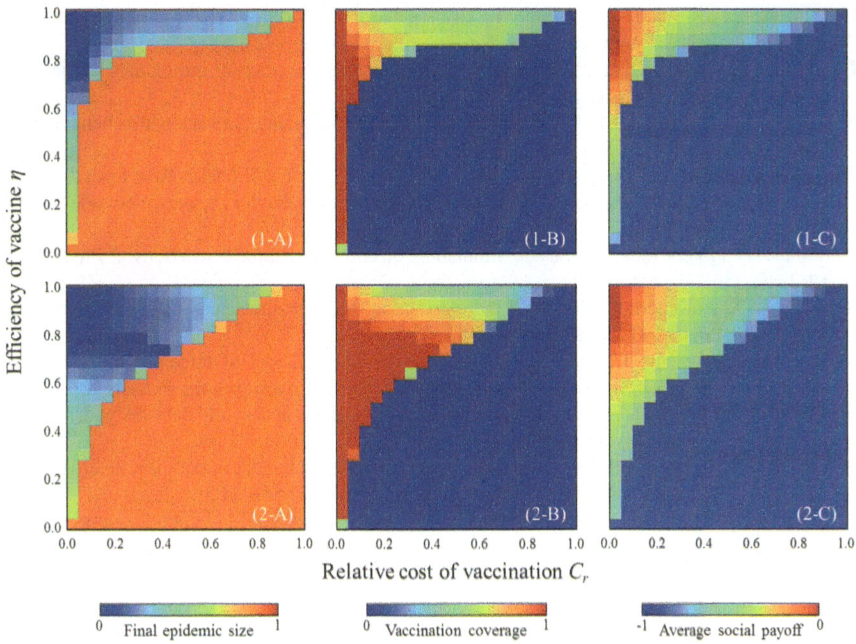

Fig. 3.24 Final epidemic size (left-hand panels; *-A), VC (central panels; *-B), and ASP (right-hand panels; *-C) for strategy-updating rules IB-RA (upper panels; 1-*) and SB-RA (lower panels; 2-*). An efficiency model is applied. We assumed a BA-SF network with $N = 10^4$, $\langle k \rangle = 8$, $\beta_e = 0.19$, and $\gamma = 1/3$

Summing up this section, the framework backed by a theoretical vaccination game validated by the MAS result shows that the power-law distribution has a smaller tendency of suppressing epidemics than the Poisson degree distribution. The framework presented here clarifies the influence from the heterogeneous degree distributions as compared with none of the spatial structures; an infinite and well-mixed population. The framework can adjust to whether the effectiveness model or the efficiency model is employed, and to which the strategy-updating rule is presumed.

References

Albert R, Albert-László B (2002) Statistical mechanics of complex networks. Rev Mod Phys 74:47–97

Barabási AL, Albert R (1999) Emergence of scaling in random networks. Science 286:509–512

Bollobás B (1985) Random graphs. Academic Press, London

Brauer F, Castillo-Chavez C, Feng Z (2019) Mathematical models in epidemiology. Springer, New York

Brian Arthur W (1994) Inductive reasoning and bounded rationality. Am Econ Rev 84:406–411

Cardillo A, Reyes-Suárez C, Naranjo F, Gómez-Gardeñes J (2013) Evolutionary vaccination dilemma in complex networks. Phys Rev E 88(3):032803
Challet D, Marsili M, Zhang Y-C (2005) Minority games: interacting individuals in financial markets. Oxford University Press, Oxford
Fu F, Rosenbloom DI, Wang L, Nowak MA (2011) Imitation dynamics of vaccination behaviour on social networks. Proc R Soc B 278:42–49
Fukuda E, Kokubo S, Tanimoto J, Wang Z, Hagishima A, Ikegaya N (2014) Risk assessment for infectious disease and its impact on voluntary vaccination behavior in social networks. Chaos Solitons Fractals 68:1–9
Gillespie DTJ (1977) Exact stochastic simulation of coupled chemical reactions. J Phys Chem 81:2340–2361
Kermack WO, McKendrick AG (1927) A contribution to the mathematical theory of epidemics. Proc R Soc Lond A 115(722):700–721
Li Q, Li M, Lv L, Guo C, Lu K (2017) A new prediction model of infectious diseases with vaccination strategies based on evolutionary game theory. Chaos Solitons Fractals 104:51–60
Matsuzawa R, Tanimoto J, Fukuda E (2017) Properties of a new small-world network with spatially biased random shortcuts. Physica A 486:408–415
Moreno Y, Pastor-Satorras R, Vespignani A (2002) Epidemic outbreaks in complex heterogeneous networks. Eur Phys J B Condens Matter Complex Syst 26(4):521–529
Pastor-Satorras R, Rubi M, Diaz-Guilera A (2003) Statistical mechanics of complex networks, vol 625. Springer, Berlin
Tanimoto J (2014) Mathematical analysis of environmental system. Springer, Tokyo
van der Helm PA (2010) Weber-Fechner behaviour in symmetry perception? Atten Percept Psychophys 72(7):1854–1864
Vardavas R, Breban R, Blower S (2007) Can influenza epidemics be prevented by voluntary vaccination? PLoS Comput Biol 3(5):e85
Watts DJ, Strogatz SH (1998) Collective dynamics of ‘small-world’ networks. Nature 393:440–442

Chapter 4
Plural Strategies: Intervention Game

In this chapter, we apply the concept of a vaccination game, using mathematical epidemiology to capture the spread of an epidemic and evolutionary game theory to model the time evolution of human decision making. We discuss how allowing

J. Tanimoto, *Sociophysics Approach to Epidemics*, Evolutionary Economics and Social Complexity Science 23, https://doi.org/10.1007/978-981-33-6481-3_4

multiple alternatives to committing to a vaccination changes social equilibrium and impacts the final epidemic size (FES) and average social payoff (ASP).

4.1 Alternative Provisions Featuring Different Combinations of Cost–Effect Performances

As illustrated schematically in Fig. 1.7, when coping with various epidemics in reality, several provisions are workable. Considering only the alternative provisions to which each individual can spontaneously commit (i.e., without being forcibly imposed by the government, like a quarantine), there are several ways to avoid infection. Although a vaccination might be most effective, the majority of people are not necessarily willing to commit to it. The biggest impediment is the cost–risk relation: namely the cost of a vaccination being comparable to the recognized infection cost resulting from the real disease cost and infection risk. In fact, as discussed in Chap. 2, when these two quantities are comparable, the social dilemma behind this social structure allows people to try to fee-ride on herd immunity, leading to everyone becoming defective, i.e., intentionally becoming a non-vaccinator. This may be socially problematic.

Leaving aside such an extreme case (either committing to a vaccination or not), we should consider a realistic context in which there are multiple provisions besides vaccination. Let us call these intermediate-defense measures (IDMs); they include gargling, wearing a mask, and taking an OTC medicine, among others. Each is basically less effective than vaccination for avoiding risk, but the cost is also less than that of a vaccine dose. Thus, from the cost–effect–performance standpoint, there might be some IDMs that are rather more beneficial than a vaccination, or at least comparable to it. If this is the case, the existence of such IDMs should be carefully investigated, even though the face value of their effectivity[1] is less than that of vaccination. Specifically, it is quite important to explore how co-existing IDMs besides vaccination affect FES and ASP through the time evolution of people's choice in provisions. For example, the existence of a high fraction of the population assuming a more cost-effective IDM than vaccination may cause FES and ASP to be lower than in the case of a binary choice between vaccinating or not vaccinating, since people who heavily rely on such IDMs may not commit to vaccination. In this sense, such an exploration is informative as a policymaking process by the public-health authority.

We consider the vaccination-game framework introduced in the previous chapter, but extend it to apply to a situation with multiple provisions; we hereafter call this intervention game. The approach that we take here presumes a mean-field

[1]The term "effectivity" broadly means the effect brought about by a certain provision. The mathematical characteristics of each provision can be modeled via either effectiveness or efficiency models, as discussed in Sect. 3.2.1.

approximation (MFA) and applies an effectiveness model for vaccination and an efficiency model for IDM. Moreover, we presume IB-RA, SB-RA, or DC for the strategy-updating rule. All of those elements were already introduced as model parts in Chap. 3.

Some previous studies by Iwamura et al.[2] and Ida et al.[3] have already investigated the effect of an IDM. However, their studies relied on a multi-agent-simulation (MAS) approach. Thus, the obtained knowledge seems narrow, in the sense of being like point-information, so it does not cover the entire space of model parameters. This is why we seek a comprehensive result using the theoretical approach established in Chap. 3.

4.2 Model Structure

4.2.1 *Formulation of the SVMBIR Model*[4]

We use the SIR model with susceptible (S), infected (I), and recovered (R) compartments as a base. As provisions to avoid infection, we additionally introduce the vaccinated (V), intermediate-defense-measure (M), and taking both vaccination and IDM (B) compartments. These relate to the strategy taken by an individual. The first strategy is adopting none of any provisions, i.e., defection. The second strategy is committing vaccination, which has been usually defined as cooperation. The present model further introduces the third strategy of committing to IDM, and the fourth strategy of committing to both vaccination and IDM.

There are two timescales in the model: local and global, as we discussed before. The local timescale describes what happens in a single season according to mathematical epidemiology, while the global one controls how an individual's strategy evolves over repeating seasons in the approach to social equilibrium.

Let the effectiveness of a vaccine be e $(0 \leq e \leq 1)$ and the efficiency of IDM be $(0 \leq \eta \leq 1)$, meaning how taking each of these measures can reduce the risk of being infected, of which the detailed mechanism was introduced in Sect. 3.2.1. A non-provisioned susceptible individual may become infected if they are exposed to infectious individuals at a disease-transmission rate of β [day^{-1} person^{-1}]. An individual in (S) who uses IDM may still become infectious at the rate $(1 - \eta)\beta$. Meanwhile, an infected individual recovers at the recovery rate γ [day^{-1}] and the ratio of the disease-spreading rate to the recovery rate (i.e., β/γ) is called the basic reproduction number, R_0. To this end, we establish the SVMBIR model, as shown in Fig. 4.1, and its dynamics are described below:

[2]Iwamura et al. (2016), p. 093501.

[3]Ida and Tanimoto (2018), pp. 337–341.

[4]Alam et al. (2019), pp. 408–422.

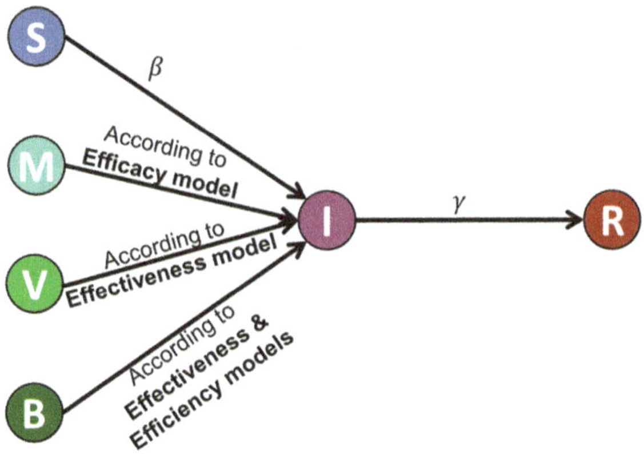

Fig. 4.1 Block-chart of the SVMBIR model

$$\frac{dS(x,y,z,t)}{dt} = -\beta S(x,y,z,t) I(x,y,z,t), \tag{4.1a}$$

$$\frac{dV(x,y,z,t)}{dt} = -\beta (V(x,y,z,t) - eV(x,y,z,0)) I(x,y,z,t), \tag{4.1b}$$

$$\frac{dM(x,y,z,t)}{dt} = -(1-\eta)\beta M(x,y,z,t) I(x,y,z,t), \tag{4.1c}$$

$$\frac{dB(x,y,z,t)}{dt} = -(1-\eta)\beta (B(x,y,z,t) - eB(x,y,z,0)) I(x,y,z,t), \tag{4.1d}$$

$$\begin{aligned}\frac{dI(x,y,z,t)}{dt} &= \beta S(x,y,z,t) I(x,y,z,t) + \beta (V(x,y,z,t) - eV(x,y,z,0)) I(x,y,z,t) \\ &+(1-\eta)\beta M(x,y,z,t) I(x,y,z,t) + (1-\eta)\beta (B(x,y,z,t) - eB(x,y,z,0)) I(x,y,z,t) \\ &-\gamma I(x,y,z,t),\end{aligned} \tag{4.1e}$$

$$\frac{dR(x,y,z,t)}{dt} = \gamma I(x,y,z,t), \tag{4.1f}$$

where x, y, and z, respectively, represent the fraction of the population taking vaccination, IDM, and both provisions at the same time. Also, for the assumed set of initial values $S(x,y,z,0) = 1\text{-}x\text{-}y\text{-}z$, $V(x,y,z,0) = x$, $M(x,y,z,0) = y$, $B(x,y,z,0) = z$, and $I(x,y,z,0) \cong 0$, we must presume the constraint:

$$\begin{aligned}&S(x,y,z,t) + V(x,y,z,t) + M(x,y,z,t) + B(x,y,z,t) + I(x,y,z,t) \\ &\quad + R(x,y,z,t) \\ &\quad = 1.\end{aligned} \tag{4.2}$$

Equation (4.1) under the constraint if Eq. (4.2) can be analytically solved by allowing the form of transcendental equation. In particular, the so-called FES and other respective fractions at equilibrium ($t = \infty$, meaning a local equilibrium at a certain season) can be expressed as

$$S(x, y, z, \infty) = (1 - \mathrm{x} - \mathrm{y} - \mathrm{z}) \exp\left[-\mathrm{R}_0 R(x, y, z, \infty)\right], \tag{4.3a}$$

$$V(x, y, z, \infty) = \mathrm{x}(e + (1 - e) \exp\left[-\mathrm{R}_0 R(x, y, z, \infty)\right]), \tag{4.3b}$$

$$M(x, y, z, \infty) = \mathrm{y}\ \exp\left[-(1 - \eta)\mathrm{R}_0 R(x, y, z, \infty)\right], \tag{4.3c}$$

$$B(x, y, z, \infty) = \mathrm{z}(e + (1 - e) \exp\left[-(1 - \eta)\mathrm{R}_0 R(x, y, z, \infty)\right]), \tag{4.3d}$$

$$\begin{aligned} R(x, y, z, \infty) = 1 &- (1 - \mathrm{x} - \mathrm{y} - \mathrm{z}) \exp\left[-\mathrm{R}_0 R(x, y, z, \infty)\right] \\ &- \mathrm{x}(e + (1 - e) \exp\left[-\mathrm{R}_0 R(x, y, z, \infty)\right]) \\ &- \mathrm{y}\exp\left[-(1 - \eta)\mathrm{R}_0 R(x, y, z, \infty)\right] \\ &- \mathrm{z}(e + (1 - e) \exp\left[-(1 - \eta)\mathrm{R}_0 R(x, y, z, \infty)\right]). \end{aligned} \tag{4.3e}$$

Table 4.1 summarizes the fractions; V, M, B and defectors at local equilibrium depending on whether an individual is healthy or infected.

4.2.2 Payoff Structure

As we discussed in Chap. 3, in a typical model without a loop back from I to S, an epidemic season is sustained until all infected individuals recover. Meanwhile, if individuals become infected during an epidemic, the infection cost C_{i} is imposed upon them. On the other hand, individuals with any provisions who remain healthy during the course of an epidemic can fortunately avoid any cost anxiety. However, some unlucky individuals who become infected despite vaccination, IDM, or both are charged with their respective provision costs, as well as the infection cost. To figure out each individual's payoff, without loss of generality, the cost is rescaled by defining a relative cost of vaccination, namely $C_{rv} = C_v / C_{\mathrm{i}}$ $(0 \le C_v \le 1;\ C_{\mathrm{i}} = 1)$. Likewise, the relative costs of IDM (C_{rm}) and the combined strategy (C_{rb}; i.e., the sum of C_{rv} and C_{rm}) are rescaled for the payoff structure. Eventually, at the end of an epidemic season, each individual's payoff is determined by their final state. Table 4.2 encapsulates the payoffs for those who commit to particular provisions depending on whether they are healthy or infected.

Now we are able to quantify the overall expected payoff as a form of the average social payoff $\langle \pi \rangle$, the average vaccinated payoff $\langle \pi_V \rangle$, the average self-protected payoff $\langle \pi_M \rangle$, the average combined payoff (vaccinated + self-protected) $\langle \pi_B \rangle$, and the average defective payoff $\langle \pi_D \rangle$ for the adoption of respective strategies:

$$\begin{aligned} \langle \pi \rangle = &-C_{rv} x(1 + (1 - e) \exp\left[-\mathrm{R}_0 R(x, y, z, \infty)\right]) - (C_{rv} + 1)x(1 - e) \\ &\times (1 - \exp\left[-\mathrm{R}_0 R(x, y, z, \infty)\right]) \\ &- C_{rm} y \exp\left[-(1 - \eta)\mathrm{R}_0 R(x, y, z, \infty)\right] \\ &- (C_{rm} + 1)y(1 - \exp\left[-(1 - \eta)\mathrm{R}_0 R(x, y, z, \infty)\right]) \end{aligned}$$

Table 4.1 Fractions of eight types of individuals

Strategy/State	Healthy	Infected
Vaccinated	$x(e+(1-e)\exp[-R_0R(x,y,z,\infty)])$	$x(1-e)(1-\exp[-R_0R(x,y,z,\infty)])$
IDM	$y\cdot\exp[-(1-\eta)R_0R(x,y,z,\infty)]$	$y(1-\exp[-(1-\eta)R_0R(x,y,z,\infty)])$
Vaccinated & IDM	$z(e+(1-e)\exp[-(1-\eta)R_0R(x,y,z,\infty)])$	$z(1-e)(1-\exp[-(1-\eta)R_0R(x,y,z,\infty)])$
Defector	$(1-x-y-z)\exp(-R_0R(x,y,z,\infty))$	$(1-x-y-z)(1-\exp[-R_0R(x,y,z,\infty)])$

Table 4.2 Estimated payoff structure at the end of an epidemic season

Strategy/State	Healthy	Infected
Vaccinated	$-C_{rv}$	$-C_{rv}-1$
IDM	$-C_{rm}$	$-C_{rm}-1$
Vaccinated & IDM	$-C_{rb}$	$-C_{rb}-1$
Defector	0	-1

$$-C_{rb}z(1+(1-e)\exp[-(1-\eta)\mathrm{R}_0R(x,y,z,\infty)])$$
$$-(C_{rb}+1)z(1-e)(1-\exp[-(1-\eta)\mathrm{R}_0R(x,y,z,\infty)])$$
$$-(1-x-y-z)(1-\exp[-\mathrm{R}_0R(x,y,z,\infty)]), \tag{4.4}$$

$$\langle\pi_V\rangle=-C_{rv}(1+(1-e)\exp[-\mathrm{R}_0R(x,y,z,\infty)])-(C_{rv}+1)(1-e)\times(1-\exp[-\mathrm{R}_0R(x,y,z,\infty)]), \tag{4.5}$$

$$\langle\pi_M\rangle=-C_{rm}\exp[-(1-\eta)\mathrm{R}_0R(x,y,z,\infty)]-(C_{rm}+1)\times(1-\exp[-(1-\eta)\mathrm{R}_0R(x,y,z,\infty)]), \tag{4.6}$$

$$\langle\pi_B\rangle=-C_{rb}(1+(1-e)\exp[-(1-\eta)\mathrm{R}_0R(x,y,z,\infty)])$$
$$-(C_{rb}+1)(1-e)(1-\exp[-(1-\eta)\mathrm{R}_0R(x,y,z,\infty)]), \tag{4.7}$$

$$\langle\pi_D\rangle=-(1-\exp[-\mathrm{R}_0R(x,y,z,\infty)]). \tag{4.8}$$

4.2.3 Strategy-Updating and Global Dynamics

Normally in the standard framework for a vaccination game, when an epidemic season ends, an individual is allowed to alter their strategy as to whether to implement a provision by reflecting on what happened in the previous epidemic season. In the present study, i.e., an intervention game, we follow same concept that was discussed in Sects. 3.2.2 and 3.2.3.

There are three strategy-updating rules: individual-based risk assessment (IB-RA), strategy-based risk assessment (SB-RA), and direct commitment (DC). We assume MFA and set the noise parameter in the Fermi function for each state probability function κ to 0.1.

4.2.3.1 Individual-Based Risk Assessment (IB-RA)

In the present framework, there are eight classes of individual in terms of cost burden: (i) a healthy vaccinator (HV) who pays $-C_{rv}$; (ii) an infected vaccinator (IV) who pays $-C_{rv} - 1$; (iii) a healthy individual with IDM (HM) who pays $-C_{rm}$; (iv) an infected individual with IDM (IM) who pays $-C_{rm} - 1$; (v) a healthy individual with both IDM and a vaccination (HB) who pays $-C_{rb}$; (vi) an infected individual with both (IB) who pays $-C_{rb} - 1$; (vii) a successful free rider (SFR) who pays nothing; and (viii) a failed free rider (FFR) who pays -1. The individual can choose one of four strategies, i.e., vaccination, self-protection by IDM, both, and neither. Thus, the transition probabilities in the IB-RA rule are covered by the following forty-eight (=(8 − 2) × 8) cases:

$$P(\text{HV} \leftarrow \text{HM}) = \frac{1}{1 + \exp\left[-(-C_{rm} - (-C_{rv}))/\kappa\right]}, \tag{4.9a}$$

$$P(\text{HV} \leftarrow \text{IM}) = \frac{1}{1 + \exp\left[-\frac{(-C_{rm}-1-(-C_{rv}))}{\kappa}\right]}, \tag{4.9b}$$

$$P(\text{HV} \leftarrow \text{HB}) = \frac{1}{1 + \exp\left[-(-C_{rb} - (-C_{rv}))/\kappa\right]}, \tag{4.9c}$$

$$P(\text{HV} \leftarrow \text{IB}) = \frac{1}{1 + \exp\left[-(-C_{rb} - 1 - (-C_{rv}))/\kappa\right]}, \tag{4.9d}$$

$$P(\text{HV} \leftarrow \text{SFR}) = \frac{1}{1 + \exp\left[-(0 - (-C_{rv}))/\kappa\right]}, \tag{4.9e}$$

$$P(\text{HV} \leftarrow \text{FFR}) = \frac{1}{1 + \exp\left[-(-1 - (-C_{rv}))/\kappa\right]}, \tag{4.9f}$$

$$P(\text{IV} \leftarrow \text{HM}) = \frac{1}{1 + \exp\left[-(-C_{rm} - (-C_{rv} - 1))/\kappa\right]}, \tag{4.9g}$$

$$P(\text{IV} \leftarrow \text{IM}) = \frac{1}{1 + \exp\left[-(-C_{rm} - 1 - (-C_{rv} - 1))/\kappa\right]}, \tag{4.9h}$$

$$P(\text{IV} \leftarrow \text{HB}) = \frac{1}{1 + \exp\left[-(-C_{rb} - (-C_{rv} - 1))/\kappa\right]}, \tag{4.9i}$$

$$P(IV \leftarrow IB) = \frac{1}{1 + exp\left[-(-C_{rb} - 1 - (-C_{rv} - 1))/\kappa\right]}, \tag{4.8j}$$

$$P(\text{IV} \leftarrow \text{SFR}) = \frac{1}{1 + \exp\left[-(0 - (-C_{rv} - 1))/\kappa\right]}, \tag{4.9k}$$

$$P(\text{IV} \leftarrow \text{FFR}) = \frac{1}{1 + \exp\left[-(-1 - (-C_{rv} - 1))/\kappa\right]}, \tag{4.9l}$$

$$P(\mathrm{HM} \leftarrow \mathrm{HV}) = \frac{1}{1 + \exp\left[-(-C_{rv} - (-C_{rm}))/\kappa\right]}, \tag{4.9m}$$

$$P(\mathrm{HM} \leftarrow \mathrm{IV}) = \frac{1}{1 + \exp\left[-(-C_{rv} - 1 - (-C_{rm}))/\kappa\right]}, \tag{4.9n}$$

$$P(\mathrm{HM} \leftarrow \mathrm{HB}) = \frac{1}{1 + \exp\left[-(-C_{rb} - (-C_{rm}))/\kappa\right]}, \tag{4.9o}$$

$$P(\mathrm{HM} \leftarrow \mathrm{IB}) = \frac{1}{1 + \exp\left[-(-C_{rb} - 1 - (-C_{rm}))/\kappa\right]}, \tag{4.9p}$$

$$P(\mathrm{HM} \leftarrow \mathrm{SFR}) = \frac{1}{1 + \exp\left[-(0 - (-C_{rm}))/\kappa\right]}, \tag{4.9q}$$

$$P(\mathrm{HM} \leftarrow \mathrm{FFR}) = \frac{1}{1 + \exp\left[-(-1 - (-C_{rm}))/\kappa\right]}, \tag{4.9r}$$

$$P(\mathrm{IM} \leftarrow \mathrm{HV}) = \frac{1}{1 + \exp\left[-(-C_{rv} - (-C_{rm} - 1))/\kappa\right]}, \tag{4.9s}$$

$$P(\mathrm{IM} \leftarrow \mathrm{IV}) = \frac{1}{1 + \exp\left[-(-C_{rv} - 1 - (-C_{rm} - 1))/\kappa\right]}, \tag{4.9t}$$

$$P(\mathrm{IM} \leftarrow \mathrm{HB}) = \frac{1}{1 + \exp\left[-(-C_{rb} - (-C_{rm} - 1))/\kappa\right]}, \tag{4.9u}$$

$$P(\mathrm{IM} \leftarrow \mathrm{IB}) = \frac{1}{1 + \exp\left[-(-C_{rb} - 1 - (-C_{rm} - 1))/\kappa\right]}, \tag{4.9v}$$

$$P(\mathrm{IM} \leftarrow \mathrm{SFR}) = \frac{1}{1 + \exp\left[-(0 - (-C_{rm} - 1))/\kappa\right]}, \tag{4.9w}$$

$$P(\mathrm{IM} \leftarrow \mathrm{FFR}) = \frac{1}{1 + \exp\left[-(-1 - (-C_{rm} - 1))/\kappa\right]}, \tag{4.9x}$$

$$P(\mathrm{HB} \leftarrow \mathrm{HV}) = \frac{1}{1 + \exp\left[-(-C_{rv} - (-C_{rb}))/\kappa\right]}, \tag{4.9y}$$

$$P(\mathrm{HB} \leftarrow \mathrm{IV}) = \frac{1}{1 + \exp\left[-(-C_{rv} - 1 - (-C_{rb}))/\kappa\right]}, \tag{4.9z}$$

$$P(\mathrm{HB} \leftarrow \mathrm{HM}) = \frac{1}{1 + \exp\left[-(-C_{rm} - (-C_{rb}))/\kappa\right]}, \tag{4.9aa}$$

$$P(\mathrm{HB} \leftarrow \mathrm{IM}) = \frac{1}{1 + \exp\left[-(-C_{rm} - 1 - (-C_{rb}))/\kappa\right]}, \tag{4.9ab}$$

$$P(\mathrm{HB} \leftarrow \mathrm{SFR}) = \frac{1}{1 + \exp\left[-(0 - (-C_{rb}))/\kappa\right]}, \tag{4.9ac}$$

$$P(\text{HB} \leftarrow \text{FFR}) = \frac{1}{1 + \exp\left[-(-1 - (-C_{rb}))/\kappa\right]}, \tag{4.9ad}$$

$$P(\text{IB} \leftarrow \text{HV}) = \frac{1}{1 + \exp\left[-(-C_{rv} - (-C_{rb} - 1))/\kappa\right]}, \tag{4.9ae}$$

$$P(\text{IB} \leftarrow \text{IV}) = \frac{1}{1 + \exp\left[-(-C_{rv} - 1 - (-C_{rb} - 1))/\kappa\right]}, \tag{4.9af}$$

$$P(\text{IB} \leftarrow \text{HM}) = \frac{1}{1 + \exp\left[-(-C_{rm} - (-C_{rb} - 1))/\kappa\right]}, \tag{4.9ag}$$

$$P(\text{IB} \leftarrow \text{IM}) = \frac{1}{1 + \exp\left[-(-C_{rm} - 1 - (-C_{rb} - 1))/\kappa\right]}, \tag{4.9ah}$$

$$P(\text{IB} \leftarrow \text{SFR}) = \frac{1}{1 + \exp\left[-(0 - (-C_{rb} - 1))/\kappa\right]}, \tag{4.9ai}$$

$$P(\text{IB} \leftarrow \text{FFR}) = \frac{1}{1 + \exp\left[-(-1 - (-C_{rb} - 1))/\kappa\right]}, \tag{4.9aj}$$

$$P(\text{SFR} \leftarrow \text{HV}) = \frac{1}{1 + \exp\left[-(-C_{rv} - (0))/\kappa\right]}, \tag{4.9ak}$$

$$P(\text{SFR} \leftarrow \text{IV}) = \frac{1}{1 + \exp\left[-(-C_{rv} - 1 - (0))/\kappa\right]}, \tag{4.9al}$$

$$P(\text{SFR} \leftarrow \text{HM}) = \frac{1}{1 + \exp\left[-(-C_{rm} - (0))/\kappa\right]}, \tag{4.9am}$$

$$P(\text{SFR} \leftarrow \text{IM}) = \frac{1}{1 + \exp\left[-(-C_{rm} - 1 - (0))/\kappa\right]}, \tag{4.9an}$$

$$P(\text{SFR} \leftarrow \text{HB}) = \frac{1}{1 + \exp\left[-(-C_{rb} - (0))/\kappa\right]}, \tag{4.9ao}$$

$$P(\text{SFR} \leftarrow \text{IB}) = \frac{1}{1 + \exp\left[-(-C_{rb} - 1 - (0))/\kappa\right]}, \tag{4.9ap}$$

$$P(\text{FFR} \leftarrow \text{HV}) = \frac{1}{1 + \exp\left[-(-C_{rv} - (-1))/\kappa\right]}, \tag{4.9aq}$$

$$P(\text{FFR} \leftarrow \text{IV}) = \frac{1}{1 + \exp\left[-(-C_{rv} - 1 - (-1))/\kappa\right]}, \tag{4.9ar}$$

$$P(\text{FFR} \leftarrow \text{HM}) = \frac{1}{1 + \exp\left[-(-C_{rm} - (-1))/\kappa\right]}, \tag{4.9as}$$

$$P(\text{FFR} \leftarrow \text{IM}) = \frac{1}{1 + \exp\left[-(-C_{rm} - 1 - (-1))/\kappa\right]}, \tag{4.9at}$$

$$P(\text{FFR} \leftarrow \text{HB}) = \frac{1}{1 + \exp\left[-(-C_{rb} - (-1))/\kappa\right]}, \tag{4.9au}$$

$$P(\text{FFR} \leftarrow \text{IB}) = \frac{1}{1 + \exp\left[-(-C_{rb} - 1 - (-1))/\kappa\right]}. \tag{4.9av}$$

At the end of each epidemic season, each individual is allowed to update their strategy in accordance with last season's payoff. Hence, increasing or decreasing x, y, and z is inevitable. The global dynamics equations obeying IB-RA are given below:

$$\begin{aligned}
\frac{dx}{dt} &= xy(1 + (1-e)\exp[-\mathrm{R}_0 R(x,y,z,\infty)])\exp[-(1-\eta)\mathrm{R}_0 R(x,y,z,\infty)] \\
&\quad \times (P(\text{HM} \leftarrow \text{HV}) - P(\text{HV} \leftarrow \text{HM})) \\
&\quad + xy(1 + (1-e)\exp[-\mathrm{R}_0 R(x,y,z,\infty)])(1 - \exp[-(1-\eta)\mathrm{R}_0 R(x,y,z,\infty)]) \\
&\quad \times (P(\text{IM} \leftarrow \text{HV}) - P(\text{HV} \leftarrow \text{IM})) \\
&\quad + xz(1 + (1-e)\exp[-\mathrm{R}_0 R(x,y,z,\infty)]) \\
&\quad \times (1 + (1-e)\exp[-(1-\eta)\mathrm{R}_0 R(x,y,z,\infty)])(P(\text{HB} \leftarrow \text{HV}) - P(\text{HV} \leftarrow \text{HB})) \\
&\quad + xz(1 + (1-e)\exp[-\mathrm{R}_0 R(x,y,z,\infty)])(1 - \exp[-(1-\eta)\mathrm{R}_0 R(x,y,z,\infty)]) \\
&\quad \times (P(\text{IB} \leftarrow \text{HV}) - P(\text{HV} \leftarrow \text{IB})) \\
&\quad + x(1-x-y-z)(1 + (1-e)\exp[-\mathrm{R}_0 R(x,y,z,\infty)])\exp[-\mathrm{R}_0 R(x,y,z,\infty)] \\
&\quad \times (P(\text{SFR} \leftarrow \text{HV}) - P(\text{HV} \leftarrow \text{SFR})) \\
&\quad + x(1-x-y-z)(1 + (1-e)\exp[-\mathrm{R}_0 R(x,y,z,\infty)]) \\
&\quad \times (1 - \exp[-\mathrm{R}_0 R(x,y,z,\infty)])(P(\text{FFR} \leftarrow \text{HV}) - P(\text{HV} \leftarrow \text{FFR})) \\
&\quad + xy(1-e)(1 - \exp[-\mathrm{R}_0 R(x,y,z,\infty)])\exp[-(1-\eta)\mathrm{R}_0 R(x,y,z,\infty)] \\
&\quad \times (P(\text{HM} \leftarrow \text{IV}) - P(\text{IV} \leftarrow \text{HM})) \\
&\quad + xy(1-e)(1 - \exp[-\mathrm{R}_0 R(x,y,z,\infty)])(1 - \exp[-(1-\eta)\mathrm{R}_0 R(x,y,z,\infty)]) \\
&\quad \times (P(\text{IM} \leftarrow \text{IV}) - P(\text{IV} \leftarrow \text{IM})) \\
&\quad + xz(1-e)(1 - \exp[-\mathrm{R}_0 R(x,y,z,\infty)]) \\
&\quad \times (1 + (1-e)\exp[-(1-\eta)\mathrm{R}_0 R(x,y,z,\infty)])(P(\text{HB} \leftarrow \text{IV}) - P(\text{IV} \leftarrow \text{HB})) \\
&\quad + xz(1-e)(1 - \exp[-\mathrm{R}_0 R(x,y,z,\infty)])(1 - \exp[-(1-\eta)\mathrm{R}_0 R(x,y,z,\infty)]) \\
&\quad \times (P(\text{IB} \leftarrow \text{IV}) - P(\text{IV} \leftarrow \text{IB})) \\
&\quad + x(1-x-y-z)(1-e)(1 - \exp[-\mathrm{R}_0 R(x,y,z,\infty)])\exp[-\mathrm{R}_0 R(x,y,z,\infty)] \\
&\quad \times (P(\text{SFR} \leftarrow \text{IV}) - P(\text{IV} \leftarrow \text{SFR}))
\end{aligned}$$

$$+x(1-x-y-z)(1-e)(1-\exp[-\mathrm{R}_0R(x,y,z,\infty)])$$
$$\times(1-\exp[-\mathrm{R}_0R(x,y,z,\infty)])(P(\mathrm{FFR}\leftarrow\mathrm{IV})-P(\mathrm{IV}\leftarrow\mathrm{FFR})), \quad (4.10\mathrm{a})$$

$$\frac{dy}{dt}=xy(1+(1-e)\exp[-\mathrm{R}_0R(x,y,z,\infty)])\exp[-(1-\eta)\mathrm{R}_0R(x,y,z,\infty)]$$
$$\times(P(\mathrm{HV}\leftarrow\mathrm{HM})-P(\mathrm{HM}\leftarrow\mathrm{HV}))$$

$$+xy(1-e)(1-\exp[-\mathrm{R}_0R(x,y,z,\infty)])\exp[-(1-\eta)\mathrm{R}_0R(x,y,z,\infty)]$$
$$\times(P(\mathrm{IV}\leftarrow\mathrm{HM})-P(\mathrm{HM}\leftarrow\mathrm{IV}))$$

$$+yz\exp[-(1-\eta)\mathrm{R}_0R(x,y,z,\infty)](1+(1-e)\exp[-(1-\eta)\mathrm{R}_0R(x,y,z,\infty)])$$
$$\times(P(\mathrm{HB}\leftarrow\mathrm{HM})-P(\mathrm{HM}\leftarrow\mathrm{HB}))$$

$$+yz\exp[-(1-\eta)\mathrm{R}_0R(x,y,z,\infty)](1-e)(1-\exp[-(1-\eta)\mathrm{R}_0R(x,y,z,\infty)])$$
$$\times(P(\mathrm{IB}\leftarrow\mathrm{HM})-P(\mathrm{HM}\leftarrow\mathrm{IB}))$$

$$+y(1-x-y-z)\exp[-(1-\eta)\mathrm{R}_0R(x,y,z,\infty)]\exp[-\mathrm{R}_0R(x,y,z,\infty)]$$
$$\times(P(\mathrm{SFR}\leftarrow\mathrm{HM})-P(\mathrm{HM}\leftarrow\mathrm{SFR}))$$

$$+y(1-x-y-z)\exp[-(1-\eta)\mathrm{R}_0R(x,y,z,\infty)](1-\exp[-\mathrm{R}_0R(x,y,z,\infty)])$$
$$\times(P(\mathrm{FFR}\leftarrow\mathrm{HM})-P(\mathrm{HM}\leftarrow\mathrm{FFR}))$$

$$+xy(1+(1-e)\exp[-\mathrm{R}_0R(x,y,z,\infty)])(1-\exp[-(1-\eta)\mathrm{R}_0R(x,y,z,\infty)])$$
$$\times(P(\mathrm{HV}\leftarrow\mathrm{IM})-P(\mathrm{IM}\leftarrow\mathrm{HV}))$$

$$+xy(1-e)(1-\exp[-\mathrm{R}_0R(x,y,z,\infty)])(1-\exp[-(1-\eta)\mathrm{R}_0R(x,y,z,\infty)])$$
$$\times(P(\mathrm{IV}\leftarrow\mathrm{IM})-P(\mathrm{IM}\leftarrow\mathrm{IV}))$$

$$+yz(1-\exp[-(1-\eta)\mathrm{R}_0R(x,y,z,\infty)])(1+(1-e)\exp[-(1-\eta)\mathrm{R}_0R(x,y,z,\infty)])$$
$$\times(P(\mathrm{HB}\leftarrow\mathrm{IM})-P(\mathrm{IM}\leftarrow\mathrm{HB}))$$

$$+yz(1-\exp[-(1-\eta)\mathrm{R}_0R(x,y,z,\infty)])(1-e)$$
$$\times(1-\exp[-(1-\eta)\mathrm{R}_0R(x,y,z,\infty)])(P(\mathrm{IB}\leftarrow\mathrm{IM})-P(\mathrm{IM}(\mathrm{IM}\leftarrow\mathrm{IB})))$$

$$+y(1-x-y-z)(1-\exp[-(1-\eta)\mathrm{R}_0R(x,y,z,\infty)])\exp[-\mathrm{R}_0R(x,y,z,\infty)]$$
$$\times(P(\mathrm{SFR}\leftarrow\mathrm{IM})-P(\mathrm{IM}\leftarrow\mathrm{SFR}))$$

$$+y(1-x-y-z)(1-\exp[-(1-\eta)\mathrm{R}_0R(x,y,z,\infty)])$$
$$\times(1-\exp[-\mathrm{R}_0R(x,y,z,\infty)])(P(\mathrm{FFR}\leftarrow\mathrm{IM})-P(\mathrm{IM}\leftarrow\mathrm{FFR})), \quad (4.10\mathrm{b})$$

$$\frac{dz}{dt}=\mathrm{xz}(1+(1-e)\exp[-\mathrm{R}_0R(x,y,z,\infty)])$$
$$\times(1+(1-e)\exp[-(1-\eta)\mathrm{R}_0R(x,y,z,\infty)])$$
$$\times(P(\mathrm{HV}\leftarrow\mathrm{HB})-P(\mathrm{HB}\leftarrow\mathrm{HV}))$$

$$+\mathrm{xz}(1-e)(1-\exp[-\mathrm{R}_0R(x,y,z,\infty)])$$
$$\times(1+(1-e)\,exp[-(1-\eta)\mathrm{R}_0R(x,y,z,\infty)])(P(\mathrm{IV}\leftarrow\mathrm{HB})-P(\mathrm{HB}\leftarrow\mathrm{IV}))$$

$$+\mathrm{yz}\ \exp\left[-(1-\eta)\mathrm{R}_0R(x,y,z,\infty)\right]\left(1+(1-e)\exp\left[-(1-\eta)\mathrm{R}_0R(x,y,z,\infty)\right]\right)$$
$$\times\left(P(\mathrm{HM}\leftarrow\mathrm{HB})-P(\mathrm{HB}\leftarrow\mathrm{HM})\right)$$

$$+\mathrm{yz}\ \left(1-exp\left[-(1-\eta)\mathrm{R}_0R(x,y,z,\infty)\right]\right)$$
$$\times\left(1+(1-e)\exp\left[-(1-\eta)\mathrm{R}_0R(x,y,z,\infty)\right]\right)\left(P(\mathrm{IM}\leftarrow\mathrm{HB})-P(\mathrm{HB}\leftarrow\mathrm{IM})\right)$$

$$+\mathrm{z}(1-x-y-z)\ \left(1+(1-e)\exp\left[-(1-\eta)\mathrm{R}_0R(x,y,z,\infty)\right]\right)\exp\left[-\mathrm{R}_0R(x,y,z,\infty)\right]$$
$$\times\left(P(\mathrm{SFR}\leftarrow\mathrm{HB})-P(\mathrm{HB}\leftarrow\mathrm{SFR})\right)$$

$$+\mathrm{z}(1-x-y-z)\ \left(1+(1-e)\exp\left[-(1-\eta)\mathrm{R}_0R(x,y,z,\infty)\right]\right)$$
$$\times\left(1-\exp\left[-\mathrm{R}_0R(x,y,z,\infty)\right]\right)\left(P(\mathrm{FFR}\leftarrow\mathrm{HB})-P(\mathrm{HB}\leftarrow\mathrm{FFR})\right)$$

$$+\mathrm{xz}\left(1+(1-e)\exp\left[-\mathrm{R}_0R(x,y,z,\infty)\right]\right)(1-e)$$
$$\times\left(1-\exp\left[-(1-\eta)\mathrm{R}_0R(x,y,z,\infty)\right]\right)\left(P(\mathrm{HV}\leftarrow\mathrm{IB})-P(\mathrm{IB}\leftarrow\mathrm{HV})\right)$$

$$+\mathrm{xz}(1-e)\left(1-\exp\left[-\mathrm{R}_0R(x,y,z,\infty)\right]\right)(1-e)$$
$$\times\left(1-\exp\left[-(1-\eta)\mathrm{R}_0R(x,y,z,\infty)\right]\right)\left(P(\mathrm{IV}\leftarrow\mathrm{IB})-P(\mathrm{IB}\leftarrow\mathrm{IV})\right)$$

$$+\mathrm{yz}\ \exp\left[-(1-\eta)\mathrm{R}_0R(x,y,z,\infty)\right](1-e)\left(1-\exp\left[-(1-\eta)\mathrm{R}_0R(x,y,z,\infty)\right]\right)$$
$$\times\left(P(\mathrm{HM}\leftarrow\mathrm{IB})-P(\mathrm{IB}\leftarrow\mathrm{HM})\right)$$

$$+\mathrm{yz}\left(1-\exp\left[-(1-\eta)\mathrm{R}_0R(x,y,z,\infty)\right]\right)\ (1-e)$$
$$\times\left(1-\exp\left[-(1-\eta)\mathrm{R}_0R(x,y,z,\infty)\right]\right)\left(P(\mathrm{IM}\leftarrow\mathrm{IB})-P(\mathrm{IB}\leftarrow\mathrm{IM})\right)$$

$$+\mathrm{z}(1-x-y-z)\ (1-e)$$
$$\times\left(1-\exp\left[-(1-\eta)\mathrm{R}_0R(x,y,z,\infty)\right]\right)\exp\left[-\mathrm{R}_0R(x,y,z,\infty)\right]$$
$$\times\left(P(\mathrm{SFR}\leftarrow\mathrm{IB})-P(\mathrm{IB}\leftarrow\mathrm{SFR})\right)$$

$$+\mathrm{z}(1-x-y-z)\ (1-e)\left(1-\exp\left[-(1-\eta)\mathrm{R}_0R(x,y,z,\infty)\right]\right)$$
$$\times\left(1-\exp\left[-\mathrm{R}_0R(x,y,z,\infty)\right]\right)\left(P(\mathrm{SFR}\leftarrow\mathrm{IB})-P(\mathrm{IB}\leftarrow\mathrm{SFR})\right). \quad (4.10\mathrm{c})$$

4.2.3.2 Strategy-Based Risk Assessment (SB-RA)

To establish the state-transition-probability, SB-RA requires the following 24 (=8 × 3) cases, where "8" comes from items (i)–(viii) depicted in IB-RA above, and "3" comes from the four strategies (D (defector), C (cooperator meaning vaccinator), M (taking IDM), and B (taking V and IDM)) minus 1:

$$P(\mathrm{HV}\leftarrow\mathrm{M})=\frac{1}{1+\exp\left[-(-\langle\pi_M\rangle-(-C_{rv}))/\kappa\right]}, \quad (4.11\mathrm{a})$$

$$P(\mathrm{HV}\leftarrow\mathrm{B})=\frac{1}{1+\exp\left[-(-\langle\pi_B\rangle-(-C_{rv}))/\kappa\right]}, \quad (4.11\mathrm{b})$$

$$P(\text{HV} \leftarrow \text{D}) = \frac{1}{1 + \exp\left[-(-\langle \pi_D \rangle - (-C_{rv}))/\kappa\right]}, \tag{4.11c}$$

$$P(\text{IV} \leftarrow \text{M}) = \frac{1}{1 + \exp\left[-(-\langle \pi_M \rangle - (-C_{rv} - 1))/\kappa\right]}, \tag{4.11d}$$

$$P(\text{IV} \leftarrow \text{B}) = \frac{1}{1 + \exp\left[-(-\langle \pi_B \rangle - (-C_{rv} - 1))/\kappa\right]}, \tag{4.11e}$$

$$P(\text{IV} \leftarrow \text{D}) = \frac{1}{1 + \exp\left[-(-\langle \pi_D \rangle - (-C_{rv} - 1))/\kappa\right]}, \tag{4.11f}$$

$$P(\text{HB} \leftarrow \text{C}) = \frac{1}{1 + \exp\left[-(-\langle \pi_C \rangle - (-C_{rb}))/\kappa\right]}, \tag{4.11g}$$

$$P(\text{HB} \leftarrow \text{M}) = \frac{1}{1 + \exp\left[-(-\langle \pi_M \rangle - (-C_{rb}))/\kappa\right]}, \tag{4.11h}$$

$$P(\text{HB} \leftarrow \text{D}) = \frac{1}{1 + \exp\left[-(-\langle \pi_D \rangle - (-C_{rb}))/\kappa\right]}, \tag{4.11i}$$

$$P(\text{IB} \leftarrow \text{C}) = \frac{1}{1 + \exp\left[-(-\langle \pi_C \rangle - (-C_{rb} - 1))/\kappa\right]}, \tag{4.11j}$$

$$P(\text{IB} \leftarrow \text{M}) = \frac{1}{1 + \exp\left[-(-\langle \pi_M \rangle - (-C_{rb} - 1))/\kappa\right]}, \tag{4.11k}$$

$$P(\text{IB} \leftarrow \text{D}) = \frac{1}{1 + \exp\left[-(-\langle \pi_D \rangle - (-C_{rb} - 1))/\kappa\right]}, \tag{4.11l}$$

$$P(\text{HM} \leftarrow \text{C}) = \frac{1}{1 + \exp\left[-(-\langle \pi_C \rangle - (-C_{rb}))/\kappa\right]}, \tag{4.11m}$$

$$P(\text{HM} \leftarrow \text{B}) = \frac{1}{1 + \exp\left[-(-\langle \pi_B \rangle - (-C_{rm}))/\kappa\right]}, \tag{4.11n}$$

$$P(\text{HM} \leftarrow \text{D}) = \frac{1}{1 + \exp\left[-(-\langle \pi_D \rangle - (-C_{rm}))/\kappa\right]}, \tag{4.11o}$$

$$P(\text{IM} \leftarrow \text{C}) = \frac{1}{1 + \exp\left[-(-\langle \pi_C \rangle - (-C_{rm} - 1))/\kappa\right]}, \tag{4.11p}$$

$$P(\text{IM} \leftarrow \text{B}) = \frac{1}{1 + \exp\left[-(-\langle \pi_B \rangle - (-C_{rm} - 1))/\kappa\right]}, \tag{4.11q}$$

$$P(\text{IM} \leftarrow \text{D}) = \frac{1}{1 + \exp\left[-(-\langle \pi_D \rangle - (-C_{rm} - 1))/\kappa\right]}, \tag{4.11r}$$

$$P(\text{SFR} \leftarrow \text{C}) = \frac{1}{1 + \exp\left[-(-\langle \pi_C \rangle - (0))/\kappa\right]}, \tag{4.11s}$$

$$P(\mathrm{SFR} \leftarrow \mathrm{M}) = \frac{1}{1 + \exp\left[-(-\langle \pi_M \rangle - (0))/\kappa\right]}, \tag{4.11t}$$

$$P(\mathrm{SFR} \leftarrow \mathrm{B}) = \frac{1}{1 + \exp\left[-(-\langle \pi_B \rangle - (0))/\kappa\right]}, \tag{4.11u}$$

$$P(\mathrm{FFR} \leftarrow \mathrm{C}) = \frac{1}{1 + \exp\left[-(-\langle \pi_C \rangle - (-1))/\kappa\right]}, \tag{4.11v}$$

$$P(\mathrm{FFR} \leftarrow \mathrm{M}) = \frac{1}{1 + \exp\left[-(-\langle \pi_M \rangle - (-1))/\kappa\right]}, \tag{4.11w}$$

$$P(\mathrm{FFR} \leftarrow \mathrm{B}) = \frac{1}{1 + \exp\left[-(-\langle \pi_B \rangle - (-1))/\kappa\right]}. \tag{4.11x}$$

The global dynamic equations obeying SB-RA are given as

$$\begin{aligned}
\frac{dx}{dt} &= -xy(1 + (1-e)\exp\left[-\mathrm{R}_0 R(x,y,z,\infty)\right])P(HV \leftarrow M) \\
&\quad - xz(1 + (1-e)\exp\left[-\mathrm{R}_0 R(x,y,z,\infty)\right])P(\mathrm{HV} \leftarrow \mathrm{B}) \\
&\quad - x(1-x-y-z)(1 + (1-e)\exp\left[-\mathrm{R}_0 R(x,y,z,\infty)\right])P(HV \leftarrow D) - xy(1-e) \\
&\qquad \times (1 - \exp\left[-\mathrm{R}_0 R(x,y,z,\infty)\right])P(\mathrm{IV} \leftarrow \mathrm{M}) \\
&\quad - xz(1-e)(1 - \exp\left[-\mathrm{R}_0 R(x,y,z,\infty)\right])P(IV \leftarrow B) - \mathrm{x}(1-x-y-z)(1-e) \\
&\qquad \times (1 - \exp\left[-\mathrm{R}_0 R(x,y,z,\infty)\right])P(\mathrm{IV} \leftarrow \mathrm{D}) \\
&\quad + xz(1 + (1-e)\exp\left[-(1-\eta)\mathrm{R}_0 R(x,y,z,\infty)\right])P(HB \leftarrow V) + xz(1-e) \\
&\qquad \times (1 - \exp\left[-(1-\eta)\mathrm{R}_0 R(x,y,z,\infty)\right])P(\mathrm{IB} \leftarrow \mathrm{V}) \\
&\quad + xy\exp\left[-(1-\eta)\mathrm{R}_0 R(x,y,z,\infty)\right]P(HM \leftarrow V) \\
&\quad + xy(1 - \exp\left[-(1-\eta)\mathrm{R}_0 R(x,y,z,\infty)\right])P(\mathrm{IM} \leftarrow \mathrm{V}) \\
&\quad + x(1-x-y-z)\exp\left[-\mathrm{R}_0 R(x,y,z,\infty)\right]P(\mathrm{SFR} \leftarrow V) \\
&\quad + x(1-x-y-z)(1 - \exp\left[-\mathrm{R}_0 R(x,y,z,\infty)\right])P(\mathrm{FFR} \leftarrow \mathrm{V}),
\end{aligned} \tag{4.12a}$$

$$\begin{aligned}
\frac{dy}{dt} &= -xy\exp\left[-(1-\eta)\mathrm{R}_0 R(x,y,z,\infty)\right]P(HM \leftarrow V) \\
&\quad - yz\exp\left[-(1-\eta)\mathrm{R}_0 R(x,y,z,\infty)\right]P(\mathrm{HM} \leftarrow \mathrm{B}) \\
&\quad - y(1-x-y-z)\exp\left[-(1-\eta)\mathrm{R}_0 R(x,y,z,\infty)\right]P(HM \leftarrow D) \\
&\quad - xy(1 - \exp\left[-(1-\eta)\mathrm{R}_0 R(x,y,z,\infty)\right])P(\mathrm{IM} \leftarrow \mathrm{V}) \\
&\quad - yz(1 - \exp\left[-(1-\eta)\mathrm{R}_0 R(x,y,z,\infty)\right])P(IM \leftarrow B) - y(1-x-y-z) \\
&\qquad \times (1 - \exp\left[-(1-\eta)\mathrm{R}_0 R(x,y,z,\infty)\right])P(\mathrm{IM} \leftarrow \mathrm{D})
\end{aligned}$$

$$+xy(1+(1-e)\exp[-R_0R(x,y,z,\infty)])P(HV\leftarrow M)+xy(1-e)\times(1-\exp[-R_0R(x,y,z,\infty)])P(\mathrm{IV}\leftarrow \mathrm{M})$$

$$+yz(1+(1-e)\exp[-(1-\eta)R_0R(x,y,z,\infty)])\,P(HB\leftarrow M)+yz(1-e)\times(1-\exp[-(1-\eta)R_0R(x,y,z,\infty)])P(\mathrm{IB}\leftarrow \mathrm{M})$$

$$+y(1-x-y-z)\exp[-R_0R(x,y,z,\infty)]P(\mathrm{SFR}\leftarrow M)+y(1-x-y-z)(1-\exp[-R_0R(x,y,z,\infty)])P(\mathrm{FFR}\leftarrow \mathrm{M}),\qquad(4.12\mathrm{b})$$

$$\frac{dz}{dt}=-\mathrm{xz}\,(1+(1-e)\,exp[-(1-\eta)R_0R(x,y,z,\infty)])P(HB\leftarrow V)-yz\,(1+(1-e)\,exp[-(1-\eta)R_0R(x,y,z,\infty)])P(\mathrm{HB}\leftarrow \mathrm{M})$$

$$-\mathrm{z}(1-x-y-z)(1+(1-e)\,exp[-(1-\eta)R_0R(x,y,z,\infty)])P(HB\leftarrow D)-xz(1-e)(1-exp[-(1-\eta)R_0R(x,y,z,\infty)])P(\mathrm{IB}\leftarrow \mathrm{V})$$

$$-\mathrm{yz}(1-e)(1-exp[-(1-\eta)R_0R(x,y,z,\infty)])P(IB\leftarrow M)-\mathrm{z}(1-x-y-z)\times(1-e)(1-exp[-(1-\eta)R_0R(x,y,z,\infty)])P(\mathrm{IB}\leftarrow \mathrm{D})$$

$$+\mathrm{xz}(1+(1-e)\,exp[-R_0R(x,y,z,\infty)])P(HV\leftarrow B)+\mathrm{xz}(1-e)\times(1-\exp[-R_0R(x,y,z,\infty)])P(\mathrm{IV}\leftarrow \mathrm{B})$$

$$+\mathrm{yz}\ exp[-(1-\eta)R_0R(x,y,z,\infty)]P(HM\leftarrow B)+\mathrm{yz}(1-\exp[-(1-\eta)R_0R(x,y,z,\infty)])P(\mathrm{IM}\leftarrow \mathrm{B})$$

$$+\mathrm{z}(1-x-y-z)\,exp[-R_0R(x,y,z,\infty)]P(SFR\leftarrow B)+z(1-x-y-z)(1-exp[-R_0R(x,y,z,\infty)])P(\mathrm{FFR}\leftarrow \mathrm{B}).\qquad(4.12\mathrm{c})$$

4.2.3.3 Direct Commitment (DC)

DC requires the following twelve (=4 × (4 − 1)) cases, where the "4" comes from the four strategies (D (defector), C (cooperator meaning vaccinator), M (taking IDM), and B (taking V and IDM)):

$$P(C\leftarrow M)=\frac{1}{1+\exp[-(-\langle\pi_M\rangle-\langle\pi_C\rangle)/\kappa]},\qquad(4.13\mathrm{a})$$

$$P(\mathrm{C}\leftarrow \mathrm{B})=\frac{1}{1+\exp[-(-\langle\pi_B\rangle-\langle\pi_C\rangle)/\kappa]},\qquad(4.13\mathrm{b})$$

$$P(\mathrm{C}\leftarrow \mathrm{D})=\frac{1}{1+\exp[-(-\langle\pi_D\rangle-\langle\pi_C\rangle)/\kappa]},\qquad(4.13\mathrm{c})$$

$$P(\mathrm{M}\leftarrow \mathrm{C})=\frac{1}{1+\exp[-(-\langle\pi_C\rangle-\langle\pi_M\rangle)/\kappa]},\qquad(4.13\mathrm{d})$$

$$P(\text{M} \leftarrow \text{B}) = \frac{1}{1 + \exp\left[-(-\langle \pi_B \rangle - \langle \pi_M \rangle)/\kappa\right]}, \tag{4.13e}$$

$$P(\text{M} \leftarrow \text{D}) = \frac{1}{1 + \exp\left[-(-\langle \pi_D \rangle - \langle \pi_M \rangle)/\kappa\right]}, \tag{4.13f}$$

$$P(\text{B} \leftarrow \text{C}) = \frac{1}{1 + \exp\left[-(-\langle \pi_C \rangle - \langle \pi_B \rangle)/\kappa\right]}, \tag{4.13g}$$

$$P(\text{B} \leftarrow \text{M}) = \frac{1}{1 + \exp\left[-(-\langle \pi_M \rangle - \langle \pi_B \rangle)/\kappa\right]}, \tag{4.13h}$$

$$P(\text{B} \leftarrow \text{D}) = \frac{1}{1 + \exp\left[-(-\langle \pi_D \rangle - \langle \pi_B \rangle)/\kappa\right]}, \tag{4.13i}$$

$$P(\text{M} \leftarrow \text{C}) = \frac{1}{1 + \exp\left[-(-\langle \pi_C \rangle - \langle \pi_M \rangle)/\kappa\right]}, \tag{4.13j}$$

$$P(\text{M} \leftarrow \text{B}) = \frac{1}{1 + \exp\left[-(-\langle \pi_B \rangle - \langle \pi_M \rangle)/\kappa\right]}, \tag{4.13k}$$

$$P(\text{M} \leftarrow \text{D}) = \frac{1}{1 + \exp\left[-(-\langle \pi_D \rangle - \langle \pi_M \rangle)/\kappa\right]}. \tag{4.13l}$$

The global dynamical equations obeying DC are given by

$$\frac{dx}{dt} = xy(P(\text{M} \leftarrow \text{V}) - P(\text{V} \leftarrow \text{M})) + xz(P(\text{B} \leftarrow \text{V}) - P(\text{V} \leftarrow \text{B})) + x(1 - x - y - z)(P(\text{D} \leftarrow \text{V}) - P(\text{V} \leftarrow \text{D})), \tag{4.14a}$$

$$\frac{dy}{dt} = xy(P(\text{V} \leftarrow \text{M}) - P(\text{M} \leftarrow \text{V})) + yz(P(\text{B} \leftarrow \text{M}) - P(\text{M} \leftarrow \text{B})) + y(1 - x - y - z)(P(\text{D} \leftarrow \text{M}) - P(\text{M} \leftarrow \text{D})), \tag{4.14b}$$

$$\frac{dz}{dt} = xz(P(\text{V} \leftarrow \text{B}) - P(\text{B} \leftarrow \text{V})) + yz(P(\text{M} \leftarrow \text{B}) - P(\text{B} \leftarrow \text{M})) + z(1 - x - y - z)(P(\text{D} \leftarrow \text{B}) - P(\text{B} \leftarrow \text{D})). \tag{4.14c}$$

4.3 Result and Discussion

By numerically solving the model above, we obtain the results below. We presume throughout that $R_0 = 2.5$. For all heat maps below, the x- and y-axes, respectively, indicate the costs of IDM and vaccination. Since we presume that the vaccination cost is larger than that of IDM, each heat map is shown as an upper triangle.

Figure 4.2 provides the FES while varying the parameter settings for the three strategy-updating rules IB-RA (left panels, A-*), SB-RA (middle panels, B-*), and

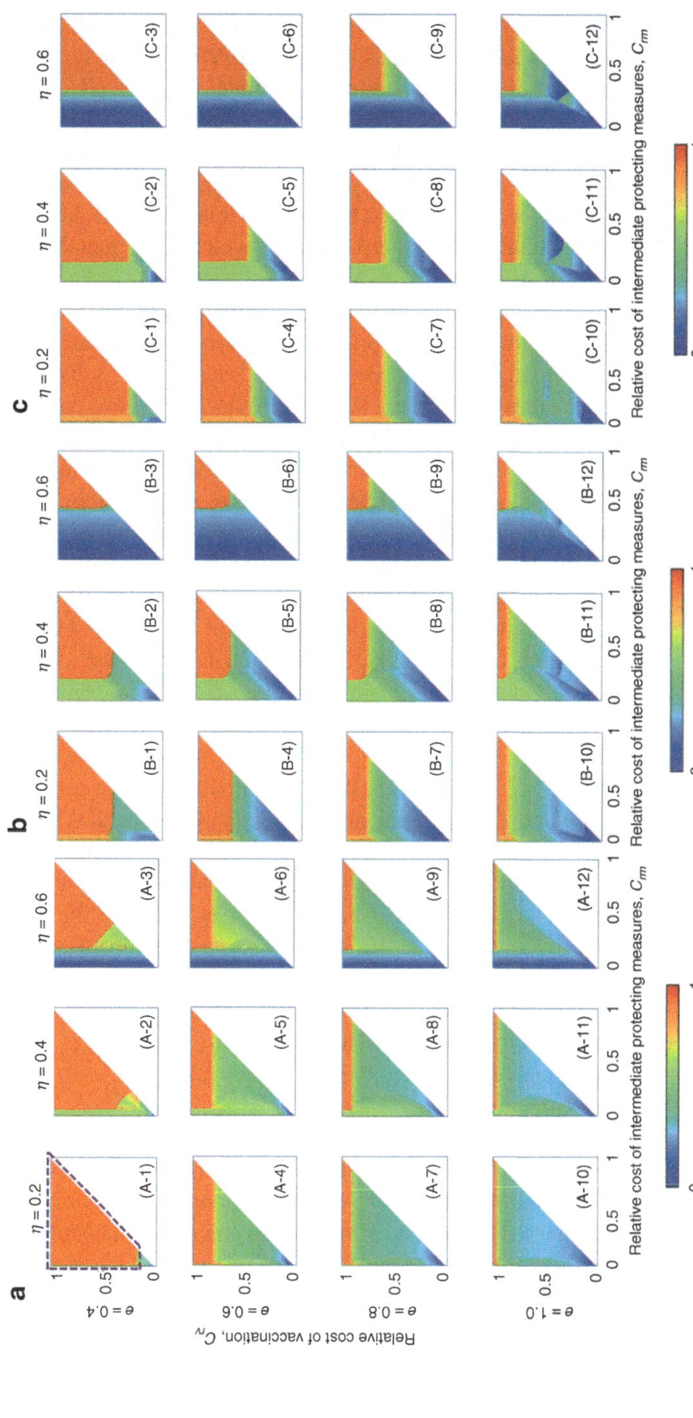

Fig. 4.2 Final epidemic size; FES (left panels (A-*) for IB-RA, middle panels (B-*) for SB-RA and right panels (C-*) for DC strategy-updating rules , respectively)

DC (left panels, C-*). Likewise, Fig. 4.3 indicates the ASP, for exactly the same parameter settings and strategy-updating rules. Throughout our discussion, we use the effectiveness parameter $e = \{0.4,0.6,0.8,1.0\}$ and the efficiency $\eta = \{0.2,0.4,0.6\}$. Figures 4.3, 4.4, 4.5 and 4.6, respectively, display the fractions for strategies of vaccination (x), IDM (y), and taking both (z).

The regions colored light red in the plot for FES (Fig. 4.2), dark blue in the plots for vaccination rate (Fig. 4.4), IDM rate (Fig. 4.5), and both (Fig. 4.6), and light blue in the plot for the ASP (Fig. 4.3) indicate a pandemic taking place. The mechanism working behind this is that, in such regions, the low reliability of both the vaccination and the IDM lead most individuals to take neither, which consequently leads to full-scale spread of the disease.

Since people rationally evaluate which strategy (including defection) is beneficial for them, it would be natural that they primarily choose vaccination when the effectiveness is reasonably high and the cost is low. Likewise, it is conceivable that, regardless of vaccination cost, when IDM is provided at a low cost and its efficiency is reasonably high (e.g., $\eta = 0.6$), many individuals will take IDM rather than vaccination. With regard to different strategy updates, it is worthwhile to note that an individual's tendency to take IDM obeys the IB-RA as the strategy-updating rule is quite meager (see the dotted rectangle in Fig. 4.5 (A-3)) vis-à-vis the two other update rules. SB-RA shows highest, and the DC case follows after SB-RA (Fig. 4.5 (B-3) and (C-3)). Comparing the definitions of the three update rules, IB-RA relies on the most detailed classification that is 48, as explained at Eqs. (4.9a–4.9av). One justification is that somehow coarse-grained information helps in for searching a globally optimal solution. On the other hand, too much detailed information with many classifications, as in the IB-RA scheme, may become trapped within a locally maximal solution.

Let us carefully observe Figs. 4.4 and 4.5. Panels (A-1) show that, over almost the entire parameter region, individuals take neither vaccination nor IDM. This is because the effectivities of the two provisions are meager ($e = 0.4$ and $\eta = 0.2$). Neither provision attracts individuals who rationally evaluate the risk of infection and benefit from provisions. Consequently, the disease breaks out (see the trapezoidal area in Fig. 4.2 (A-1)). Leaving aside this specific case (panel (A-1)), we should note that a reasonable fraction (not blue) in either the vaccination (Fig. 4.4) or IDM (Fig. 4.5) plots can be observed, except in the right part of the upper triangle in each heat map that implies that both vaccination and IDM costs, C_{rv} and C_{rm}, are quite high. In particular, unless both costs are quite high, people rationally take one of the two provisions to avoid infection.

With respect to the fourth strategy of simultaneously taking both vaccination and IDM (B), we point out several interesting discoveries below. The first is that the cases with the lowest vaccination reliability ($e = 0.4$) only allow the fourth strategy to survive. Moreover, either SB-RA or DC is imposed as strategy-update rules; in the region of lower vaccination cost and extremely low IDM cost, the vast majority of people take the fourth strategy (see the solid red-line boxes in panels (B-1), (B-2), (C-1), and (C-2) in Fig. 4.6). This seems conceivable because, despite the fact that vaccination may be an individual's first choice to avoid disease (even if they are not

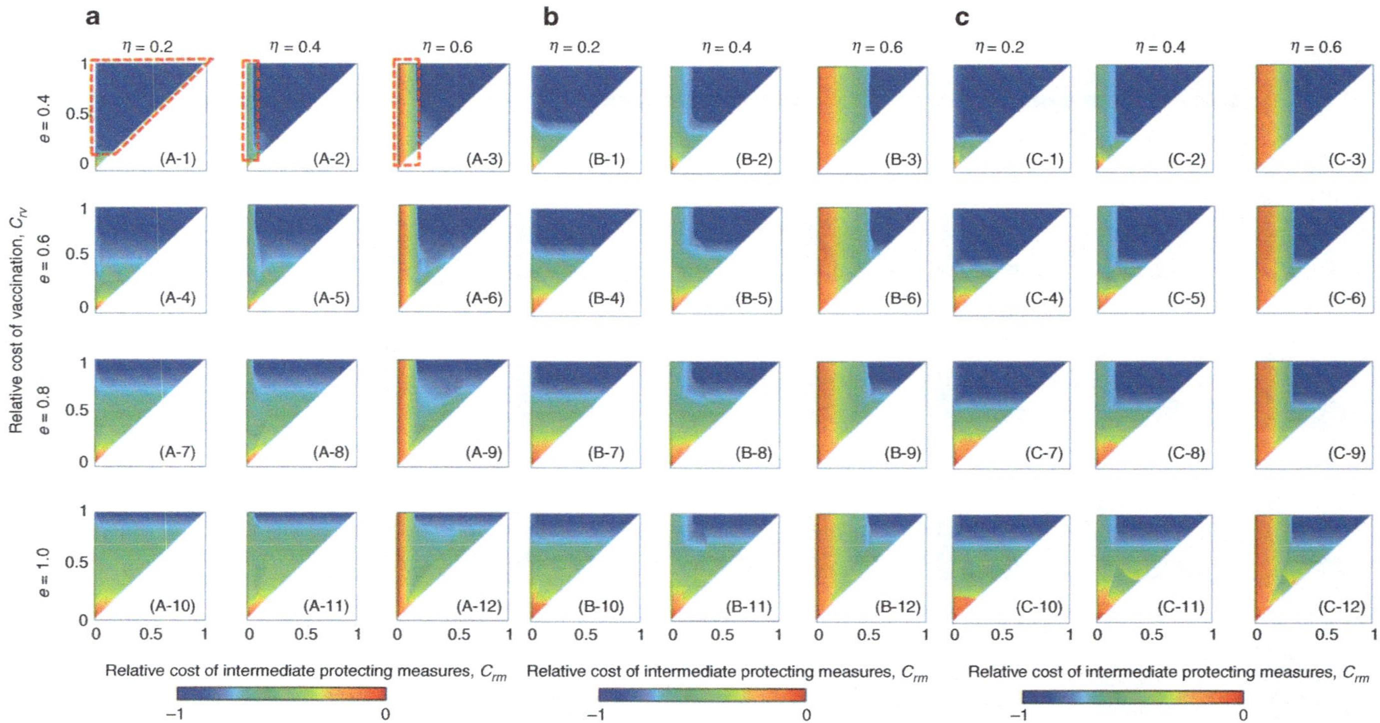

Fig. 4.3 Average social payoff; ASP (left panels (A-*) for IB-RA, middle panels (B-*) for SB-RA and right panels (C-*) for DC strategy-updating rules, respectively)

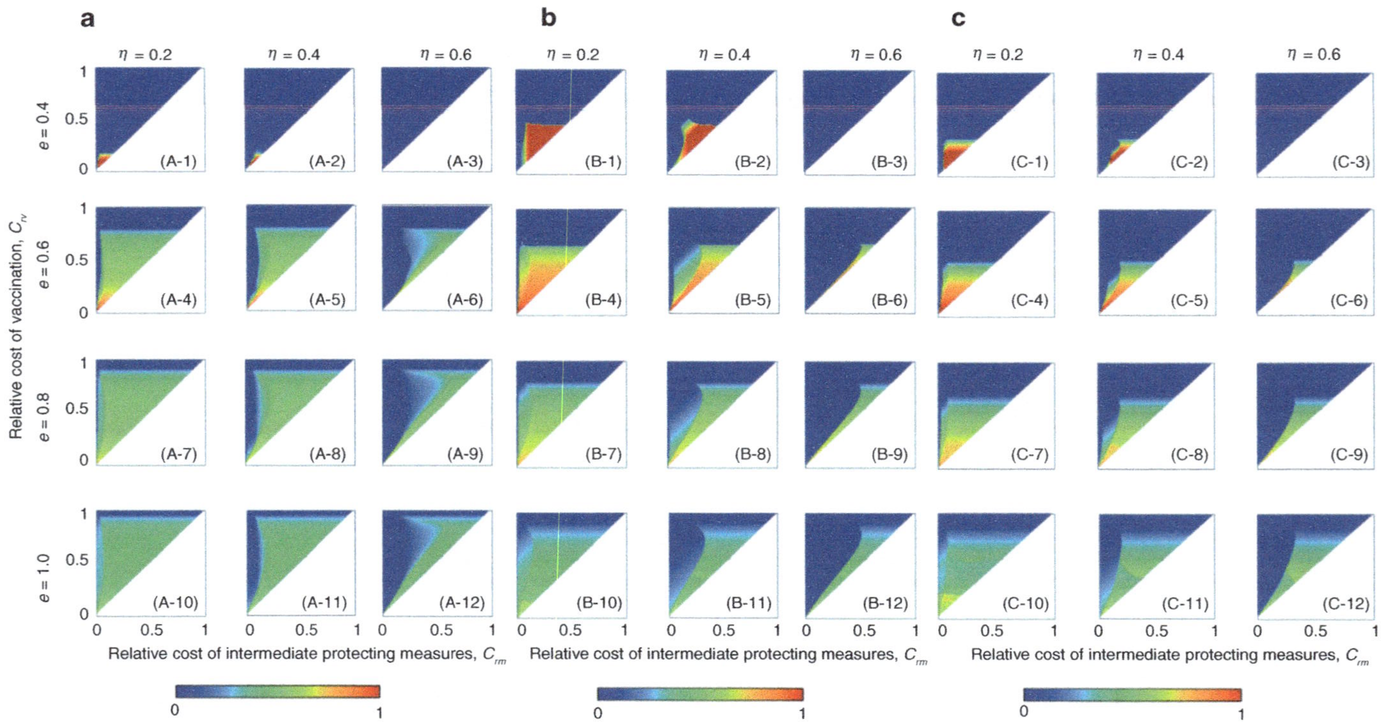

Fig. 4.4 Fraction of the population taking vaccination; fraction of V (cooperators) (left panels (A-*) for IB-RA, middle panels (B-*) for SB-RA, and right panels (C-*) for DC strategy-updating rules respectively)

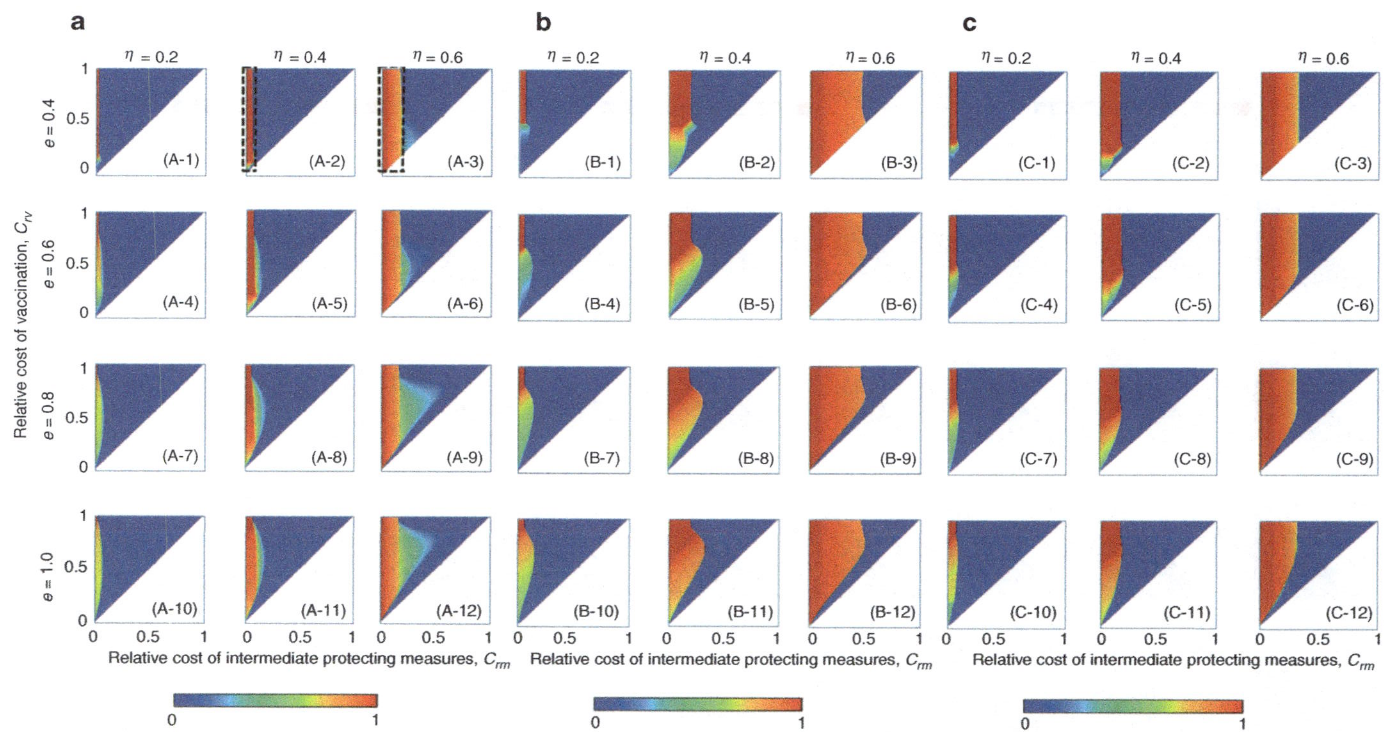

Fig. 4.5 Fraction of the population taking intermediate-defense measures (IDM); fraction of M (left panels (A-*) for IB-RA, middle panels (B-*) for SB-RA, and right panels (C-*) for DC strategy-updating rules, respectively)

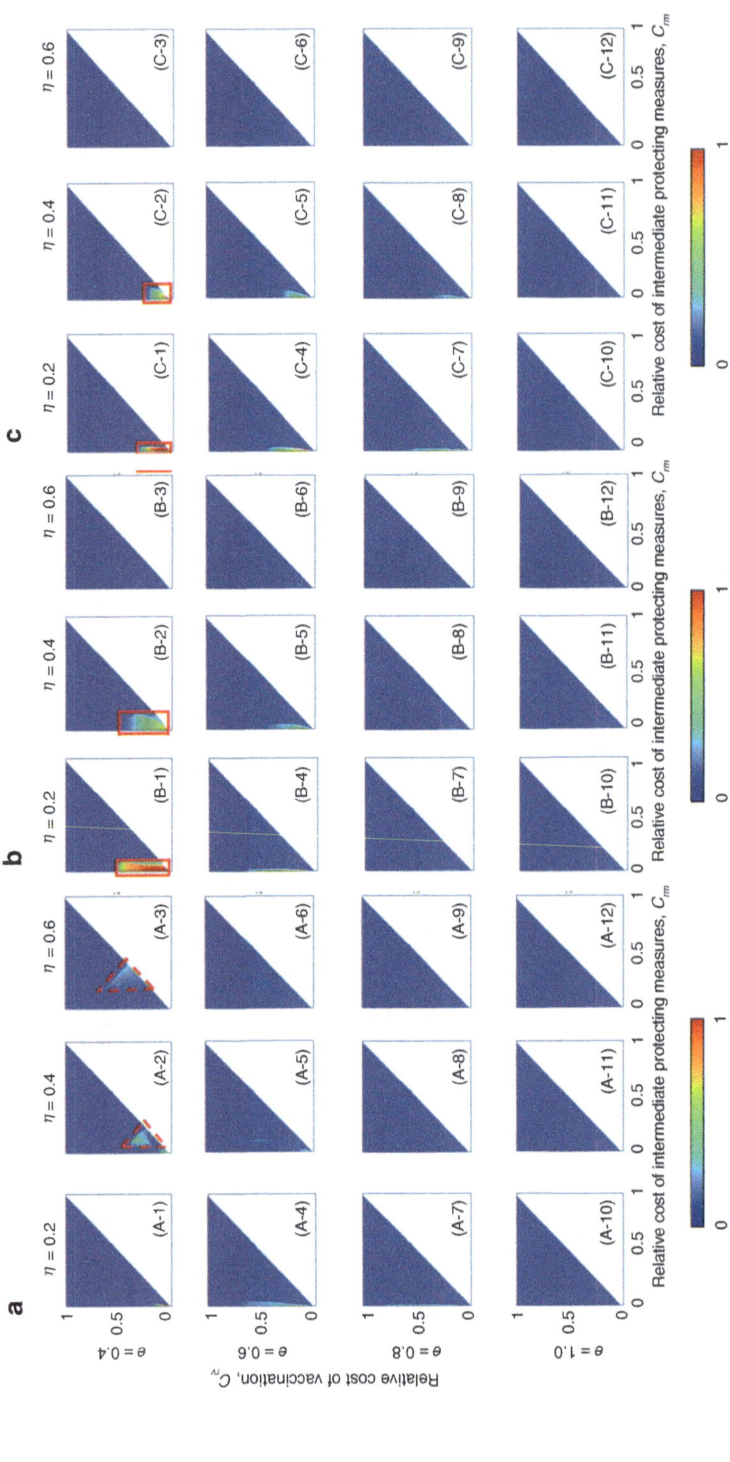

Fig. 4.6 Fraction of the population taking both V and IDM; fraction of B (left panels (A-*) for IB-RA, middle panels (B-*) for SB-RA and right panels (C-*) for DC strategy-updating rules)

able to acquire immunity, there is a possibility e that they can acquire perfect immunity), a lower e pushes individuals who are committing vaccination to take IDM as an additional measure. Taking both measures is fairly conceived as a more reliable alternative than taking one of these two. Thus, from a public-health standpoint, when the vaccine has a low reliability, the delivery of an additional provision (i.e., IDM) to the public helps to compensate for vaccination's drawbacks, which is sufficient to attract people who behave in a basically rational way. By contrast, when IB-RA is presumed, we note that the fourth strategy can survive in the region of moderate C_{rm} and C_{rv}, which is highlighted by broken-line triangles in panels (A-2) and (A-3) of Fig. 4.6. This happens because an extremely low C_{rm} leads most individuals to take IDM, as confirmed in the counterpart panels (A-2) and (A-3) of Fig. 4.5, highlighted by broken-line rectangle-boxes. This implies that, as a strategy-updating rule, IB-RA misleads individuals, since the ASP of that particular parameter region (highlighted by broken-line rectangle-boxes in Fig. 4.3 (A-2) and (A-3)) is somewhat less than in the counterpart regions for SB-RA and DC (see panels (B-2) & (B-3) and (C-2) & (C-3) of Fig. 4.3). This may be caused by IB-RA's specific drawback for too much detailed information, as mentioned above.

From all of these discussions, we believe that the introduction of a third strategy (taking IDM) and a fourth strategy (taking both vaccination and IDM) is beneficial in suppressing the spread of an epidemic at a considerably large scale from social-application point of view.

In this chapter, we extended the vaccination-game concept introduced in Chap. 3 to define the "intervention game." The modeling procedure, the coupling of a mathematical epidemiological framework with an evolutionary game model, and the obtained results are deliberately discussed.

We can conclude that the intervention game allows more alternatives than vaccination and is thus a powerful tool for quantitatively evaluating more complex situations to emulate real society.

References

Alam M, Kuga K, Tanimoto J (2019) Three-strategy and four-strategy model of vaccination game introducing an intermediate protecting measure. Appl Math Comput 346:408–422

Ida Y, Tanimoto J (2018) Effects of noise-perturbing intermediate defense measures in voluntary vaccination games. Chaos Solitons Fractals 106:337–341

Iwamura Y, Tanimoto J, Fukukda E (2016) Effect of intermediate defense measures in voluntary vaccination games. J Statistical Mech Theory Exp:093501

Chapter 5
Quarantine and Isolation

The intervention-game framework introduced in Chap. 4 included provisions for vaccination and intermediate defense measures (IDM), which were defined as voluntary strategies decided upon by individuals. This Chapter introduces another framework for interventions that are forcefully imposed upon individuals, regardless of their intentions. These include quarantine and isolation. We show a theoretical

J. Tanimoto, *Sociophysics Approach to Epidemics*, Evolutionary Economics and Social Complexity Science 23, https://doi.org/10.1007/978-981-33-6481-3_5

framework based on the intervention-game concept and discuss how the introduction of quarantine and isolation impacts the final epidemic size (FES) and average social payoff (ASP).

5.1 Social Background; Quarantine or Isolation?

In general, vaccination and IDM can be regarded as more or less spontaneous, simply because they are self-costly and self-beneficial. The point of how one's decision impacts others is less addressed in the literature. Although commitment to those provisions helps to establish a herd immunity, the primary effect, i.e., benefit, comes to the person who commits. However, if the vigor of a spreading disease becomes critical, as we are currently experiencing with COVID-19, public authorities and governments cannot help but to interfere to save their own population as individual-based provisions are no longer sufficient. Thus, provisions no longer focus on an individual, but on the mass of individuals in the population. For this purpose, authorities sometimes resort to enforcing compulsory provisions.

There are two ideas: quarantine and isolation. Hereafter, we use "quarantine" to refer to a state that is more compulsory than isolation. In quarantine, the authorities force people who are not symptomatic but might be infected to be segregated. By contrast, isolation is only applied to people who are symptomatic. The word "quarantine" originally corresponded to a period of 40 days,[1] which is the length of time that arriving ships suspected of black-plague infection were constrained from intercourse in Mediterranean ports in the fourteenth century. The terms quarantine and isolation have been heavily highlighted in the media in the wake of COVID-19, and would be familiar to lay people. Although the removal of even a small number of infected individuals from the general population is beneficial from the standpoint of public health, it impinges upon individual rights and freedom; thus, in democratic countries, it is believed that applying quarantine on a large scale without due cause would not be possible. However, COVID-19 has altered this traditional picture. Western democratic countries in Europe and the USA have introduced quite strong intervention provisions, even including legal penalties like lock-down to dam the voracious momentum of COVID-19. The only successful example of a democracy not employing such a strong intervention but still managing to control the epidemic is Japan, where the government basically "asks" people and industry to introduce "self-isolation," rather than compulsory quarantine. One reason for their success is that Japanese people's compliancy causes them to interpret the government's "asking" as "demands" or even "decrees."

The mechanism for quarantine and isolation to control the spread of infectious diseases is quite simple, as it forcefully reduces possible contacts between

[1]"Quarantine" is a loan-word from Italian, where it means "segregation"; this in turn comes from the Latin word "quadrāgintā" meaning "40".

susceptible (S) and infected (I) individuals. Before modern medicine and medical science were established, such a social provision of forcefully segregating infected people was the only workable way to cope with dreadful communicable diseases; this is shown, for instance, by biblical passages that refer to the ostracism of lepers.

This chapter concerns another intervention game in which both quarantine and isolation co-exist alongside pre-emptive vaccination. Thus, intervention provisions are threefold: vaccination, quarantine, and isolation. The key point of our theoretical framework is that vaccination is introduced as an individual's voluntary strategy and is therefore affected by a decision-making process backed by the evolutionary game theory, while both quarantine and isolation are defined as globally imposed provisions that are irrespective of an individual's intention.

A fundamental question coming to mind is whether the simultaneous introduction of these two different interventions affects the social dynamics observed in terms of FES, vaccination coverage (VC), or ASP. Another question is whether basic public-health measures, such as isolation (i.e., the removal of only symptomatic individuals from the general population) and quarantine (i.e., the removal of individuals who have had contact with an infectious individual but are not displaying symptoms) are likely to control the spread of the disease. For cases in which these public-health measures are good enough, the next likely question is whether both isolation and quarantine should be used or whether one of the two can be sufficient. For example, even in retrospect, it is not clear whether isolation or quarantine had the greater impact in stopping the spread of SARS, or whether both control measures were essential.[2] Still another interesting question is whether the introduction of such non-voluntary provisions as quarantine and isolation cooperatively bolster the suppressing effect of committing to a voluntary vaccination.[3]

We will answer these questions by establishing a new model for the intervention game.

5.2 Model Structure

5.2.1 *Formulation of the SVEIR Model*[4]

We use the SEIR model as a baseline for the current intervention-game mode. The compartment E implies "infected but not infectious"; thus, those individuals in E can

[2]There have been several studies on quarantine and isolation in the specific case of SARS. For example:

Gumel (2004), pp. 2223–2232.

[3]There have been several pioneering studies on this point. One of those suggested that the combined use of quarantine and isolation policies with an imperfect vaccination strategy can lead to complete elimination of disease, even when a low efficacy level of the universal strategy is considered.

Safi and Gumel (2011), pp. 3044–3070.

[4]Alam et al. (2020), p. 033502.

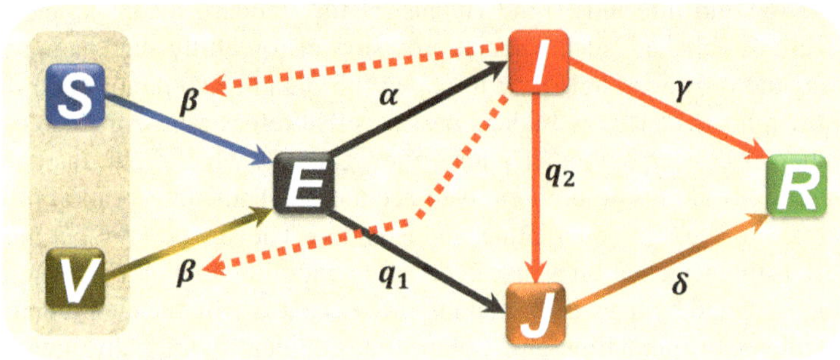

Fig. 5.1 Schematic diagram of the proposed epidemic model: Susceptible (S) individuals get infected with a disease-transmission rate β, which also applies to the fraction of vaccinated individuals who failed to induce perfect immunity. The vaccinated individuals who get immunity and who do not are fully determined by the vaccine effectiveness, denoted by e. Here, the solid arrows indicate the state transfer taking place from one compartment to other with relevant rates. The red dotted arrows show the influence on state transfer from either susceptible (S) to exposed (E) compartments, or from vaccinated (V) to exposed (E) compartments, as triggered by contact with infectious individuals. Once the individuals in S or V become exposed (E), a fraction of them become infectious (as well as infected) with a progression rate α, and are immediately shifted to the infected state (I). The remaining fraction of exposed individuals who remain asymptomatic are picked up with quarantine rate q_1 and thereby placed into compartment (J). Disease-bearers are considered to be infected (I), in that a fraction of infected individuals may be forcefully sent to compartment (J) with isolation rate q_2, while the remaining fraction recovers at a rate γ. The fraction of individuals who are taken to J by imposing either of the policies (quarantine or isolation or joint) eventually recover with a revised recovery rate δ. When an individual enters R, they are removed from further consideration on the local timescale at which the epidemic progression takes place

be picked up and transferred to the compartment J, meaning quarantine, in accordance with the "precautionary principle" to preserve public welfare. By contrast, to represent isolation, those who are at I can be transferred to J, meaning isolation. Hence, compartment J is defined as a haunt that implies both quarantine and isolation.

In practice, we rely upon the basic concepts of the widely used SVEIR (Susceptible $\rightarrow$ Vaccinated $\rightarrow$ Exposed $\rightarrow$ Infected $\rightarrow$ Recovered) epidemic model, because, as with what we discussed in the previous chapters, we presume that pre-emptive "vaccination" is a voluntary provision. To adjust to the specific situation, our new predictive scheme (depicted in Fig. 5.1) includes an additional compartment "J" (as mentioned above) to represent quarantine and isolation. Our proposed model also considers the population to be infinite and ideally well-mixed without having any spatial structure, in keeping with the mean-field approximation (MFA).

To this end, the epidemic model contains compartments: susceptible (S), vaccinated (V), exposed (E), infected (I), recovered (R), and one additional compartment (J), meaning a joint compartment which serves as a universal compartment when quarantine and isolation policies are taken simultaneously or separately.

A susceptible person is an uninfected individual who can be infected through active contact with an infectious person. An exposed person, initially introduced as susceptible or vaccinated, later on encounters an infectious person and gets infected but usually remains asymptomatic; by contrast, an infected person is symptomatic. A recovered person is someone who gets immunized from the disease, and we assume can no longer be infected. A quarantined person is infected but asymptomatic and is forcefully removed from contact with the mass of people (thus, they are transferred from E to J). On the other hand, an isolated person is definitely both infected and infectious and is removed from contact with the general population by being admitted to a hospital or any rehabilitation center, as represented by transferring from I to J. The most important feature of the present model is that both quarantine and isolation are assigned to the same compartment J, despite the original meaning of those two different provisions being preserved. When quarantine is taken as the only control policy, J serves as a single compartment called quarantine (Q). Similarly, when solely isolation is considered, J behaves like a single compartment for isolation (Is). Our theoretical model assumes that susceptible (S) and vaccinated (V) individuals get exposed (E) with an infection-spreading rate β [day^{-1} person^{-1}] before becoming infectious. When the joint policy is imposed, a fraction of exposed (E) individuals who were already becoming symptomatic are immediately transferred to the infected (I) state with progression rate α, and the remaining asymptomatic exposed individuals are then quarantined and thereby forcefully moved to J at a constant quarantine rate, q_1. On the other hand, a certain fraction of infected individuals is isolated and consequently transferred to J with a constant isolation rate, q_2, while the rest of the infected individuals gradually reach the recovery state R with a disease-recovery rate γ [day^{-1}]. Furthermore, the people in J can also recover with a revised recovery rate, δ [day^{-1}].

Our mathematical model substantively shows the interplay between voluntary vaccination (active provision) and the imposed quarantine-isolation (passive provision) policy during an epidemic outbreak. As we have presumed so far, vaccination is costly when introduced. Since we presume that pre-emptive vaccination is imperfect and voluntary, we would therefore like to address the possibility of vaccine failure. In particular, a vaccine may not work equally well for all vaccinators (i.e., the fraction of vaccinators for which the vaccine yields an immune response is termed the effectiveness of vaccination, denoted by e ($0 \leq e \leq 1$), and kept fixed for repeated seasons). Another key property is the VC, expressed as a positive-valued variable $x (0 \leq x \leq 1)$, which evolves over repeated seasons in accordance with the predictions of evolutionary game theory, as explained later. Consequently, we can classify all pre-emptive vaccinators into two categories: immune individuals, who receive perfect immunity with a probability e, and non-immune individuals, who fail to obtain vaccine-induced immunity with a complementary probability of $1 - e$. Throughout this study, the ratio of the infection-spreading rate, β, and the disease-recovery rate, γ, is treated as the basic reproduction number, R_0 ($= \beta/\gamma$).

The evolution of individuals through the SVEIR model with quarantine, isolation, and joint policy in a single season is modeled by the following dynamical equations:

$$\frac{dS(x,t)}{dt} = -\beta S(x,t)I(x,t), \tag{5.1a}$$

$$\frac{dV(x,t)}{dt} = -\beta(V(x,t) - eV(x,0))I(x,t), \tag{5.1b}$$

$$\begin{aligned}\frac{dE(x,t)}{dt} = {} & \beta S(x,t)I(x,t) + \beta(V(x,t) - eV(x,0))I(x,t) - \alpha E(x,t) \\ & - q_1 E(x,t),\end{aligned} \tag{5.1c}$$

$$\frac{dI(x,t)}{dt} = \alpha E(x,t) - q_2 I(x,t) - \gamma I(x,t), \tag{5.1d}$$

$$\frac{dJ(x,t)}{dt} = q_1 E(x,t) + q_2 I(x,t) - \delta J(x,t), \tag{5.1e}$$

$$\frac{dR(x,t)}{dt} = \gamma I(x,t) + \delta J(x,t). \tag{5.1f}$$

For simplicity, the entire population is kept at fixed as:

$$S(t) + V(t) + E(t) + I(t) + J(t) + R(t) = 1. \tag{5.2}$$

The initial conditions for the system of Eqs. (5.1)–(5.2) take the form:

$$S(0) \geq 0, V(0) \geq 0, E(0) \geq 0, I(0) \geq 0, J(0) \geq 0, R(0) \geq 0. \tag{5.3}$$

As we previously assumed, committing to vaccination is costly and being infected also entails an illness cost that is greater than the vaccination cost. In the current evolutionary framework, we can classify individuals into four states in terms of their health condition and cost burden: (i) a healthy vaccinator (HV) who pays only the vaccination cost and remains healthy in an epidemic season; (ii) an infected vaccinator (IV) who commits vaccination but unfortunately becomes infected and must therefore bear the vaccination cost as well as the infection cost; (iii) a successful free rider (SFR), who does not incur any cost and fortunately can survive without being infected; and (iv) a failed free rider (FFR), who relies utterly upon herd immunity and tries to free ride without taking any protective provisions but eventually becomes infected, thereby bearing the infection cost itself.

Presuming a positive VC, x, we can observe the level of each compartment when t (local timescale; meaning the time in a single season) becomes infinitely large ($t \to \infty$). Thus, we obtain:

$$HV(x,\infty) = V(x,\infty), \tag{5.4a}$$

$$SFR(x,\infty) = S(x,\infty). \tag{5.4b}$$

Meanwhile, we are not able to solve the set of Eqs. (5.1a, 5.1b, 5.1c, 5.1d, 5.1e, 5.1f, 5.2 and 5.3) in a fully analytical way; thus, we introduce the concept of flux

"$\varphi_{A \to B}$," indicating the total number of individuals passing through state (compartment) A to B. In practice, we define three flux terms: $\varphi_{V \to I}(x, \infty)$, $\varphi_{S \to I}(x, \infty)$, and $\varphi_{E \to I}(x, \infty)$, to express the transferring flux from vaccinated to infected, susceptible to infected, and exposed to infected, respectively. We can calculate the fraction of infected vaccinators as the conditional expectation of vaccinated individuals who get exposed and are subsequently infectious. The details of this mechanism can be illustrated as a product of the conditional probability of vaccinated individuals who are exposed. The fraction of those exposed individuals who eventually become infectious is given by

$$IV(x, \infty) = \frac{\left(\int\limits_{t=0}^{\infty} \varphi_{V \to E}(x,t)dt\right)}{\int_{t=0}^{\infty}(\varphi_{S \to E}(x,t) + \varphi_{V \to E}(x,t))dt} * \left(\int\limits_{t=0}^{\infty} \varphi_{E \to I}(x,t)dt\right). \tag{5.4c}$$

In a similar fashion, the fraction of non-vaccinators who are infected (i.e., the fraction of failed free riders) can be obtained through

$$FFR(x, \infty) = \frac{\left(\int\limits_{t=0}^{\infty} \varphi_{S \to E}(x,t)dt\right)}{\int_{t=0}^{\infty}(\varphi_{S \to E}(x,t) + \varphi_{V \to E}(x,t))dt} * \left(\int\limits_{t=0}^{\infty} \varphi_{E \to I}(x,t)dt\right). \tag{5.4d}$$

Finally, the corresponding fractions of the four types of individuals existing in the population are encapsulated in Table 5.1.

5.2.2 *Payoff Structure*

We should follow the Table 3.1, because there is only a single costly strategy, which is vaccination defined as an active provision. We recall the relative cost of vaccination, $C_r = C_v/C_i$ $(0 \le C_r \le 1)$, where the costs for vaccination and infection are denoted by C_v and C_i, respectively. As we discussed in the previous sub-section, at the end of a single season, we can use the game-theoretic approach to classify all individuals who initially chose either vaccination or free riding (defection) into the four classes depending upon their final health status and whether they are healthy or infected at the equilibrium point. Therefore, we depict the present payoff structure in Table 5.2, just as in Table 3.1.

Table 5.1 Fractions of four types of individuals

Strategy/State	Healthy	Infected
Vaccinated (Cooperator; V)	$HV(x, \infty)$	$IV(x, \infty)$
Non-vaccinated; Defector (Free Rider; FR)	$SFR(x, \infty)$	$FFR(x, \infty)$

Table 5.2 Estimated payoff structure at the end of each epidemic season

Strategy/State	Healthy	Infected
Vaccinated (V)	$-C_r$	$-C_r - 1$
Non-vaccinated (NV)	0	-1

An individual's decision about vaccinating depends utterly upon the trade-off between the prevention costs and the perceived risks involved. Thus, their decision might evolve with global-scale time (repeating seasons) based on the epidemic incidence observed in the population. Therefore, we can formulate the overall expected payoff by means of the average social payoff $\langle\pi\rangle$, the average payoff of vaccinated individuals $\langle\pi_V\rangle$, and that of non-vaccinated (defective) individuals $\langle\pi_{NV}\rangle$, which are given below:

$$\langle\pi\rangle = -C_r\,HV(x,\infty) - (C_r+1)IV(x,\infty) - FFR(x,\infty), \tag{5.5a}$$

$$\langle\pi_V\rangle = (-C_r\,HV(x,\infty) - (C_r+1)IV(x,\infty))/x, \tag{5.5b}$$

$$\langle\pi_{NV}\rangle = (-FFR(x,\infty))/(1-x). \tag{5.5c}$$

5.2.3 *Strategy Updating and Global Dynamics*

In the present model, we only apply individual-based risk assessment (IB-RA), introduced in Sect. 3.2.2.[5] As described in Eq. (3.25), there are eight state-transition-probability functions, $P(\mathrm{HV} \leftarrow \mathrm{SFR}) = \frac{1}{1+\exp[-(0-(-C_r))/\kappa]}$, $P(\mathrm{HV} \leftarrow \mathrm{FFR}) = \frac{1}{1+\exp[-(-1-(-C_r))/\kappa]}$, $P(\mathrm{IV} \leftarrow \mathrm{SFR}) = \frac{1}{1+\exp[-(0-(-C_r-1))/\kappa]}$, $P(\mathrm{IV} \leftarrow \mathrm{FFR}) = \frac{1}{1+\exp[-(-1-(-C_r-1))/\kappa]}$, $P(\mathrm{SFR} \leftarrow \mathrm{HV}) = \frac{1}{1+\exp[-(-C_r-(0))/\kappa]}$, $P(\mathrm{SFR} \leftarrow \mathrm{IV}) = \frac{1}{1+\exp[-(-C_r-1-(0))/\kappa]}$, $P(\mathrm{FFR} \leftarrow \mathrm{HV}) = \frac{1}{1+\exp[-(-C_r-(-1))/\kappa]}$, and $P(\mathrm{FFR} \leftarrow \mathrm{IV}) = \frac{1}{1+\exp[-(-C_r-1-(-1))/\kappa]}$, throughout which we presume a noise parameter in the Fermi function, κ, of 0.1.

At the end of each epidemic season, everyone can update their strategy depending upon the last season's payoff. Hence, with increasing or decreasing VC, x is inevitable. Here, the independent variable, t, indicates the global timescale, which in other words means the number of elapsed seasons. Since we consider IB-RA strategy-updating rule for decision making in each subsequent season, the dynamic equation following this particular rule can be expressed as follows:

[5] Needless to say, you can apply SB-RA instead of IB-RA.

$$\frac{dx}{dt} = HV(x,\infty)SFR(x,\infty)(P(SFR \leftarrow HV) - P(HV \leftarrow SFR))$$
$$+HV(x,\infty)\ FFR(x,\infty)(P(FFR \leftarrow HV) - P(HV \leftarrow FFR))$$
$$+SFR(x,\infty)IV(x,\infty)(P(SFR \leftarrow IV) - P(IV \leftarrow SFR))$$
$$+FFR(x,\infty) * IV(x,\infty)(P(FFR \leftarrow IV) - P(IV \leftarrow FFR)). \quad (5.6)$$

5.3 Result and Discussion

5.3.1 Local Dynamics in a Single Season

Let us confirm what happens in a single season to show how well quarantine and isolation policies manage to suppress the spread of disease. The result here is not influenced by evolutionary game theory, which implies a fixed VC of 0.46 at the beginning of a season. We presume the set of parameters to be $\beta = 0.833, \gamma = \frac{1}{3}, \alpha = 0.25,\ \ \delta = 0.33,$ and $e = 0.5$, where the value of β comes from the basic reproduction number, $R_0 = 2.5$, considered for the case of a seasonal-influenza-like flu.[6] Another two crucially important parameters, q_1 and q_2, which are the pick-up rates for quarantine and isolation policies, respectively, have been introduced here to quantify the levels of quarantine and isolation policies. If both parameters take non-zero positive values, then both quarantine and isolation policies co-exist and we call this the "joint policy." Figure 5.2 compares the no-policy, quarantine-policy, isolation policy, and joint-policy cases by displaying the fraction of people in each compartment at the equilibrium state. Note that the compartment denoted "isolated" refers to "J," which contains individuals not only resulting from isolation, but also from quarantine policies. In all four cases, a monotonically increasing tendency is observed in the fraction of recovered individuals, which is triggered mostly by the monotonic decrement of the fraction of susceptible individuals; this is quite conceivable. Moreover, a decreasing tendency is also observed in the vaccinated fraction, which eventually reaches a stable equilibrium. This phenomenon can be justified by the fact that vaccines do not bring perfect immunity ($e = 0.5$ is presumed); thus, the population initially labeled "V" monotonically decreases as some of them are infected (transferring from V to E). This is also conceivable, since it confirms that the model works fairly closely to how it was designed. Figure 5.3 highlights the fraction of infection for four respective scenarios. Obviously, the promptness of an infection dying out due to the joint policy justifies its outperformance among the four scenarios. From the quantitative viewpoint of an infection spreading, quarantine and isolation policies are next implemented

[6]Fukuda et al. (2014), pp. 1–9.

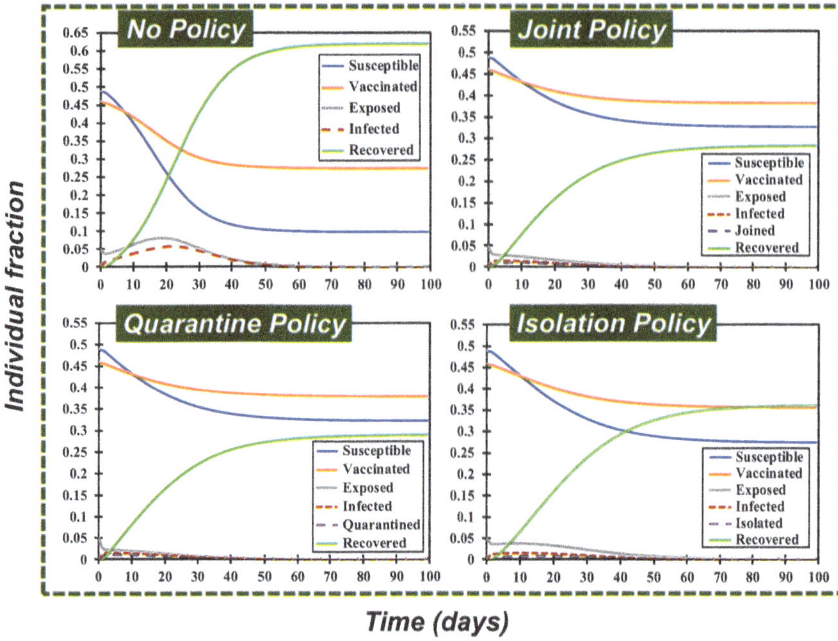

Fig. 5.2 Time-evolutionary line graphs of four different schemes based on their policy selection. For the no-policy, joint-policy, isolation policy, and quarantine-policy cases (considered in a clockwise direction), we depict the fractions of individuals existing in each compartment at local equilibrium. The highest fraction of exposed individuals is observed in the case of the no-policy condition. The exposed fraction dies out much earlier when either joint or quarantine policy is imposed. In all four cases, the baseline values of the parameters are considered as $\beta = 0.8333$, $\gamma = 0.3333$, $\alpha = 0.25$, $\delta = 0.33$, and $e = 0.5$. For simplicity, we choose, $q_1 = q_2 = 0$ for the no – policy condition, $q_1 = 0.1$, $q_2 = 0.1$ for the joint – policy condition, $q_1 = 0.2$, $q_2 = 0$ for the quarantine – policy condition, and $q_1 = 0.0$, $q_2 = 0.2$ for the isolation policy condition

according to the joint policy. We observe that the quarantine policy outperforms the isolation policy, while the joint policy (i.e., adopting both at the same time) performs understandingly well in keeping the infection size smaller.

5.3.2 Social Equilibrium from Global Dynamics

In Fig. 5.4, varying the basic reproduction number R_0 and progression rate α, we present a set of two-dimensional full-phase heat maps to quantify the FES under a wide range of parametric conditions used for the quarantine-isolation policy. Each 2D heat map displays FES along with vaccination cost, C_r, ranging from 0 to 1 along the x-axis, as well as the vaccine effectiveness e, ranging 0 to 1 along the y-axis. For

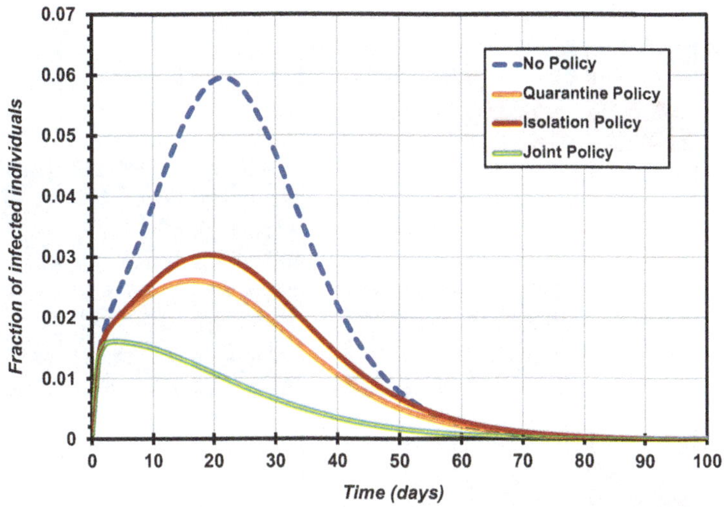

Fig. 5.3 The time evolution of the infection densities observed for different policies: the trajectories of the infected fractions obtained from four different policy conditions are utterly controlled by two pick-up rates, namely the quarantine rate q_1 and the isolation rate q_2. While constructing the figure, we take $q_1 = q_2 = 0$ for no − policy, $q_1 = 0.2$, $q_2 = 0$ for the quarantine policy, $q_1 = 0.0$, $q_2 = 0.2$ for the isolation policy, and $q_1 = q_2 = 0.1$ for the joint policy. The baseline values for each parameter, $\beta = 0.8333$, $\gamma = 0.3333$, $\alpha = 0.25$, $\delta = 0.33$, are considered as positive constants. Also, the vaccine efficacy throughout the season is considered to be $e = 0.5$. The infected fraction observed at its pick point is around 0.06 for the no-policy case. When other policies are taken, the infected fraction drops significantly. We observe that the quarantine policy outperforms the isolation policy while the joint policy (i.e., adopting both at the same time) performs understandingly well in keeping the infection size smaller

a comprehensive understanding, we introduce two control parameters, namely, q_1 and q_2, which represent the contributions of quarantine and isolation policies, respectively. Additionally, we include six blocks labeled a, b, c, d, e, and f, each having nine panels to explain the comparative impact of quarantine and isolation policies besides pre-emptive vaccination at varying combinations of basic reproduction number R_0 in the row-wise direction and the infection-progression rate α in the column-wise direction. For the sake of discussion, we presume two different values of $R_0 = \{4.5, 2.0\}$ and three different $\alpha = \{0.10, 0.25, 0.40\}$ to represent all evolutionary outcomes. Being shifted from the standard value of the basic reproduction number ($R_0 = 2.5$), which came from the case of the flu in many previous studies and was presumed in Figs. 5.2 and 5.3 we investigate two extreme cases: $R_0 = 4.5$ (i.e., the infection-spreading rate β is higher and the recovery rate γ is lower than usual in flu cases) and $R_0 = 2.0$ (i.e., the infection-spreading rate β is lower and the recovery rate γ is a little bit higher). In a nutshell, $R_0 = 4.5$ is considered to be a worse situation than the standard flu, whereas $R_0 = 2.0$ can be treated as more moderate. Moreover, implementing a pair of control parameters $q_i = \{0.0, 0.1, 0.2\}_{i\,=\,1,\,2}$ in two different directions, we can classify nine panels

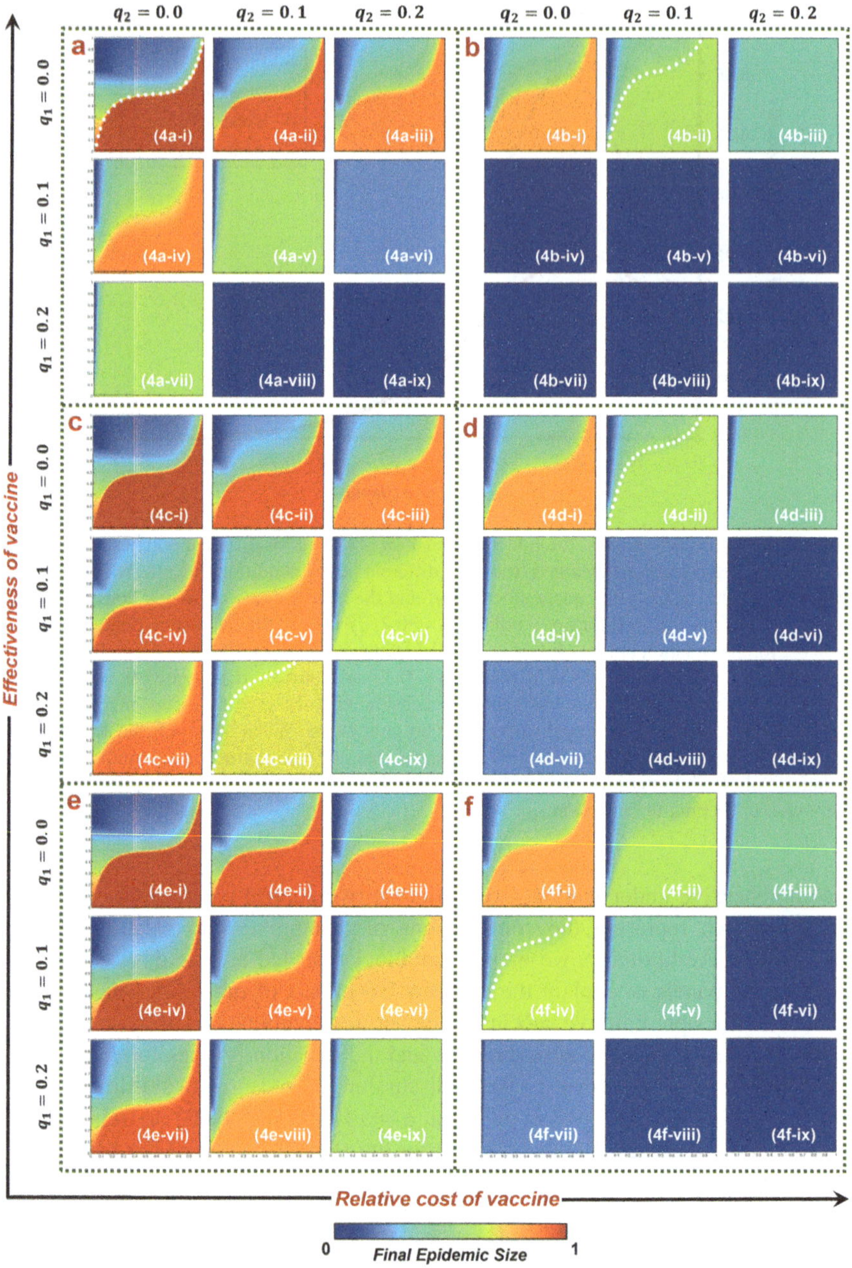

Fig. 5.4 Final epidemic size (FES) observed under varying vaccine cost and efficacy. First, we distinguish the entire parameter space into six blocks labeled **a**, **b**, **c**, **d**, **e**, and **f** based on two different basic reproduction numbers ($R_0 = 4.5, 2.0$) in a row – wise direction and three different progression rates ($\alpha = 0.10, 0.25, 0.40$) in a column – wise direction. Also, while applying control policies, we use two different pick – up rates; $q_1(=0,0.1,0.2)$ for a quarantine policy and $q_2(=0, 0.1, 0.2)$ for an isolation policy. For example, block "**a**" is solely designed for $R_0 = 4.5$ and $\alpha = 0.10$, which contains nine panels, each labeled with (4a – ∗), depending upon the values of

quantitatively inside each of the six blocks. Therefore, on the basis of parameter settings, we can distinguish four different policy regions as follows: (i) no-policy ($q_1 = 0$, $q_2 = 0$); (ii) quarantine policy ($q_1 \neq 0$, $q_2 = 0$); (iii) isolation policy ($q_1 = 0$, $q_2 \neq 0$); and (iv) joint policy ($q_1 \neq 0$, $q_2 \neq 0$). Panels (4*-i) show the FES when no-policy is initiated by the governing authorities (the default situation). Accordingly, panels (4*-vii), (4*-iii), and (4*-ix), respectively, indicate the full-scale implementation of quarantine, isolation, and joint policies considered under varying conditions.

An in-depth analysis of the basic reproduction number and progression rate reveals a few interesting points that must be shared with readers here. First, if we presume that $R_0 = 4.5$ and $\alpha = 0.10$, we can see from Fig. 5.4 that the quarantine policy (panel (4a-vii)) more effectively works to suppress disease than the isolation policy (panel (4a-iii)), while both perform better than the default case (panel (4a-i)). The introduction of control policies at a certain level can sometimes push up the critical line (for instance, the white dotted line in panel (4a-ii)) between the controlled and endemic phases (see, for example, panels (4b-ii), (4c-viii), (4d-ii), (4f-iv)) when varying R_0 and α are presumed.

Interestingly, when $R_0 = 4.5$ and α are set to 0.25 and 0.40 (depicted in blocks c and e, respectively), no such pushing takes place as long as a pure-quarantine policy is imposed (i.e., fixing $q_2 = 0$). Keeping the parameter settings unchanged for R_0 and α, we observe that the pure-isolation policy reduces FES compared to the pure-quarantine policy (see panel (4c-iii) vs. (4c-vii) and panel (4e-iii) vs. (4e-vii)). In both cases, the isolation policy offers better disease attenuation, as measured in terms of the existing size of the epidemic, than the quarantine policy, especially in the endemic regions below the critical line. This is a quite interesting finding. One possible reason behind this is that the higher progression rate (when $\alpha > 0.10$) might increase the flow of individuals from an exposed situation to an infected state, rather than shifting them to a quarantined state. This in return enables the government to capture infected and infectious people (I) efficiently through an isolation policy, which ensures public safety. Therefore, an isolation policy seems to work better than quarantine in this condition. In Fig. 5.4, panel (4c-iii) clearly suggests the supremacy of an isolation policy in producing an impressive result compared to a quarantine policy (4c-vii), thereby contributing as a dominant factor when a joint policy is taken (4c-ix). In other words, we can say that the increasing progression rate, α, gradually diminishes the contribution of the quarantine policy. This might be important when establishing the strategy for a public-health policy.

Meanwhile, when a smaller basic reproduction number ($R_0 = 2.0$) is presumed, the promptness of the epidemic outbreak is weakened, and thus moderate policies are expected from the government. As we can see from panels (4b-*), a reasonable level of quarantine can bring back the epidemic-free situation (when $\alpha = 0.10$). Even

Fig. 5.4 (continued) pick − up rates q_1 and q_2. The construction of the remaining blocks is analogous to the way block "**a**" has been designed. Here, the white dotted line indicates the critical line between the controlled and endemic phases

when the progression rate (α) is a little bit higher than our prior consideration (e.g., $\alpha = 0.25, 0.40$), a reasonable contribution from either of the policies is fair enough (panels (4d-vi), (4d-viii), (4f-vi), (4f-viii)) to ensure a pleasant situation in terms of epidemic size.

Figure 5.5 shows the VC under different parametric conditions and varying R_0 and α, as presented in the same manner as Fig. 5.4. Below the critical line, there is no hope for individuals to be vaccinated due to the meager effectiveness and higher vaccination cost throughout that region. However, in the no-policy cases (e.g., panels (5a-i), (5c-i), (5e-i)), VC converges toward its maximum limit compared to other policy cases. This is because neither quarantine nor isolation policies encourage people to vaccinate as long as a reasonably higher effectiveness and lower cost are ensured. When a full-scale quarantine or isolation policy (panel (5*-ix)) is taken, VC immediately declines to zero, but the corresponding FES observed in Fig. 5.4 is not bad. Meanwhile, for $R_0 = 2.0$, it is observed that an intermediate quarantine policy determined by the pick-up rate $q_1 = 0.1$ is adequate to maintain a small FES (see panel (4b-iv)), as long as the progression rate α is set to 0.10 although VC is zero (see panel (5b-iv)). That means that the spread of the epidemic can be kept under control by only quarantining a limited number of exposed individuals. In this sense, quarantine can be a powerful provision to suppress the epidemic. We also observe that, in such a situation, an isolation policy is relatively less effective than quarantine. Therefore, results coming from the dual-policy cases (see panels (4b-v), (4b-ix)) are mostly influenced by the quarantine policy. On the other hand, the incredible dominance with complete remission of the disease triggered by either quarantine or isolation policies is no longer observed when the progression rate, α, is set to 0.25 to 0.40 (see panels (4d-*) and (4f-*)).

Although it sounds trivial, implementing two control policies (quarantine and isolation) at full scale is the most efficient way to suppress the spread of a disease. Therefore, apart from pre-emptive vaccination, the public-health authority must prepare sufficient resources and support, as well as formulate a set of comprehensive quarantine-isolation policies to address this pressing issue. At times when a quarantine-isolation policy works very well, the spread of a disease can be controlled even under a low or zero vaccination rate. Looking at the panels ((4a-ix), (4c-ix), (4e-ix)) for FES and the corresponding panels ((5a-ix), (5c-ix), (5e-ix)) for VC, we can justify what we mentioned earlier. It also presents a positive impression of how these two types of preventive measures (pre- and post-provisions, or active provision and passive provisions using our earlier terminology) complement each other as a means of containing or preventing infectious diseases.

The importance of introducing a quarantine-isolation policy can be well understood from the phase diagrams drawn for ASP in Fig. 5.6. Although the holistic tendency observed at ASP is quite analogous to FES, it shows subtle differences above the critical line. The question then arises: why is this color-gradient change observed in the ASP heat map but not in the FES heat map? This might be an interesting phenomenon and the exact reasons must be noted. These reasons are twofold: a subtle change in FES can create an enlarged difference in ASP, and a fractional difference in vaccines contributes to producing a better social payoff.

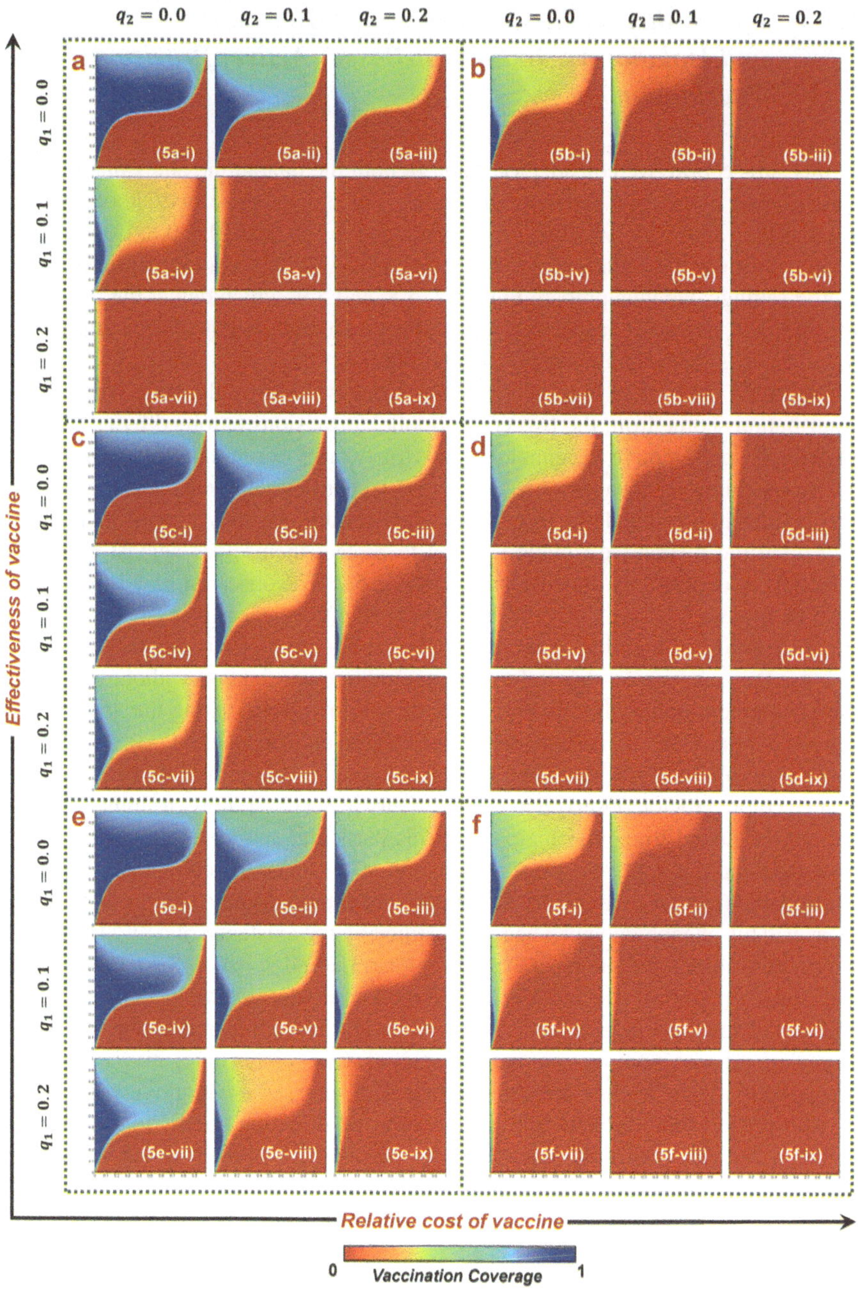

Fig. 5.5 Vaccination coverage, VC, observed under varying vaccine cost and efficacy. First, we distinguish the entire parameter space into six blocks labeled **a**, **b**, **c**, **d**, **e,** and **f** based on two different basic reproduction numbers ($R_0 = 4.5, 2.0$) in the row – wise direction and three different progression rates ($\alpha = 0.10, 0.25, 0.40$) in the column – wise direction. Also, while applying control policies, we use two different pick – up rates, namely $q_1(=0,\ 0.1,\ 0.2)$ for quarantine policy and $q_2(=0,\ 0.1,\ 0.2)$ for isolation policy. For example, block **a** is solely designed

5.3.3 Public-Based (Passive) Provision: Quarantine and Isolation vs. Individual-Based (Active) Provision: Vaccination

In this sub-section, we consider variation in both the quarantine and isolation rates (i.e., q_1 and q_2) in two different directions. For this purpose, we choose several parametric values for vaccine effectiveness ($e = 0.2, 0.4, 0.6, 0.8, 1.0$) and the relative cost of vaccination ($C_r = 0.2, 0.4, 0.6, 0.8$). With the aforementioned parametric settings, we thus prepare Figs. 5.7, 5.8, and 5.9, which depict the heat maps for FES, VC, and ASP, respectively. Each heat map, both x-axis for quarantine rate; q_1, and y-axis for isolation rate; q_2, range 0 to 0.7.

In Fig. 5.7, a broken line colored white along 45° shows the state where equal contributions come from the two control-policy provisions enforced by authorities. Essentially, pre-emptive vaccination is not trustworthy for common people as long as its reliability is meager (e.g., $e = 0.4$), regardless of what relative cost of vaccination is presumed. In fact, under such conditions, they cannot help but to rely upon the quarantine-isolation policy to protect themselves against threats of infection, as depicted by dotted triangles in Fig. 5.7. The observed FES is quite high because of poor VC (see Fig. 5.8). Yet, when the effectiveness is reasonably high, a certain fraction of individuals prefer to take pre-emptive provisions (see Fig. 5.8), because they can protect themselves by pre-emptive vaccination without relying on the government policy. Intuitively, when we go downwards along the 45° line, we can see a monotonic increase tendency in FES. Except for panels ((7-a), (7-f), (7-k), (7-p), (7-g), (7-l), (7-q)), this monotonic increase takes place until it reaches a ring-shaped colored region observed in Fig. 5.7. Only panel (7-b) shows a slight deviation around the dotted triangular region. For this particular setting of the effectiveness ($e = 0.4$) and the relative cost of vaccination ($C_r = 0.2$), the epidemic size is still controllable (marked with a curved triangle), even though the isolation rate is quite meager. This phenomenon can be fully justified by the existence of a tiny fraction of vaccinators around the same parametric region observed in panel (8-b). Beneath this ring-shaped region (down the 45° line), a monotonic decreasing tendency is found where both pick-up rates q_1 and q_2 are relatively low compared to other regions. Complete reliance upon government policies always creates a negative impression among individuals, leading them to put their faith solely in pre-emptive vaccination.

Along the white broken line in the upward direction, both quarantine and isolation policies contribute equally to suppressing the spread of the disease. In

---

Fig. 5.5 (continued) for $R_0 = 4.5$ and $\alpha = 0.10$, which contains nine panels, each labeled with (5a $-$ $*$), depending on the values of pick $-$ up rates q_1 and q_2. The constructions of the remaining blocks are analogous to the way block **a** has been designed. At a glance we can see that the vaccination coverage is a little higher in most cases when $R_0 = 4.5$ (see blocks **a, c, e**) than when $R_0 = 2.0$ (see blocks **b**, **d**, **f**)

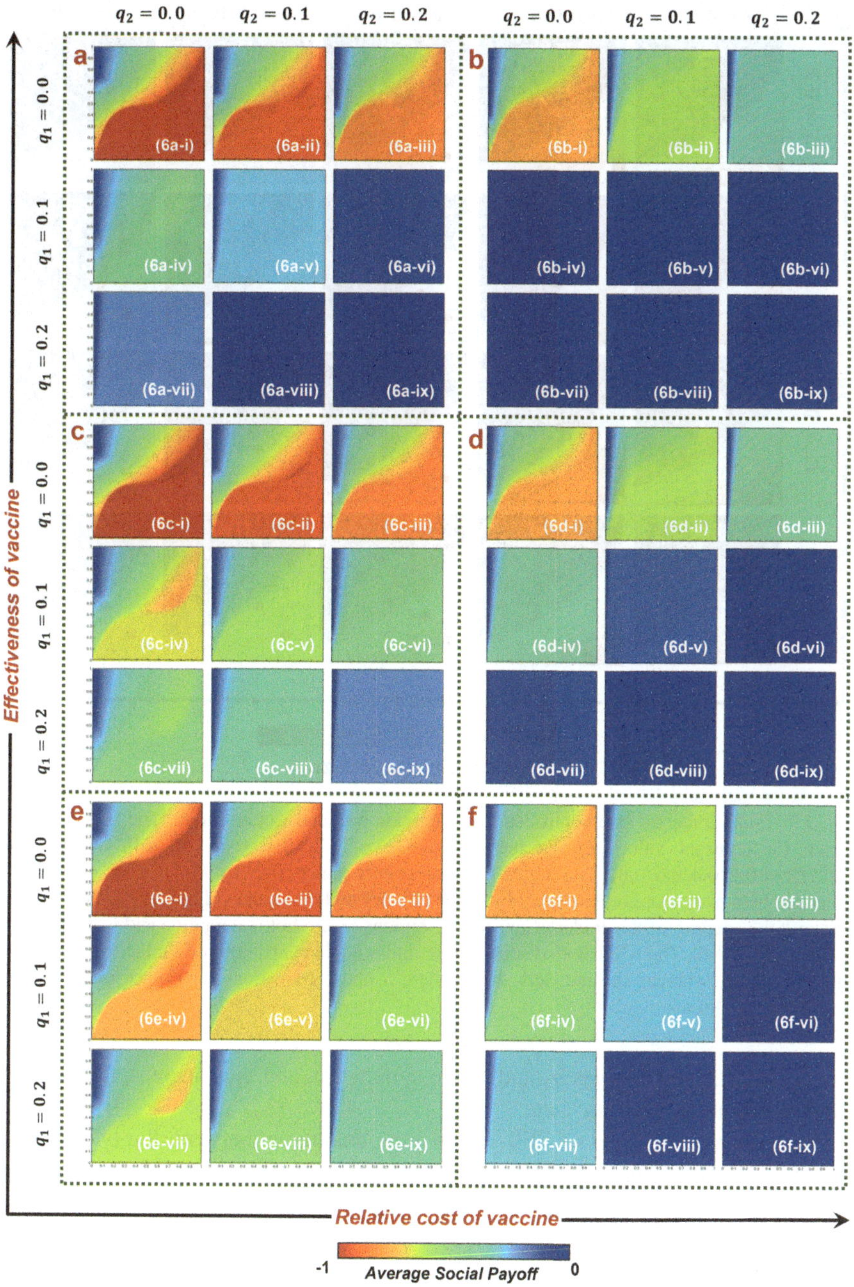

Fig. 5.6 Average social payoff (ASP) observed at varying vaccine costs and efficacies. We distinguish the entire parameter space into six blocks labeled **a**, **b**, **c**, **d**, **e**, and **f** based on two different basic reproduction numbers ($R_0 = 4.5, 2.0$) in the row – wise direction and three different progression rates ($\alpha = 0.10, 0.25, 0.40$) in the column – wise direction. Also, while applying control policies, we use two different pick – up rates, namely q_1(=0, 0.1, 0.2) for quarantine policy and q_2(=0, 0.1, 0.2) for isolation policy. For example, block **a** is solely designed

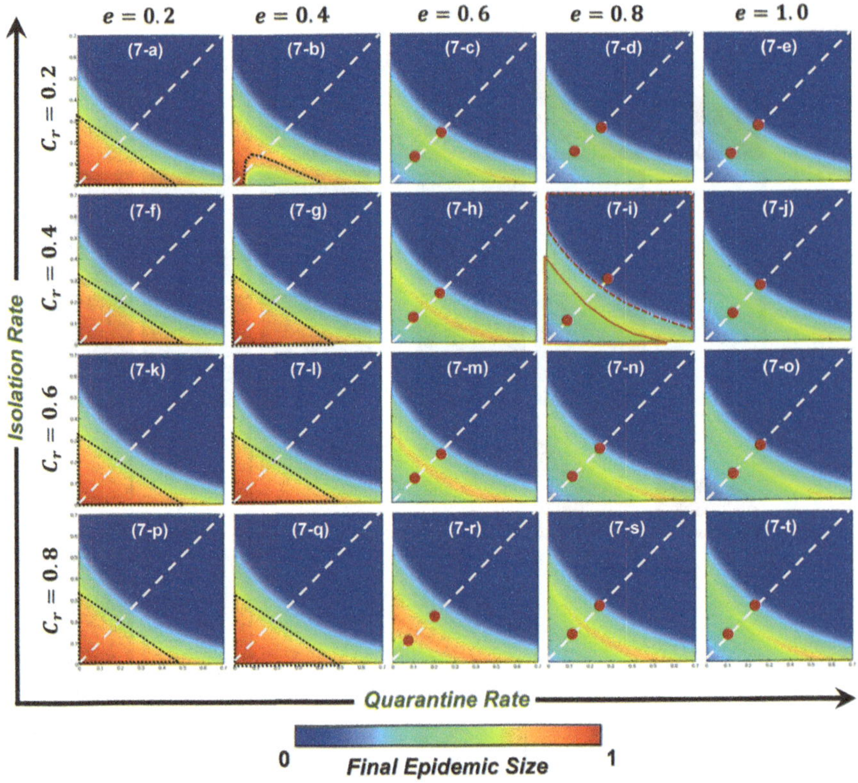

Fig. 5.7 Final epidemic size controlled by quarantine rate and isolation rate. The heat maps are drawn for constant values of vaccine efficacy ($e = 0.2, 0.4, 0.6, 0.8, 1.0$) and relative cost ($C_r = 0.2, 0.4, 0.6, 0.8$). This figure contains twenty panels (labeled (7-*)), each of which is generated by a particular combination of vaccine efficacy and cost values. The black dotted triangle indicates the severity of the epidemic existing in regions where both quarantine and isolation rates are relatively small. The white dotted straight line shows equal contributions coming from isolation and quarantine, whereas the red dots show where a phase transition takes place under certain parametric conditions

that direction, pick-up rates q_1 and q_2 are symmetric, and their increasing tendency ensures a smaller epidemic size.

Interestingly, the level of FES seems quite analogous between the two regions (below and above the ring marked with two red dots along the 45° line), which indicates that individual-based pre-emptive provisions (vaccination) and public-

---

Fig. 5.6 (continued) for $R_0 = 4.5$ and $\alpha = 0.10$, which contains nine panels, each labeled with $(5a - *)$, depending upon the values of pick − up rates q_1 and q_2. The constructions of the remaining blocks are analogous to the way block **a** has been designed. At a glance we can see that the average payoff is higher for most cases in which $R_0 = 2.0$ (see blocks **b**, **d**, **f**) than those in which $R_0 = 4.5$ (see blocks **a**, **c**, **e**)

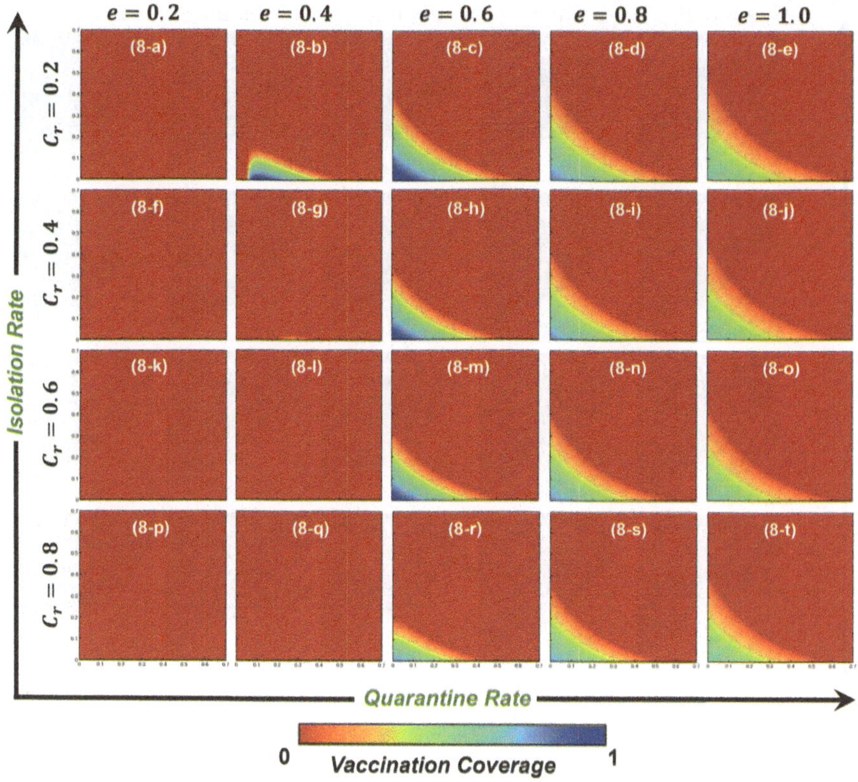

Fig. 5.8 Vaccination coverage controlled by the quarantine and isolation rates. The heat maps are drawn for constant values of vaccine efficacy (e = 0.2, 0.4, 0.6, 0.8, 1.0) and relative cost (C_r = 0.2, 0.4, 0.6, 0.8). This figure contains twenty panels (labeled with (8-*)), each of which is generated by a particular combination of vaccine efficacy and cost values

based late provisions (quarantine/isolation) exert an equal influence. Therefore, we can separate the phase diagrams (heat maps) into two major regions based on the instinct contributions coming from either of the two types of provisions. To justify our hypothesis through 2D heat maps, let us consider the example of panel (7-i), where the solid orange line shows the contribution of pre-emptive vaccination and the dotted red line indicates the contribution brought about by the quarantine-isolation policy. Thus, the contributory characteristics shown by the pre-emptive provision and control policies are immensely beneficial for each individual living in society. We can also verify our hypothesis through Fig. 5.9, which provides us with a more general idea of how a dual contribution driven by two types of provisions can significantly enhance the ASP. One obvious occurrence with vaccination is that, when reasonably reliable vaccines are offered at a cheap price, they strongly attract people despite the fact that fully relying on a public-based government policy can contribute significantly to producing a better social payoff (see, for example, panels (8-c), (8-d), (8-h), (8-i), (8-n), where the social payoff is radically improved).

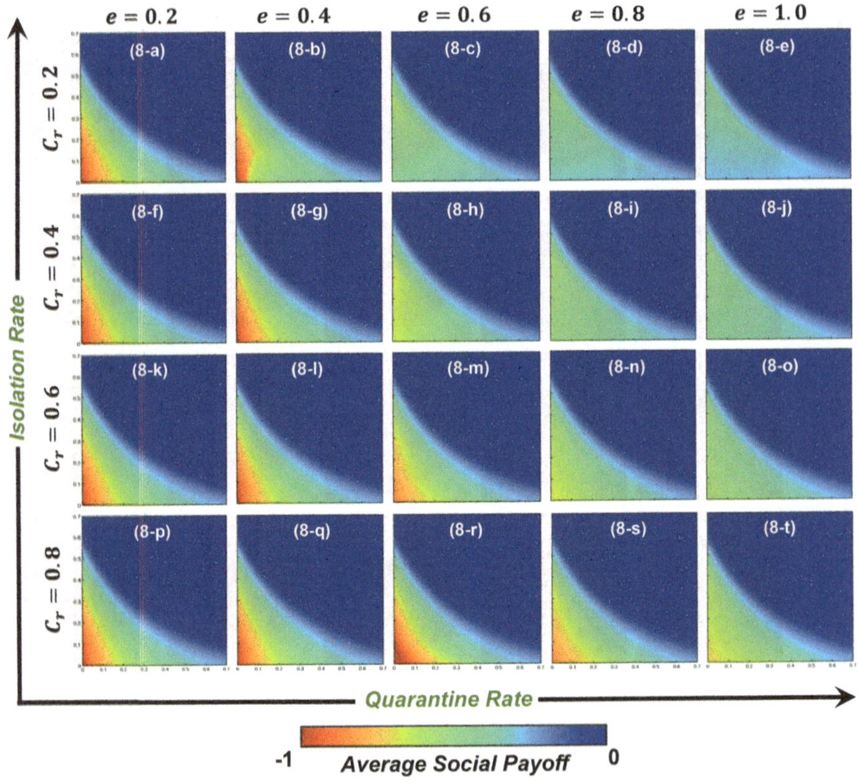

Fig. 5.9 Average social payoff controlled by the quarantine and isolation rates. The heat maps are drawn for constant values of vaccine efficacy (e = 0.2, 0.4, 0.6, 0.8, 1.0) and relative cost (C_r = 0.2, 0.4, 0.6, 0.8). This figure contains twenty panels (labeled with (9-*)), each of which is generated by a particular combination of vaccine efficacy and cost values

5.3.4 Passive Provision Rather Compensates the Shadow by Active Provision Than Mutually Competing

In this chapter, we considered pre-emptive vaccination and forced control policies, two major protection approaches against pandemics and severe epidemics. We then systematically studied the positive impacts triggered by a quarantine-isolation policy while modeling the mathematical epidemiology aided by the theory of vaccination games. The central theme of this study has been to answer the questions of when, why, and under what conditions do authorities have the strongest justification for implementing quarantine and isolation policies to prevent infectious diseases. Our proposed model deeply elaborates the delicate impact of these control policies through the SVEIJR epidemic model, as well as their appropriateness under a wide range of parametric conditions to suppress the spread of disease among individuals who primarily took vaccination as a pre-emptive provision. However, due to the transient effect of vaccination and the widespread occurrence of epidemic breakout, there is an emergent need to seek more sustainable approaches to protect the global community from viral diseases. This study emphasized the importance of using quarantine or isolation policies to prevent the spread of infection. Particularly at times when the majority of people prefer not take vaccination due to its meager efficacy or higher cost, these control policies can somehow offer some degree of reassurance. Through numerical simulation, it is obvious that adopting a quarantine-isolation policy can also relax any bad situation in the hopeless region. We also showed which policy performs best under what conditions. Our theoretical analysis suggests that a joint policy should be taken when the disease-spreading rate is higher. Nonetheless, the government should pay more attention to securing a healthy state in society by taking either of the control policies. The important question remains: how should the pick-up rates be designed? We justified this fact by varying the pick-up rate parameters q_1 (designed for quarantine policy) and q_2 (dedicated to isolation policy) in two different directions while generating phase diagrams for FES, VC, and ASP throughout this study. In other words, the relative contributions coming from either pre-emptive vaccination or late-control policies are equally important for maintaining the FES at a controllable state.

Using the MFA technique and the IB-RA strategy-updating rule, our model can successfully address the importance of dual provision techniques to the audience which has never been used before. In this study, we have deliberately skipped all visual datasets using the strategy-based risk-assessment (SB-RA) update rule introduced in Chap. 3. Basically the same general tendency was observed even when SB-RA was applied instead of IB-RA, other than suggesting some difference arising inherently from the substantial difference between IB-RA and SB-RA, as discussed in Chap. 4 (Sect. 4.3).

5.3.5 Comprehensive Discussion

This chapter has demonstrated that the mathematical-epidemiological model is quite useful for depicting various situations to adjust the respective circumstances brought about by an epidemic and the social provisions followed to contain it. Quarantine and isolation, which have been introduced for COVID-19 since no protective medical provisions like vaccines or antiviral treatments are available, can be embedded into the baseline ODE model. Note that those two provisions are treated as independent of an individual's decision. That is why the focal compartments I and J do not connect with the framework of the evolutionary game. Hence, it can be said that the flexibility of that part of the mathematical-epidemiological model can be preserved.

References

Alam M, Kabir KMA, Tanimoto J (2020) Based on mathematical epidemiology and evolutionary game theory, which is more effective: quarantine or isolation policy? J Statist Mech Theory Exp:033502

Fukuda E, Kokubo S, Tanimoto J, Wang Z, Hagishima A, Ikegaya N (2014) Risk assessment for infectious disease and its impact on voluntary vaccination behavior in social networks. Chaos Solitons Fractals 68:1–9

Gumel AB, Ruan S, Day T, Watmough J, Brauer F, Van Den Driessche P, Gabrielson D, Bowman C, Alexander ME, Ardal S, Wu J, Sahai BM (2004) Modelling strategies for controlling SARS outbreaks. Proc Royal Society B 271:2223–2232

Safi MA, Gumel AB (2011) Mathematical analysis of a disease transmission model with quarantine, isolation and an imperfect vaccine. Comp Math Appl 62(10):3044–3070

Chapter 6
Media Information Effect Hampering the Spread of Disease

The media effect is expected to deliver appropriate information to the people, allowing them to prepare some provisions for the spread of a disease in advance. As discussed in Chap. 4, gargling, wearing a mask, or avoiding crowds to maintain a "social distance" may help to considerably reduce the transmission rate. The effect from such social preventive measures can be enhanced by the media through rumor, information obtained through personal networks, SNS, and, of course, mass media.

In this chapter, we quantify the media effect to lighten the disease impact. Shared information is diffused through a human network from one individual to another according to the physics of diffusion. Establishing a theoretical model based on the SIR process and taking account of this media effect, we discuss how its introduction

J. Tanimoto, *Sociophysics Approach to Epidemics*, Evolutionary Economics and Social Complexity Science 23, https://doi.org/10.1007/978-981-33-6481-3_6

mitigates the final epidemic size (FES) and improves the vaccination coverage (VC) as well as the average social payoff (ASP).

6.1 Positive Effect of Media Helps to Suppress the Spread of an Epidemic

There are various opinions concerning the pros and cons of the media effect in the wake of COVID-19. Media supports people's groundless terror by providing rootless information and dubious news, which should be called agitation. This is a negative effect from the media that can worsen a situation. In some cases, a certain authority or nation may intentionally control information by covering-up facts for their own benefit, rather than offering correct information to the worldwide public. However, it is also true that media effectively works to deliver useful information to the people, letting them prepare for oncoming waves of an epidemic in the near future. Such information drives people to be aware of risk and allows them to take some intermediate defense measures (IDMs), such as gargling or wearing a mask, as we discussed in Chap. 4, while people without such useful information are at a relative disadvantage. Information may be brought by the media, but also by neighbors through word-of-mouth. If this is the case, the diffusion of information to implement a mathematical model should be not depicted only by discounting the transmission rate of an epidemic, but should also consider how information propagates on an underlying network connecting individuals.

Spreading "awareness" of a disease can play a significant role in reducing the infection, as has been observed by many previous studies of complex population networks.[1] In order to address the effect of awareness upon an epidemic's dynamics, this chapter considers the voluntary-vaccination dilemma whose template was introduced in Chap. 3, and has applied in both Chaps. 4 and 5, to characterize the decision-making process in a population.

This chapter integrates a vaccination game with different types of strategy-updating rules for vaccination in a susceptible-infected-recovered (SVIR) process, coupled with an unaware–aware (UA) situation (hereafter: SVIR-UA model) in a heterogeneous network. Unlike usual vaccination games on well-mixed populations based on MFA, we mainly concentrate upon two different types of degree distributions: the Poisson (Erdős–Rényi random (E-R random) graph) and power-law (Barabási–Albert scale-free (BA-SF) network) distributions. This model deals with imperfect vaccination, through which a vaccine only offers perfect immunity with a

[1]There have been many previous studies. Some representative ones are:
Salathe and Bonhoeffer (2008), pp. 1505–1508.
Gómez et al. (2013), p. 028701.
Granell et al. (2014), p. 012808.
Guo et al. (2015), p. 012822.
Wang et al. (2016), p. 29259.

probability of *e* (effectiveness), and otherwise fails to offer immunity. More importantly, the proposed model treats the situation in which information is defined by the binary states of awareness (A) and unawareness (U), allowing an individual to avoid infection by reducing the disease-transmission rate. To do this, we assume that the portion of aware individuals (information-advantageous) may withdraw or stay safe from crowds or undesirable contamination. Thus, in the present study, we assume that an individual entering an aware state does not require any cost, which is different from what the IDM measures presumed.

6.2 Model Structure

6.2.1 Formulation of the SVIR-UA Model[2]

We use the SVIR model as a base and apply the methodology for considering a spatial structure among individuals, as introduced in Sect. 3.4. Hence, let $US_k(\mathrm{t})$, $AS_k(\mathrm{t})$, $UV_k(\mathrm{t})$, $AV_k(\mathrm{t})$, $UI_k(\mathrm{t})$, $AI_k(\mathrm{t})$, and $R_k(\mathrm{t})$ be the densities of the unaware susceptible, aware susceptible, unaware vaccinated, aware vaccinated, unaware infected, aware infected, and recovered nodes of connectivity (degree) k at time t in a population. These satisfy the following normalization condition:

$$US_k(\mathrm{t}) + AS_k(\mathrm{t}) + UV_k(\mathrm{t}) + AV_k(\mathrm{t}) + UI_k(\mathrm{t}) + AI_k(\mathrm{t}) + R_k(\mathrm{t}) = 1. \quad (6.1)$$

Figure 6.1 shows all compartments and state transfer in the present model. We presume that there are two spatial structures (say, two-layer networks): a physical network and a virtual network. In a physical network, an epidemic spreads, while in a virtual network, the information (actually “awareness”) is diffused. The disease state is threefold: susceptible, infected, and recovered, with state transfer represented by black arrows. The information state is binary: one is either aware (A) or unaware (U), with state transfer being drawn by a red arrow. The connection between unaware susceptible (US) and aware susceptible (AS), or unaware vaccinated (UV) and aware vaccinated (AV) depicts the virtual interaction, suggesting that information spreads from an unaware state to an aware one. On the other hand, the transitions from US to unaware infected (UI), aware susceptible (AS) to aware infected (AI), UV to UI, and aware vaccinated (AV) to aware infected (AI) show physical interaction, suggesting that the infection spreads. The unaware-susceptible and unaware-vaccinated compartments become UI with the disease-transmission rate β [day^{-1} person^{-1}], with both unaware-susceptible and unaware-vaccinated individuals becoming aware immediately after infection. Aware-infected individuals decay into the removed class at the recovery rate γ [day^{-1}]. We presume a rate of information spreading from unaware to aware α [day^{-1} person^{-1}]. The aware-

[2]Kabir et al. (2020), p. 109548.

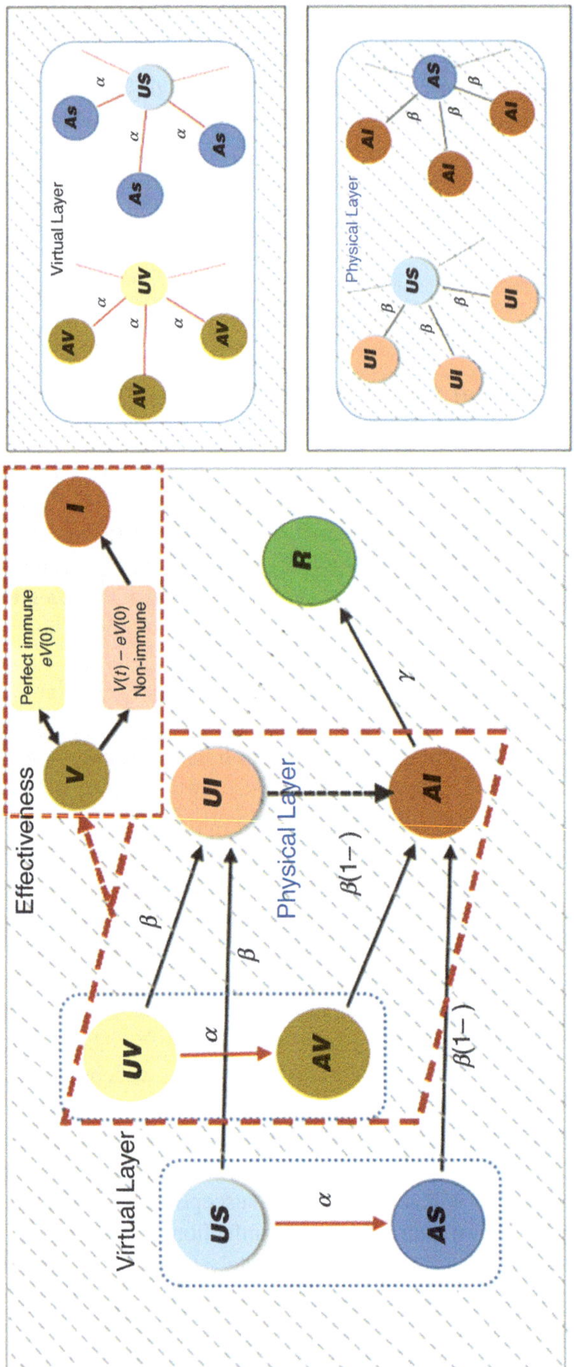

Fig. 6.1 The structure of the SIR/V-UA network model in which vaccination dynamics are displayed as two layers: a virtual layer and a physical layer

susceptible and aware-vaccinated individuals become infected at a rate of $(1-\eta)\,\beta$, where η is the discount rate of infection with the help of information, implying that the individual takes a costless IDM. The case with $\alpha = 0$ (no information spreading) recovers the default setting, which is the networked SIR/V with the effectiveness model introduced in Sect. 3.4.3.

The vaccinated population fraction is divided into two groups: perfectly immune and non-immune. The effectiveness of the vaccine is denoted as e $(0 \le e \le 1)$ and the efficiency of IDM is η $(0 \le \eta \le 1)$, which accounts for the media effect that ensures that individuals protect themselves from an epidemic in advance. The set of differential equations of the SVIR-UA model under a networked population can be given as

$$\frac{dUS_k(t)}{dt} = -\beta k\; US_k(t)\Theta_1(t)(1-\alpha k\Theta_2(t)) - \alpha k\; US_k\Theta_2(t)(1-\beta k\Theta_1(t)), \tag{6.2a}$$

$$\frac{dAS_k(t)}{dt} = \alpha k\; US_k\Theta_2(t)(1-\beta k\;\Theta_1(t)) - (1-\eta)\beta k AS_k(t)\;\Theta_1(t), \tag{6.2b}$$

$$\begin{aligned}\frac{dUV_k(t)}{dt} &= -\beta k(UV_k(t) - eUV_k(0))\Theta_1(t)\;(1-\alpha k\Theta_2(t)) - \alpha k UV_k\Theta_2(t)\\ &\quad\times(t)(1-\beta k\;\Theta_1(t)),\end{aligned} \tag{6.2c}$$

$$\begin{aligned}\frac{dAV_k(t)}{dt} &= -(1-\eta)\beta k(AV_k(t) - eAV_k(0))\Theta_1(t) + \alpha k UV_k(t)\Theta_2(t)\\ &\quad\times(1-\beta k\Theta_1(t)),\end{aligned} \tag{6.2d}$$

$$\begin{aligned}\frac{dUI_k(t)}{dt} &= \beta k(UV_k(t) - eUV_k(0))\Theta_1(t)(1-\alpha k\Theta_2(t)) + \beta k US_k(t)\Theta_1(t)\\ &\quad\times(1-\alpha k\Theta_2(t)) - UI_k(t),\end{aligned} \tag{6.2e}$$

$$\begin{aligned}\frac{dAI_k(t)}{dt} &= (1-\eta)\beta k AS_k(t)\Theta_1(t) + (1-\eta)\beta k(AV_k(t) - eAV_k(0))\Theta_1(t)\\ &\quad+ UI_k(t) - \gamma AI_k(t),\end{aligned} \tag{6.2f}$$

$$\frac{dR_k(t)}{dt} = \gamma AI_k(t), \tag{6.2g}$$

where $\Theta(t)$ refers to the mean probability of infected people, which is illustrated by the spatial structure for both the aware and unaware states. Herein, $\Theta_1(t)$ and $\Theta_2(t)$ illustrate the mean probabilities for given link points to an infected node (physical interaction), and the links are associated with unaware-to-aware transitions (virtual interactions), respectively, which takes account of the spatial structure through the degree distribution, of which basic concept was introduced at Eq. (3.36) (introduced for the governing equation; Eq. (3.43)) in Sect. 3.4.3:

$$\Theta_1(t) = \frac{\sum_k (k-1)P(k)(UI_k(t) + AI_k(t))}{\langle k \rangle}, \tag{6.3a}$$

$$\Theta_2(t) = \frac{\sum_k (k-1)P(k)(AS_k(t) + AI_k(t) + AV_k(t))}{\langle k \rangle}. \tag{6.3b}$$

We must carefully note the substantial difference between the present model (Eqs. 6.2a, 6.2b, 6.2c, 6.2d, 6.2e, 6.2f, 6.2g) and the networked SIR/V with an effectiveness model presented in Eq. (3.43). Let us confirm these one by one. On the right-hand side of Eq. (6.2a), $(1 - \alpha k\Theta_2(t))$ is newly multiplied by $\beta k\ US_k(t)\Theta_1(t)$, which is its counterpart as the right-hand term in the first equation of Eq. (3.43) $(\lambda k\ S_k(t)\Theta(t))$. Why does the present model (two-layer spatial structure by the SVIR-UA process) require $(1 - \alpha k\Theta_2(t))$ to be further multiplied? This is explained below. The term $\beta k\ US_k(t)\Theta_1(t)$ indicates the extent to which US_k is infected by contacts with infectious individuals in the physical network (denoted by subscript 1). The focal term should represent newly infected individuals (transferred from S to I) who are not aware. Mathematically, the condition of "and not being aware" requires multiplication by $(1 - \alpha k\Theta_2(t))$ (where subscript 2 indicates the virtual network). Likewise, the term $\alpha k\ US_k\Theta_2(t)$ indicates the extent to which US_k is aware through contacts with aware individuals in the virtual network. Since the focal term must represent the individuals newly becoming aware (from U to A) without being infected, the condition of "and not being infected" mathematically requires multiplication by $(1 - \beta k\Theta_1(t))$. Let us move to the right-hand terms of Eq. (6.2b). The second term, $(1 - \eta)\beta kAS_k(t)\ \Theta_1(t)$, must represent individuals becoming newly infected (from S to I) while being aware. Above, we presumed that "both US and unaware-vaccinated individuals become aware immediately after infected," meaning that the state-transfer probability is 1, so that the condition of "and being aware" requires nothing to be multiplied. In Eqs. (6.2c) to (6.2g), the same concepts as above are applied to obtain the respective terms. Another quite important point to be carefully confirmed is that the last term on the right-hand side in Eq. (6.2e) does not implement $-\gamma UI_k(t)$ (and the right-hand side of Eq. (6.2g) does not have $+\gamma UI_k(t)$ in return), but $-UI_k(t)$ and in return the third term on the right-hand side of Eq. (6.2f) appears as $+UI_k(t)$. This is because we can assume that an UI individual must immediately become aware, and never recovers in an unaware state.

As we will precisely define later, committing to a vaccination is costly, and being infected also entails an illness cost that is greater than the vaccination cost. Likewise from Sect. 5.2.1, in the current evolutionary framework, we can classify individuals into four states in terms of their health condition and cost burden: (i) a healthy vaccinator (HV) who pays only the vaccination cost and remains healthy in an epidemic season; (ii) an infected vaccinator (IV) who commits to a vaccination but unfortunately becomes infected and must therefore bear the vaccination cost as well as the infection cost; (iii) a successful free rider (SFR) is a non-vaccinator who does not incur any cost and fortunately can survive without being infected; and (iv) a failed free rider (FFR) is also a non-vaccinator who relies utterly upon the herd

Table 6.1 Fractions of four types of individuals

Strategy/State	Healthy	Infected
Vaccinated (Cooperator; V)	$V_k(x,\infty)$	$\varphi_{AV_k \to AI_k}(x,\infty) + \varphi_{UV_k \to UI_k}(x,\infty)$
Non-vaccinated; Defector (Free Rider; FR)	$S_k(x,\infty)$	$\varphi_{AS_k \to AI_k}(x,\infty) + \varphi_{US_k \to UI_k}(x,\infty)$

Table 6.2 Estimated payoff structure at the end of each epidemic season

Strategy/State	Healthy	Infected
Vaccinated (V; C)	$-C_r$	$-C_r - 1$
Non-vaccinated (NV; D)	0	-1

immunity state and tries to free ride without taking any protective provisions but eventually becomes infected, thereby bearing the infection cost themselves.

Presuming a positive vaccination coverage; VC, x, we can observe the level of each compartment when t (the local timescale, meaning the time in a single season) goes to infinity ($t \to \infty$). Thus, we obtain:

$$HV_k(x,\infty) = V_k(x,\infty), \tag{6.4a}$$

$$SFR_k(x,\infty) = S_k(x,\infty). \tag{6.4b}$$

To evaluate the other two fractions IV and FFR, we introduce the flux notation $\varphi_{A \to B}$, which indicates the total number of individuals changing the state from A to B (one state to another). At this point, the flux between the aware-vaccinated and aware-infected compartments is denoted by $\varphi_{AV_k \to AI_k}(x,\infty)$, that from the unaware-vaccinated to unaware-infected compartments is $\varphi_{UV_k \to UI_k}(x,\infty)$, that from the aware-susceptible to aware-infected compartments is $\varphi_{AS_k \to AI_k}(x,\infty)$, and that from the unaware-susceptible to unaware-infected compartments is $\varphi_{US_k \to UI_k}(x,\infty)$. Thus, we recover Table 6.1.

6.2.2 *Payoff Structure*

We should follow Table 3.1, because there is only a single costly strategy, which is vaccination (defined as an active provision). We recall the relative cost of vaccination, $C_r = C_v / C_i$ ($0 \le C_r \le 1$), where the costs for vaccination and infection are denoted by C_v and C_i, respectively. As we discussed in the previous sub-section, at the end of a single season, using the game-theoretic approach, we can classify all individuals who initially chose either vaccination or free riding (defection) into the four classes depending upon their final health status, whether they are healthy or infected at the equilibrium point. Therefore, we depict the present-payoff structure, Table 6.2, as in Table 3.1.

In order to construct the expected payoff, the ASP $\langle \pi \rangle$, cooperative payoff $\langle \pi_C \rangle$ (vaccinated), and defective payoff $\langle \pi_D \rangle$ (non-vaccinated) are formulated as follows:

$$\langle\pi\rangle = -C_r\sum\nolimits_k x_k - (C_r+1)\sum\nolimits_k(1-x_k), \tag{6.5a}$$

$$\langle\pi_C\rangle = \Big\{-C_r\frac{1}{\langle k\rangle}\sum\nolimits_k kp(k)V_k(x,\infty) - (C_r+1)\frac{1}{\langle k\rangle}\sum\nolimits_k kp(k)\big(\varphi_{AV_k\to AI_k}(x,\infty)$$

$$+\varphi_{UV_k\to UI_k}(x,\infty)\big)\Big\}/\sum\nolimits_k x_k, \tag{6.5b}$$

$$\langle\pi_D\rangle = \Big\{-\frac{1}{\langle k\rangle}\sum\nolimits_k kp(k)\big(\varphi_{AS_k\to AI_k}(x,\infty)+\varphi_{US_k\to UI_k}(x,\infty)\big)\Big\}/\sum\nolimits_k(1-x_k). \tag{6.5c}$$

6.2.3 *Strategy Updating and Global Dynamics*

In the present model, we apply individual-based risk assessment (IB-RA) and strategy-based risk assessment (SB-RA), as introduced in Sect. 3.2.2. We use the MFA parameter and set the noise parameter in the Fermi function in each state probability function κ to 0.1.

6.2.3.1 Individual-Based Risk Assessment (IB-RA)

As described in Eq. (3.25), there are eight state-transition-probability functions; $P(\mathrm{HV}\leftarrow\mathrm{SFR}) = \frac{1}{1+\exp[-(0-(-C_r))/\kappa]}$, $P(\mathrm{HV}\leftarrow\mathrm{FFR}) = \frac{1}{1+\exp[-(-1-(-C_r))/\kappa]}$, $P(\mathrm{IV}\leftarrow\mathrm{SFR}) = \frac{1}{1+\exp[-(0-(-C_r-1))/\kappa]}$, $P(\mathrm{IV}\leftarrow\mathrm{FFR}) = \frac{1}{1+\exp[-(-1-(-C_r-1))/\kappa]}$, $P(\mathrm{SFR}\leftarrow\mathrm{HV}) = \frac{1}{1+\exp[-(-C_r-(0))/\kappa]}$, $P(\mathrm{SFR}\leftarrow\mathrm{IV}) = \frac{1}{1+\exp[-(-C_r-1-(0))/\kappa]}$, $P(\mathrm{FFR}\leftarrow\mathrm{HV}) = \frac{1}{1+\exp[-(-C_r-(-1))/\kappa]}$, and $P(\mathrm{FFR}\leftarrow\mathrm{IV}) = \frac{1}{1+\exp[-(-C_r-1-(-1))/\kappa]}$.

At the end of each epidemic season, everyone can update their strategy depending upon the last season's payoff. Hence, increases or decreases to VC, x, is inevitable. Here, the independent variable, t, indicates the global timescale, which means the number of elapsed seasons. Since we consider the IB-RA strategy-updating rule for decision-making in each subsequent season, the dynamic equation following this particular rule can therefore be expressed as follows:

$$\frac{dx_k}{dt} = -V_k(x,\infty)\frac{1}{\langle k\rangle}\sum\nolimits_k kp(k)AS_k(x,\infty)\mathrm{P}(\mathrm{HV}\leftarrow\mathrm{SFR})$$

$$-V_k(x,\infty)\frac{1}{\langle k\rangle}\sum\nolimits_k kp(k)\big(\varphi_{AS_k\to AI_k}(x,\infty)+\varphi_{US_k\to UI_k}(x,\infty)\big)\mathrm{P}(\mathrm{HV}\leftarrow\mathrm{FFR}))$$

$$-\left(\varphi_{AV_k \to AI_k}(x,\infty) + \varphi_{UV_k \to UI_k}(x,\infty)\right)\frac{1}{\langle k \rangle}\sum\nolimits_k kp(k)S_k(x,\infty)\mathrm{P}(\mathrm{IV} \leftarrow \mathrm{SFR})$$

$$-\left(\varphi_{AV_k \to AI_k}(x,\infty) + \varphi_{UV_k \to UI_k}(x,\infty)\right)\frac{1}{\langle k \rangle}\sum\nolimits_k kp(k)$$

$$\times\left(\varphi_{AS_k \to AI_k}(x,\infty) + \varphi_{US_k \to UI_k}(x,\infty)\right)\mathrm{P}(\mathrm{IV} \leftarrow \mathrm{FFR})$$

$$+S_k(x,\infty)\frac{1}{\langle k \rangle}\sum\nolimits_k kp(k)V_k(x,\infty)\mathrm{P}(\mathrm{SFR} \leftarrow \mathrm{HV})$$

$$+\left(\varphi_{AS_k \to AI_k}(x,\infty) + \varphi_{US_k \to UI_k}(x,\infty)\right)\frac{1}{\langle k \rangle}\sum\nolimits_k kp(k)V_k(x,\infty)\mathrm{P}(\mathrm{FFR} \leftarrow \mathrm{HV})$$

$$+S_k(x,\infty)\frac{1}{\langle k \rangle}\sum\nolimits_k kp(k)\left(\varphi_{AV_k \to AI_k}(x,\infty) + \varphi_{UV_k \to UI_k}(x,\infty)\right)\mathrm{P}(\mathrm{SFR} \leftarrow \mathrm{IV})$$

$$+\left(\varphi_{AS_k \to AI_k}(x,\infty) + \varphi_{US_k \to UI_k}(x,\infty)\right)\frac{1}{\langle k \rangle}\sum\nolimits_k kp(k)$$

$$\times\left(\varphi_{AV_k \to AI_k}(x,\infty) + \varphi_{UV_k \to UI_k}(x,\infty)\right)\mathrm{P}(\mathrm{FFR} \leftarrow \mathrm{IV}), \tag{6.6}$$

where x_k represents the fraction of vaccinators (cooperators: C). The product of the two classes, as an example, $V_k(x,\infty)$ and $\frac{1}{\langle k \rangle}\sum_k kp(k)AS_k(x,\infty)$ are probabilities that any two portions of individuals of the two classes participate for imitation (if possible).

6.2.3.2 Strategy-Based Risk Assessment (SB-RA)

As described in Eq. (3.27), there are four state-transition-probability functions: $P(\mathrm{HV} \leftarrow \mathrm{NV}) = \frac{1}{1+\exp\left[-(<\pi_D>-(-C_r))/\kappa\right]}$, $P(\mathrm{IV} \leftarrow \mathrm{NV}) = \frac{1}{1+\exp\left[-(<\pi_D>-(-C_r-1))/\kappa\right]}$, $P(\mathrm{SFR} \leftarrow \mathrm{V}) = \frac{1}{1+\exp\left[-(<\pi_C>-0)/\kappa\right]}$, and $P(\mathrm{FFR} \leftarrow \mathrm{V}) = \frac{1}{1+\exp\left[-(<\pi_C>-(-1))/\kappa\right]}$.

Following the IB-RA case above, we can establish the dynamic equation following SB-RA as follows:

$$\frac{dx_k}{dt} = -x_k V_k(x,\infty)\frac{1}{\langle k \rangle}\sum\nolimits_k kp(k)(1-x_k)P(HV \leftarrow NV)$$

$$-x_k\left(\varphi_{AV_k \to AI_k}(x,\infty) + \varphi_{UV_k \to UI_k}(x,\infty)\right)\frac{1}{\langle k \rangle}\sum\nolimits_k kp(k)(1-x_k)P(IV \leftarrow NV)$$

$$+(1-x_k)S_k(x,\infty)\frac{1}{\langle k \rangle}\sum\nolimits_k kp(k)x_k P(SFR \leftarrow V)$$

$$+(1 - x_k)\left(\varphi_{AS_k \to AI_k}(x, \infty) + \varphi_{US_k \to UI_k}(x, \infty)\right)$$
$$\times \frac{1}{\langle k \rangle} \sum\nolimits_k kp(k) x_k P(FFR \leftarrow V). \quad (6.7)$$

6.2.4 Spatial Structure

We must give an explicit form to the degree distribution for $P(k)$. In this model, we apply the same topology for the physical network (meaning subscript 1 in Eq. (6.3a)) and the virtual network (meaning the subscript 2 in Eq. (6.3b)). Following Sect. 3. 4.1, we presume two typical degree distributions; the Poisson and power-law degree distributions, which are, respectively, associated with the Erdős–Rényi random-graph (E-R random) and the Barabási–Albert scale-free (BA-SF) networks. In the Poisson degree distribution $P(k) = \exp(-\langle k \rangle)\langle k \rangle^k / k!$, most of the modes have a connectivity k close to the mean value $\langle k \rangle = \sum_k kP(k)$. Although BA-SF's power law obeys; $P(k) \sim k^{-3}$, we use the approximation presumed in Sect. 3.4.1: $P(k) = A/k(k + 1)(k + 2)$.

6.2.5 Initial Condition and Numerical Procedure

To solve the above-stated sets of equations for each season as well as the repeating seasons numerically, an explicit finite-difference method is used. Throughout, the minimum and maximum degrees are assumed to be $k_{\min} = 3$ and $k_{\max} = 100$, respectively.

Initially, we presumed initial values of $UV_k(x, 0) = x$, $US_k(x, 0) = 1 - x$, $UI_k(x, 0) = 0$, $AS_k(x, 0) = 0$, $R_k(x, 0) = 0$, and $AI_k(x, 0) \approx 0$ to start a new season ($x = 0.5$, & $AI_k(x, 0) = 0.0001$).

6.3 Results and Discussion

Throughout our simulations, we presume that $R_0 = \frac{\beta}{\gamma} = 2.5$. By varying the topology of spatial structure (where "well-mixed" means that there is no spatial structure, E-R random and BA-SF, and the average degree is set to $\langle k \rangle = 4$, $\langle k \rangle = 8$ and $\langle k \rangle = 12$), we explore the FES, VC, and ASP when presuming either risk assessment (IB-RA) or strategy-based risk assessment (SB-RA) for the strategy-update rule. All visual results below are presented as a 2D heat map, where the relative vaccination cost C_r (effectiveness of vaccination, e) is presented along the X-axis (Y-axis) and ranges

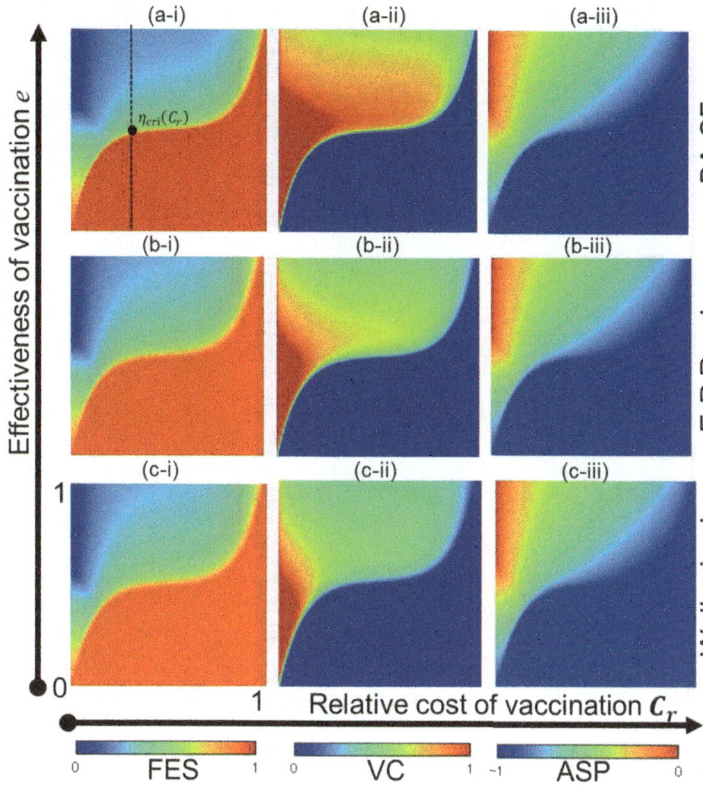

Fig. 6.2 The first column shows the final epidemic size (*-i), the second column shows the vaccination coverage (*-ii), and the last column shows the average social payoff (*-iii). The power-law degree distribution (a-*), Poisson degree distribution (b-*), and well-mixed-population case (c-*) are depicted in the upper, middle, and lower panels, respectively, for the IB-RA updating rule. Here, the information-spreading rate is considered to be $\alpha = 0.0$, and the other parameters are set as $\beta = 0.83$, $\gamma = 0.333$, and $\eta = 0.3$

from 0 to 1. In each figure, panels (*-i), (*-ii) and (*-iii), respectively, show the FES ranging for 0 to 1; VC also ranges from 0 to 1 and the ASP ranges from -1 to 0.[3]

Figure 6.2 shows the default setting where $\alpha = 0$ is presumed, which implies there is no information effect. The top, middle, and bottom panels show different topologies. For BA-SF and E-R random, we presume $\langle k \rangle = 8$. As strategy-update rule, IB-RA is presumed. For the well-mixed case, panels (c-i), (c-ii), and (c-iii) recover what we discussed in Chap. 3; more precisely, panels (1-A), (1-B), and (1-C) in Fig. 3.13 (or panels (3-A), (3-B) and (3-C) in Fig. 3.19).

[3]Theoretically ASP can be less than -1. The possible minimum is -2, where an individual pays the vaccination cost, fails to obtain immunity, and is unfortunately infected (paying an additional disease cost).

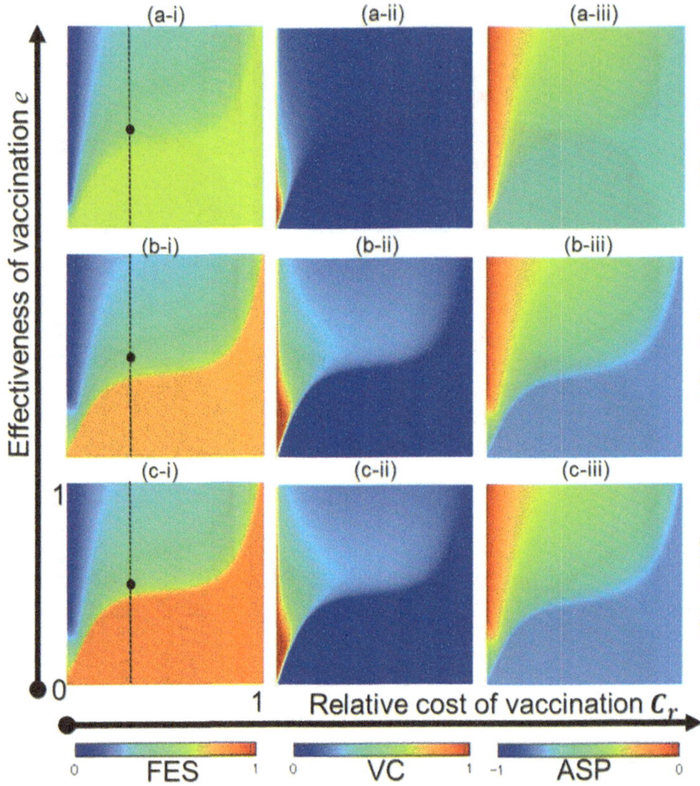

Fig. 6.3 The power-law degree distributions (BA-SF) for IB-RA are presented for varying final epidemic size (*-i), vaccination coverage (*-ii), and average social payoff (*-iii). Three different average degrees are depicted in figures (a-*), (b-*) and (c-*), respectively. Here, the information-spreading rate is $\alpha = 0.4$ and the other parameters are $\beta = 0.83$, $\gamma = 0.333$, and $\eta = 0.3$

Figure 6.5 is the counterpart to Fig. 6.2 when presuming SB-RA as an update rule. Thus, for the well-mixed case, panels (c-i), (c-ii), and (c-iii) recover what we discussed in Chap. 3; more precisely, panels (2-A), (2-B), and (2-C) in Fig. 3.13 (or panels (3-A), (3-B), and (3-C) in Fig. 3.21).

Returning to Fig. 6.2, the same general tendency among FES and ASP seems common for different topologies. More precisely, the monotone region of red indicating a pandemic phase in (a-i) (BA-SF) is redder than those in (b-i) (E-R random) and (c-i) (well-mixed), implying that the pandemic in BA-SF is more severe than under the other two topologies. This is likely because hub individuals work as so-called super-spreaders in the case of an SF network, due to a much more homogeneous degree distribution than the other two topologies. One thing that is worthwhile to note is that outside of the blue, epidemic-controlled region in (a-ii), there is a higher VC (more red) than in the other two topologies. This makes sense because, in an SF network, inherent vulnerability against an epidemic drives individuals to commit to vaccination more.

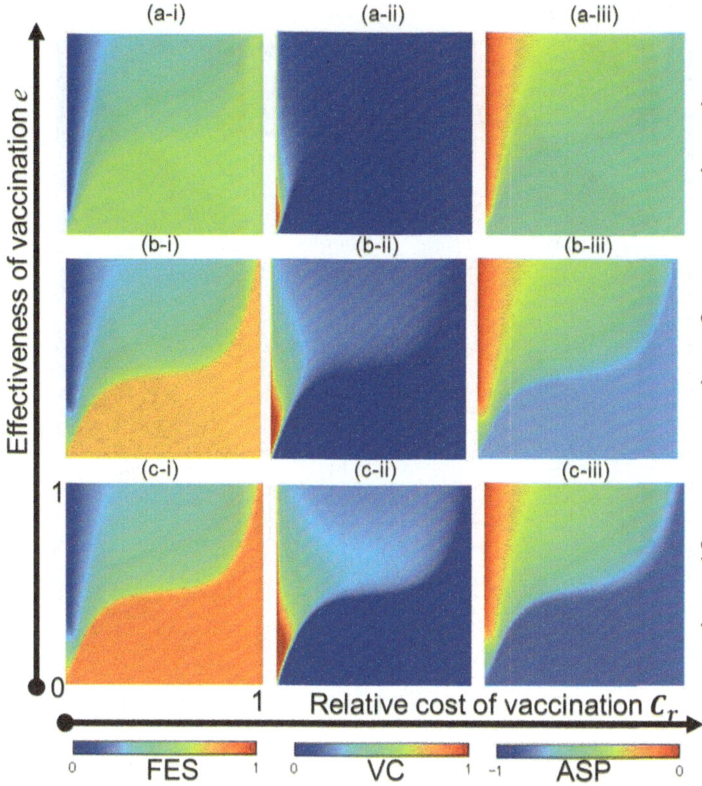

Fig. 6.4 The Poisson degree distributions (ER-RG) of IB-RA are presented for varying final epidemic size (*-i), vaccination coverage (*-ii), and average social payoff (*-iii). Three different average degrees are depicted in figures (a-*), (b-*), and (c-*), respectively. Here, the information-spreading rate is $\alpha = 0.4$, and the other parameters are set as $\beta = 0.83$, $\gamma = 0.333$, and $\eta = 0.3$

Let us consider Fig. 6.2 (a-i). The boundary between monotone red and the remaining area implies a phase transition between the pandemic (red) and controlled (bluer) regions. Such an abrupt transition of a certain physical system as observed in FES is well-recognized as a first-order phase change. Thus, we can define the critical effectiveness for a certain vaccination cost, $\eta_{\text{cri}}(C_r)$, bringing this phase transition between the pandemic and the controlled phase (see the closed circle in Fig. 6.2 (a-i)).

Now, let us move on to Fig. 6.3, where BA-SF is considered and its average degree is varied in the top, middle, and bottom rows. More importantly we presume that $\alpha = 0.4$, with informational awareness, helps to suppress the spread of disease. In fact, comparing Figs. 6.3 (b-*) and 6.2 (a-*), we can confirm that the information effect is remarkable. FES can be significantly reduced and ASP improved significantly. More importantly, this does not come from a higher VC, but results from the awareness effect through η. Notably, panel (b-ii) is almost colored blue except in the region where the vaccination cost is quite low but still has reasonable effectiveness.

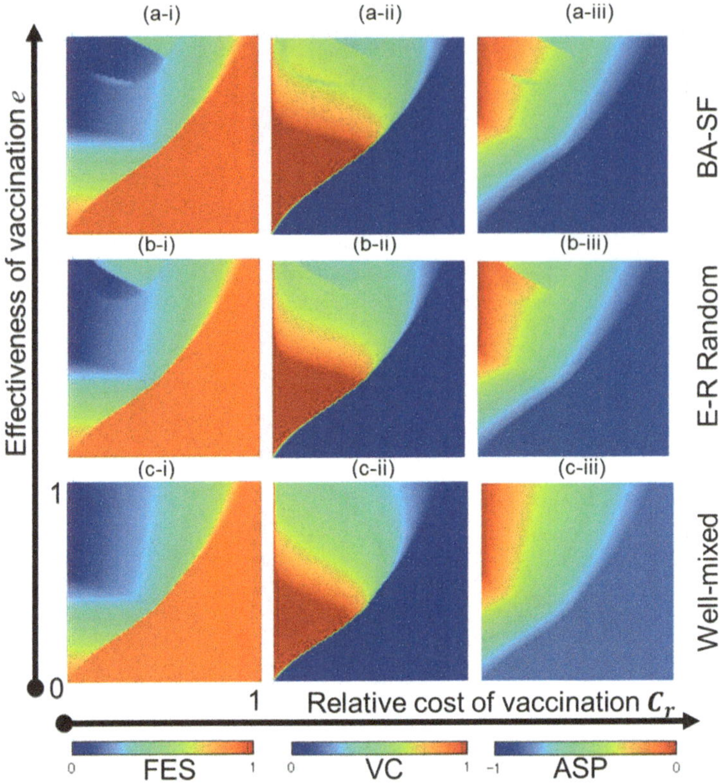

Fig. 6.5 Columns 1, 2, present the final epidemic size (*-i), vaccination coverage (*-ii), and average social payoff (*-iii), respectively. The power-law degree distribution (a-*), Poisson degree distribution (b-*), and well-mixed population (c-*) are depicted in the upper, middle, and lower panels for the SB-RA updating rule, respectively. Here, the information-spreading rate is $\alpha = 0.0$ and the other parameters are $\beta = 0.83$, $\gamma = 0.333$, and $\eta = 0.3$

We can confirm that the information effect works significantly to minimize the impact of the pandemic. Comparing rows in Fig. 6.3, we can note that this positive effect is greater in a case with a lower average degree, which might be conceivable because a lower average degree allows fewer hub individuals. Observing the closed circles plotted at the exact same position in Fig. 6.2 (a-i), we note that the critical effectiveness is decreased, which may be explicable in terms of what we have discussed above.

Figure 6.4 is the counterpart to Fig. 6.3 when presuming ER-random. We also confirm that informational of awareness helps to significantly reduce the pandemic risk. Figure 6.5 is the counterpart to Fig. 6.2, and Figs. 6.6 and 6.7 are the counterparts to Figs. 6.3 and 6.4, respectively, when SB-RA is presumed for update rule. Observing Figs. 6.5, 6.6, 6.7 and comparing them with Figs. 6.2, 6.3, 6.4, our above discussion assuming IB-RA for the update rule is almost true for the alternate update rule SA-RA.

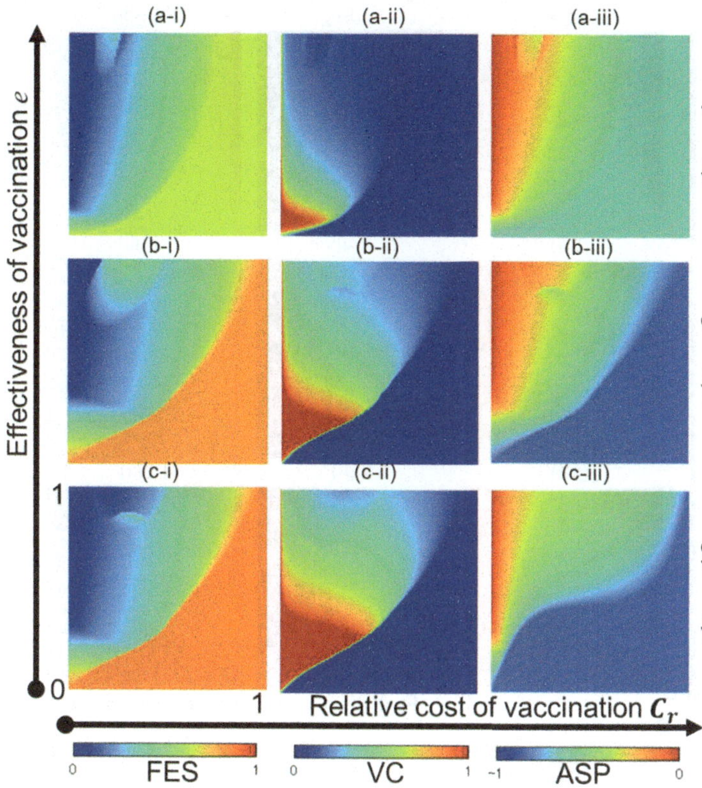

Fig. 6.6 The power-law degree distributions (BA-SF) of SB-RA are presented for the final epidemic size (*-i), vaccination coverage (*-ii), and average social payoff (*-iii). Three different average degrees are depicted in figures (a-*), (b-*), and (c-*), respectively. Here, the information-spreading rate is $\alpha = 0.4$ and the other parameters are set as $\beta = 0.83$, $\gamma = 0.333$, and $\eta = 0.3$

Let us conclude by summarizing what we learned in this Chapter.

We confirmed that the effect of awareness of a spreading infection can remarkably reduce the number of infected individuals for all updating rules, because such awareness encourages people to protect themselves.

We also presumed that both the disease and information spread on the same network. Although the most effective way to hamper a virus's spread may be through vaccination, this practice is expensive, costly to oneself, and difficult to apply to inoculate the mass of people. Lethal diseases like Dengue, Chikungunya, AIDS, Plague, and Malaria have no active vaccinations, so only awareness can effect limit their spread. The practice of safe sex, use of mosquito coils or nets, hand washing, mask wearing, and other self-protection measures informed by information play a substantial effect in the epidemic-diffusion model.

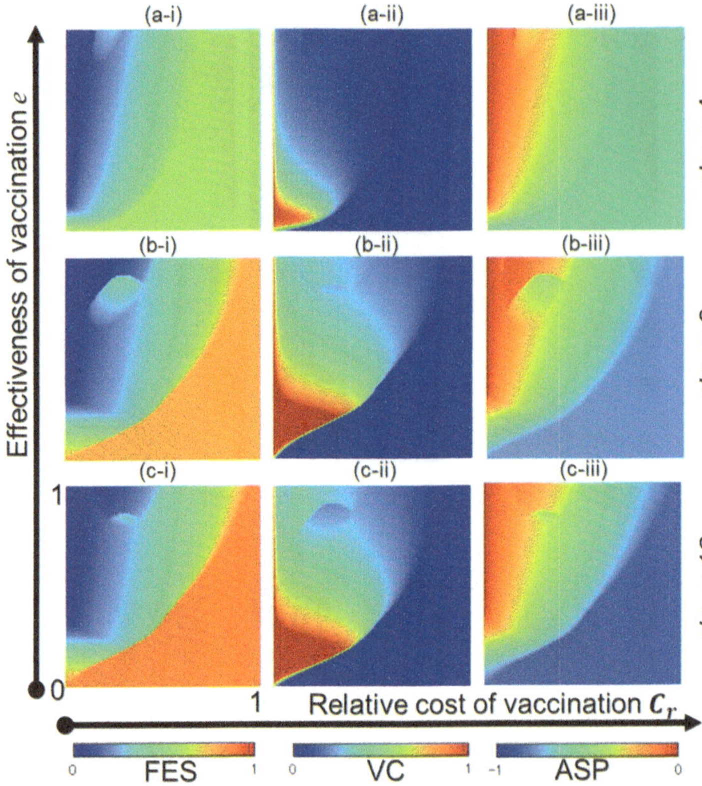

Fig. 6.7 The Poisson degree distributions (ER-RG) of SB-RA are presented for varying final epidemic size (*-i), vaccination coverage (*-ii), and average social payoff (*-iii). Three different average degrees are depicted in figures (a-*), (b-*), and (c-*), respectively. Here, the information-spreading rate is $\alpha = 0.4$, and the other parameters are $\beta = 0.83$, $\gamma = 0.333$, and $\eta = 0.3$

Some pioneering studies[4] have suggested that presuming that the physical network (on which disease spreads) is different from the virtual network (on which strategy-updating rules and information spread) will yield very different dynamical features. As future work, it would be interesting to establish another vaccination game with an improved SVIR-UA epidemic model based on what we have discussed here, referring to two different social networks for disease and awareness spreading.

[4]Fukuda et al. (2015), pp. 47–55.

References

Fukuda E, Tanimoto J, Akimoto M (2015) Influence of breaking the symmetry between disease transmission and information propagation network on stepwise decisions concerning vaccination. Chaos Solitons Fractals 80:47–55

Gómez S, Diaz-Guilera A, Gomez-Gardeñes J, Perez-Vicente CJ, Moreno Y, Arenas A (2013) Diffusion dynamics on multiplex networks. Phys Rev Lett 110(2):028701

Granell C, Gómez S, Arenas A (2014) Competing spreading processes on multiplex networks: Awareness and epidemics. Phys Rev E 90:012808

Guo Q, Jiang X, Lei Y, Li M, Ma Y, Zheng Z (2015) Two-stage effects of awareness cascade on epidemic spreading in multiplex networks. Phys Rev E 91(1-1):012822

Kabir KMA, Kuga K, Tanimoto J (2020) The impact of information spreading on epidemic vaccination game dynamics in a heterogeneous complex network- A theoretical approach. Chaos Solitons Fractals 132:109548

Salathe M, Bonhoeffer S (2008) The effect of opinion clustering on disease outbreaks. J R Soc Interface 5:1505–1508

Wang W, Liu QH, Cai S, Tang M, Lidia A, Braunstein L, Stanley HE (2016) Suppressing disease spreading by using information diffusion on multiplex networks. Sci Rep 6:29259

Chapter 7
Immunity Waning Effect

Immunity is an indispensable function of biological defense system that protects us against being infected with same pathogen again. Although some kinds of immunity last for long time, almost life span, it is, in general, degrading gradually. This chapter introduces a simplified theoretical framework embedded into an epidemiological model so that basically presumes SIRS process. Unlike our previous models, this chapter does not account the framework of evolutionary game theory based on the strategy transition probability, but relies on the so-called behavior model instead. It is because the dynamical system of this chapter premises a single long time series with no seasonal partition.

J. Tanimoto, *Sociophysics Approach to Epidemics*, Evolutionary Economics and Social Complexity Science 23, https://doi.org/10.1007/978-981-33-6481-3_7

7.1 Introduction and Background: Immunity and Its Degrading in View of Infectious Disease

Immunity is an essential means of contending against infectious diseases and harmful invaders that are actively responsible for protecting to attack from bacteria, viruses, parasites, and more. As a subject, immunology is the branch of medical and biological sciences that covers fundamental mechanisms for acquiring immunity, efficacy, and durability of protection. A person lacking a proper immune system (known as an immune system disorder) becomes sick continuously from various germs. AIDS; Acquired Immune Deficiency Syndrome, is one of such well-known diseases, which is brought by HIV (human immunodeficiency virus).

An individual can acquire immunity by (i) infection, (ii) immunization, or (iii) non-immunological factors, which are categorized by several axes as: adaptive vs innate, and active vs passive. An active immunity is acquired by exposing the body to an antigen to produce an adaptive immune response, which takes days or weeks to develop but leads to long-lasting protection. Meanwhile, a passive immunity is obtained as a protection by-products produced by an animal or human and transferred to another human (by injection), which gives an abrupt effect but short-lived immunity.

The two immunity systems this chapter concerns are formed by a previous infection (naturally obtained through being infected) and immunization. Let us call the former as naturally-obtained immunity, the latter as artificial immunity. The most common form of artificial immunity being implanted to a body is vaccination.

The durability of immunity depends on the type of disease. For example, the protection by a vaccine against pertussis, and that by pneumococcal infections as well as human papillomavirus lasts for 4–6 years. Once immunity is obtained, the protection lasts at least 10 years for the cases of diphtheria, mumps, and varicella, and that lasts for around 20 years against tetanus, polio, hepatitis B, and rubella.

In this chapter, we approach such a problem concerning durability of immunity. In other words, how waning immunity, or say, attenuating the effect of immunity influences on the epidemic size, and on the incentive to vaccination amid individuals.

Unlike what we have formulated so far which is dovetailing the epidemiological framework such as SIR process with the evolutionary game theory to describe an individual's decision of whether committing vaccination or not, we take, in this chapter, another approach that is called the behavior model[1] to depict the time evolution of people's attitude toward vaccination. It is because the situation the present model does not premise the two-layer timescale structure; the local timescale meaning time elapsing in a single season, and the global timescale meaning repeating seasons are co-existing, but does only presume the local timescale. Thus, the

[1]One of the most representative materials about the behavior model for mathematical epidemiology is;

Bauch and Bhattacharyya (2012), p. e1002452.

strategy-updating process we previously presumed, where an individual brushes up whether taking a vaccine or not for next season based on what happed in the current season, cannot be applied to the present framework. More straightforwardly speaking, both disease spreading and strategy updating take place in same timescale; that is day-by-day bases.

Another important point of discussion in the present model is how immunity waning process should be introduced. One likely solution is the introduction of age-structure to an epidemiological frame such as SIR, and SEIR process, in which some dynamical equations, in practice, equation of quantifying the compartment of infected individuals should have two independent variables: absolute time (usually denoted by t) and time elapsing after infection (usually denoted by a; age), i.e., $I(t, a)$, which inevitably needs the focal equation to be a partial differential equation (PDE) not an ordinal differential equation (ODE) anymore.[2] It would entail cumbersome procedures in mathematical treatment. In contrast in this chapter, we establish a new idea to describe "age-structure" (meaning the time after getting immunity) in a simplified way that does not require PDE but allows to stay in ODE system.

7.2 Model Structure

7.2.1 Formulation of the SV_nIR_{2n} Model[3]

Let us build a mathematical epidemic model called SV_nIR_{2n} model, where dynamics of both naturally-obtained and artificial immunity systems in view of protecting effect from those two independently attenuating. In this respect, we base a SVIRS model, which takes into account an individual's behavior toward vaccination by a variant of vaccination game with human social learning aspect, where both disease spreading and behavior of vaccination are introduced in same timescale.

Figure 7.1 shows a holistic block chart of our model structure. In this chapter, we presume: $n = 4$, since we verified that the minimal number of compartments to properly depict a decaying curve visually shown in Fig. 7.2 needs at least four. The vaccination premises imperfectly and stochastically working along the efficiency model. The total population comprises four classes, namely (i) susceptible (denoted by S as state; quantified by $S(t)$), (ii) artificial immunity (AIM; $AIM(t)$), (iii) infected (I; $I(t) = I^S(t) + I^V(t)$, where I^S and I^V are respective mean infected states transferred form susceptible and vaccinated), and (iv) naturally-obtained immunity (NIM; NIM (t)). The AIM portion comprises four different vaccination states; V_O ($V_0(t)$), V_1 ($V_1(t)$), V_2 ($V_2(t)$), and V_3 ($V_3(t)$), such that $AIM(t) = \sum_{i=0}^{n-1} V_i(t) = \sum_{i=0}^{3} V_i(t)$. Similarly, NIM defines as the sum of two recovered states from unvaccinated and

[2]One of the recent literatures dealing with age-structured epidemiological model is; Zhang and Guo (2018), pp. 214–233.

[3]Kabir and Tanimoto (2021), p. 110531.

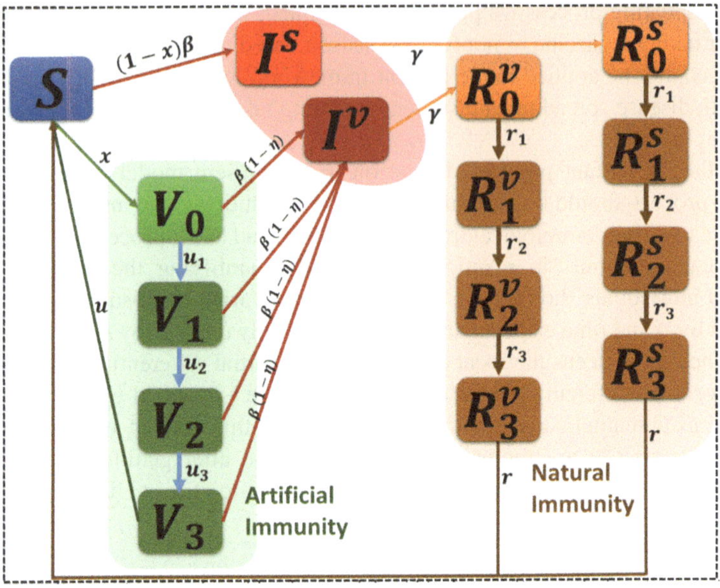

Fig. 7.1 Immunity system and epidemic model. (**a**) Immunity and its types. (**b**) Schematic representation of process described in SIVRS model, starting with susceptible (S) individuals becoming infected at rate $(1 - x)\beta$ and taking vaccination (termed artificial immunity) at rate x. Individuals fail to acquire perfect immunity because of vaccine efficiency η. The infected individuals recover (termed naturally – obtained immunity) at rate γ. Both artificial and naturally-obtained immunity return to the susceptible state after the protection duration expires

infected & recovered; quantified by $\mathrm{NIMS}(t) = \sum_{i=0}^{n-1} R_i^s(t) = \sum_{i=0}^{3} R_i^s(t)$, and vaccinated but infected & recovered; $\mathrm{NIMV}(t) = \sum_{i=0}^{n-1} R_i^v(t) = \sum_{i=0}^{3} R_i^v(t)$. Thus, NIM yields to; $\mathrm{NIM}(t) = \mathrm{NIMS}(t) + \mathrm{NIMV}(t)$. We consider the total population size as $S(t) + \mathrm{AIM}(t) + I(t) + \mathrm{NIM}(t) = 1$. In this chapter, the number of interim transferring-class; n, is presumed 4, which is generally arbitral and is better with a larger number. As abovementioned, the dynamical equations of the compartments: V, and R, are necessarily formulated as PDE in order to account the age-structure in a mathematically accurate way. In the current model, though, due to the introduction of V_i, R_i^s and R_i^v, an ODE formulation is allowed instead of PDE.

The associated transmission rate from susceptible (S) to vaccinated (V_0) is regulated by a variable x, that represents the portion of individuals from susceptible class to actively taking vaccination.[4] After vaccination (state V_0), individuals move

[4]Note that the newly introduced evolutionary variable; x, does not mean the vaccination coverage appeared in our previous frameworks; models of the vaccination game and the intervention game, but the transfer rate from S to V. This is quite important point.

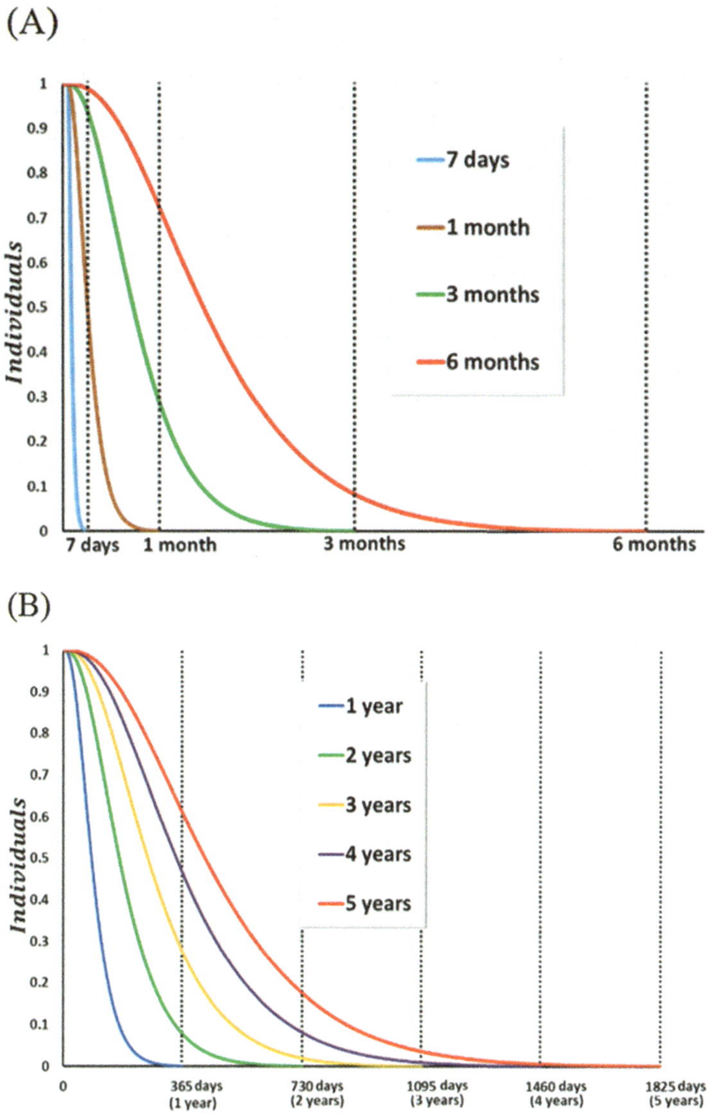

Fig. 7.2 Durability of protection. Evaluated parametric $u_i(r_i)$ values for (**a**) 7 days, 1 month, 3 months, and 6 months. (**b**) 1 year, 2 years, 3 years, 4 years, and 5 years

to the subsequent vaccinated states ($V_0 \rightarrow V_1 \rightarrow V_2 \rightarrow V_3$) with time-constant rates of u_0, u_1, and u_2 [day^{-1}], respectively, where an individual is protected by the immunity brought by a vaccination as long as staying at V_i (AIM). The vaccine loses its protective properties with time, and eventually the vaccinated become susceptible again with rate u that remains fixed at $u = 1.0$. By contrast, the susceptible people (controlled by the rate of $1-x$) are infected at the rate β [day^{-1} person^{-1}], allowing unvaccinated individuals to transfer from S to I^s. Because we presume that the vaccination does not provide artificial immunity in a perfect way, those who were vaccinated but infected are evaluated by the concept of Efficiency model, explained in Sect. 3.2.1; letting individuals transfer from V_i to I^v at the rate $(1-\eta)\beta$, where η is the vaccination efficiency indicating how the immunity helping to keep an individual from infection. Subsequently, γ [day^{-1}] defines the recovery rate of state transfer from I^s to R_0^s and I^v to R_0^v, respectively.

Quantified by the aforementioned NIMS(t) and NIMV(t), the newly recovered individuals finally acquire naturally-obtained immunity, remaining in the R_i^s or R_i^v states until their immunity expires. They return to the susceptible state at the rate $r = 1.0$ by following the state transmission from R_0^s to R_3^s $\left(R_0^s \to R_1^s \to R_2^s \to R_3^s\right)$ or R_0^v to R_3^v $\left(R_0^v \to R_1^v \to R_2^v \to R_3^v\right)$ according to the transition rates of r_0, r_1, and r_2 [day^{-1}]. After losing the protective properties of naturally-obtained immunity (NIM) originated from vaccinated (NIMV) and non-vaccinated (NIMS), the individuals move to the susceptible class again. Under these assumptions, the corresponding set of differential equations for the model is as follows:

$$\frac{dS(t)}{dt} = -(1-x)\beta S(t)[I^s(t) + I^v(t)] - xS(t) + uV_3(t) + rR_3^s + rR_3^v, \tag{7.1a}$$

$$\frac{dV_0(t)}{dt} = xS(t) - \beta(1-\eta)V_0(t)[I^s(t) + I^v(t)] - u_0V_0(t), \tag{7.1b}$$

$$\frac{dV_1(t)}{dt} = u_0V_0(t) - u_1V_1(t) - \beta(1-\eta)V_1(t)[I^s(t) + I^v(t)], \tag{7.1c}$$

$$\frac{dV_2(t)}{dt} = u_1V_1(t) - u_2V_2(t) - \beta(1-\eta)V_2(t)[I^s(t) + I^v(t)], \tag{7.1d}$$

$$\frac{dV_3(t)}{dt} = u_2V_2(t) - uV_3(t) - \beta(1-\eta)V_3(t)[I^s(t) + I^v(t)], \tag{7.1e}$$

$$\frac{dI^s(t)}{dt} = (1-x)\beta S(t)[I^s(t) + I^v(t)] - \gamma I^s(t), \tag{7.1f}$$

$$\frac{dI^v(t)}{dt} = \sum\nolimits_{i=0}^{3} \beta(1-\eta)V_i(t)[I^s(t) + I^v(t)] - \gamma I^v(t), \tag{7.1g}$$

$$\frac{dR_0^s(t)}{dt} = \gamma I^s(t) - r_0R_0^s, \tag{7.1h}$$

$$\frac{dR_1^s(t)}{dt} = r_0R_0^s - r_1R_1^s, \tag{7.1i}$$

$$\frac{dR_2^s(t)}{dt} = r_1R_1^s - r_2R_2^s, \tag{7.1j}$$

$$\frac{dR_3^s(t)}{dt} = r_2R_2^s - r\,R_3^s, \tag{7.1k}$$

$$\frac{dR_0^v(t)}{dt} = \gamma I^v(t) - r_0 R_0^v, \quad (7.1l)$$

$$\frac{dR_1^v(t)}{dt} = r_0 R_0^v - r_1 R_1^v, \quad (7.1m)$$

$$\frac{dR_2^v(t)}{dt} = r_1 R_1^v - r_2 R_2^v, \quad (7.1n)$$

$$\frac{dR_3^v(t)}{dt} = r_2 R_2^v - r R_3^v. \quad (7.1o)$$

7.2.2 Parameterization for Immunity Waning Effect

The present model, as explained in the previous sub-section, reproduces the immunity waning effect by introduction of V_i and R_i^s as well as R_i^v. Thus, we have to parameterize u_0, u_1, and u_2 [day^1] as well as r_0, r_1, and r_2 [day^{-1}].

For simplicity, we presume a universal constant value for both u_i and r_i ($i = 0$, 1, 2) to reproduce a certain realistic attenuating period. Varying u_i, we quantify AIM (t) for each of respective cases with different attenuating periods under the initial condition; $S(0) = I(0) = V_1(0) = V_2(0) = V_3(0) = 0$, $R_0^s(0) = R_1^s(0) = R_2^s(0) = R_3^s(0) = R_0^v(0) = R_1^v(0) = R_2^v(0) = R_3^v(0) = 0$, and $V_0(0) = 1$. Likewise, varying r_i, we quantify $NIM(t)$ under the initial condition; $S(0) = I(0) = V_0(0) = V_1(0) = V_2(0) = V_3(0) = 0$, $R_0^s(0) = R_1^s(0) = R_2^s(0) = R_3^s(0) = R_0^v(0) = R_1^v(0) = R_2^v(0) = R_3^v(0) = 0$, and $R_0^s(0) = R_0^v(0) = 0.5$. By solving Eqs. (7.1a, 7.1b, 7.1c, 7.1d, 7.1e, 7.1f, 7.1g, 7.1h, 7.1i, 7.1j, 7.1k, 7.1l, 7.1m, 7.1n, 7.1o), we verified u_i and r_i at which AIM(t) and NIM(t) become 0.1 % of respective $V_0(0)$ and $R_0^s(0) = R_0^v(0)$ when an elapsed time becomes each of the designed waning periods; 7 days, 1 month, 3 months. 6 months, 1 year, 2 years, 3 years, 4 years and 5 years.

To the end of this parameter identification process, we get Table 7.1. And Fig. 7.2 reproduces respective decaying curves identified the attenuating characteristics.

7.2.3 Time Evolution of Vaccination by Behavior Model

As notified beforehand, to plausibly emulate the human behavioral resulting from decision making, this model relies on the so-called behavior model unlike what we presumed in the previous frameworks applying the concept of state transition probability based on the evolutionary game theory.

For a susceptible individual, the perceived benefit, or say, payoff, depends on the difference between (i) the drawback of being vaccinated and (ii) the penalty for

Table 7.1 Estimated durability of protection periods

Times	u_i, r_i ($i = 0, 1, 2$)
7 days	8.20×10^{-1}
1 month	3.20×10^{-1}
3 months	1.17×10^{-1}
6 months	6.00×10^{-2}
1 year	3.03×10^{-2}
2 years	1.53×10^{-2}
3 years	1.02×10^{-2}
4 years	7.66×10^{-3}
5 years	6.12×10^{-3}

risking disease prevalence. The vaccine drawback is quantified in terms of the vaccination cost, which we assumed to be evaluated as $C_vV(t)$, and the infection penalty is taken simply as the product of the perceived cost and the risk of being infected, namely $C_iI(t)$. Thus, the payoff gain for changing strategy is $-C_vV(t) + C_iI(t)$, where C_v is the cost for vaccination, C_i is the cost for disease, and $V(t)$ is given by AIM(t).

This assumption contains twofold to be noted. One is that an individual only sees the surface cost of vaccination. In this sense, an individual could be called myopic, simply because he/she does not evaluate the risk of infection even if committing a vaccination due to the imperfectness of the vaccination. Such a myopic estimation and ignoring the possibility of infection even with a vaccination might be justified, because an individual tends to be influenced by just a surface value not referring to deep an analysis if observing real-world situations. As the second, the formulation of $-C_vV(t)$ implies that an individual would not be proactive for vaccination if other people willing to vaccinate, which premises that he/she expects the free ride on herd immunity. In general, the expression $-C_rV(t) + C_iI(t)$ quantifies the expected payoff gain for transferring strategies, and its sign regulates whether vaccinator or non-vaccinator is preferred.

Recalling the basic concept of what-is-called replicator equation introduced in Sect. 2.1.1, we are able to build the global dynamics of x as: $\dot{x} = m \cdot x(t) \cdot [1 - x(t)][-C_vV(t) + C_iI(t)]$, where x (t) is the rate at time t at which non-vaccinated (susceptible) individuals in the population become vaccinated, and m is the proportionality constant converting fraction of individual into transferring probability of switching strategies according to the expected gain in the payoff. Throughout the present study, we presumed $m = 0.2$. If the relative cost of vaccination is defined as $C_r = C_v/C_i$ and $C_i = 1$, then the evolutionary dynamics eventually could be:

$$\dot{x} = m \cdot x(t) \cdot [1 - x(t)][-C_rV(t) + I(t)]. \tag{7.2}$$

7.3 Result and Discussion

7.3.1 Fundamental Characteristic of Time Evolution

First off, let us confirm the model stability. Figure 7.3 displays the respective time evolutions of (A) susceptible, (B) infected (denoted by INF meaning $I^S(t) + I^V(t)$), (C) unvaccinated–recovered, defined as naturally-obtained immunity without vaccination (NIMS), (D) vaccinated–recovered, labeled as naturally-obtained immunity despite committing vaccination (NIMV) (i.e., the fraction of those who acquired naturally-obtained immunity from infection despite getting vaccinated), and (E) artificial immunity (AIM) for different durations. We set following parameters: $\beta = 0.4$, $\gamma = 0.1$, $\eta = 0.9$, and $C_r = 0.1$, which corresponds to the situation with a cheaper but less-efficient vaccine and reasonably strong epidemic ($\frac{\beta}{\gamma} = 4$).

Most important point is that all 9 cases varying the duration of immunity do converge to each stable equilibrium neither showing divergence, let alone, or fluctuation at all. Thus, as we do this time, with carefully assuming a sufficiently small time-discrete step-size, the dynamics can be inherently absorbed with a stable equilibrium state, although respective dynamical evolutions transiently show fluctuations as we could observe in Fig. 7.3. Panel (C) displays that the order of which case showing higher level of naturally-obtained immunity from susceptible compartment; NIMS is completely consistent with the order of the duration of immunity, where a longer duration shows a higher NIMS. It might be quite conceivable as the general tendency. Interestingly panel (E) shows the inverse order to what we observed in panel (C), where a longer duration shows a lower level of artificial immunity; AIM. This might be also likely because an individual protected by a longer NIMS has less incentive to vaccination. But the order observed in panel (D) showing level of naturally-obtained immunity from vaccinated compartment; NIMV is not consistent with the order of duration of immunity at all. It is because what fraction of individuals taking vaccination is influenced by not only the duration of immunity brought from either vaccination or infection but also cost of vaccination as well as inherent strength of epidemic (evaluated by $\frac{\beta}{\gamma}$), which might be dominated by complex dynamics.

One obvious thing observed in panel (A) is that too short effect of immunity that is the case of 7 days inevitably results in a specifically large epidemic size.

7.3.2 Dynamics Observed in Trajectory

To investigate the dynamical behavior, it is meaningful to observe the trajectories of infected— INF : $I(t)$, AIM: $V(t)$, NIMS: $R^S(t)$, and NIMV: $R^V(t)$—against $S(t)$ for different durations of immunity. The result is summarized in Fig. 7.4. We varied the infection rate, i.e., $\beta = 0.2$ (*-i), $\beta = 0.4$ (*-ii), and $\beta = 0.8$ (*-iii), which indicates three different levels of basic reproduction number ($R_0 = \beta/\gamma$) due to fixing to $\gamma = 0.1$.

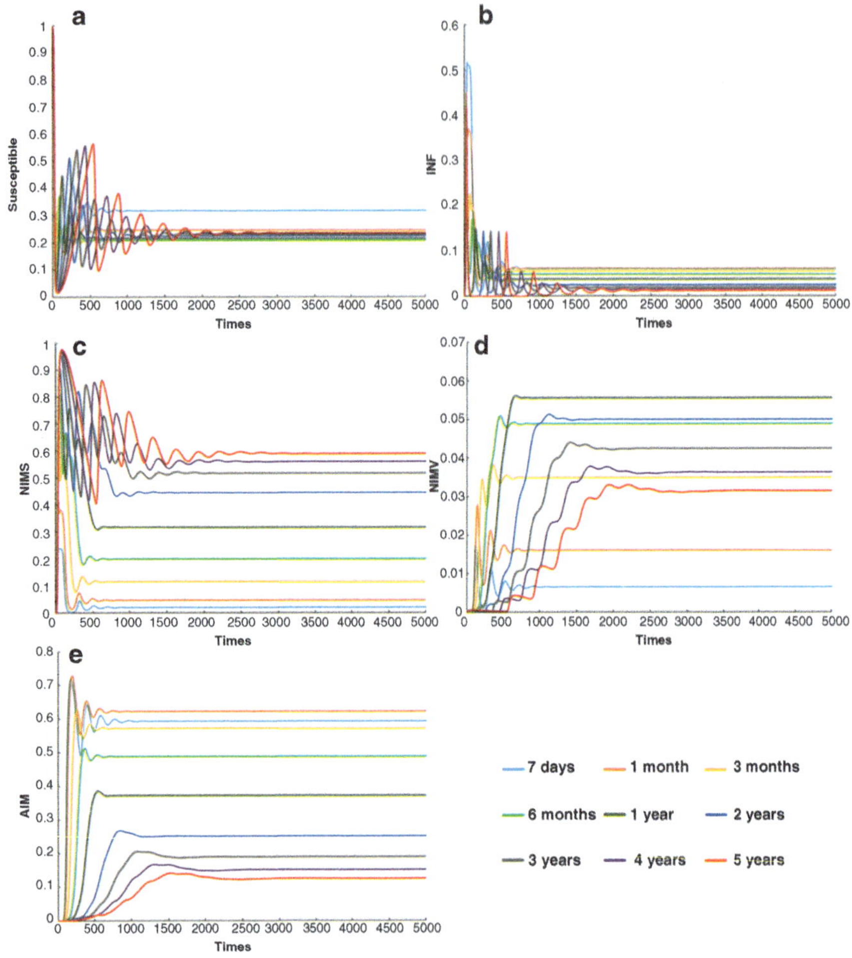

Fig. 7.3 Dynamical behavior of system with $\gamma = 0.1, \beta = 0.4, \eta = 0.9$, and $C_r = 0.1$. Time series of (**a**) susceptible $S(t)$, (**b**) infected (INF) $I^s(t) + I^v(t)$, (**c**) natural immunity from susceptible —NIMS $\left(\sum_{i=0}^{3} R_i^s(t)\right)$, (**d**) natural immunity from vaccinated—NIMV $\left(\sum_{i=0}^{3} R_i^v(t)\right)$, and (**e**) artificial immunity—AIM $\left(\sum_{i=0}^{3} V_i\right)$

We also fixed $\eta = 0.9$, and $C_r = 0.1$. Each of all trajectories starts from the zero point (i.e., the bottom right-hand corner of each graph; (X-axis, Y-axis) = (1, 0) and approaches to an equilibrium value showing a closed loop eventually.

With increase of R_0, equilibrium level of INF increases, which is a trivial thing. More importantly, with increase of the duration, an equilibrium of AIM degrades while that of NIMS increases as the general tendency. It suggests that the naturally-obtained

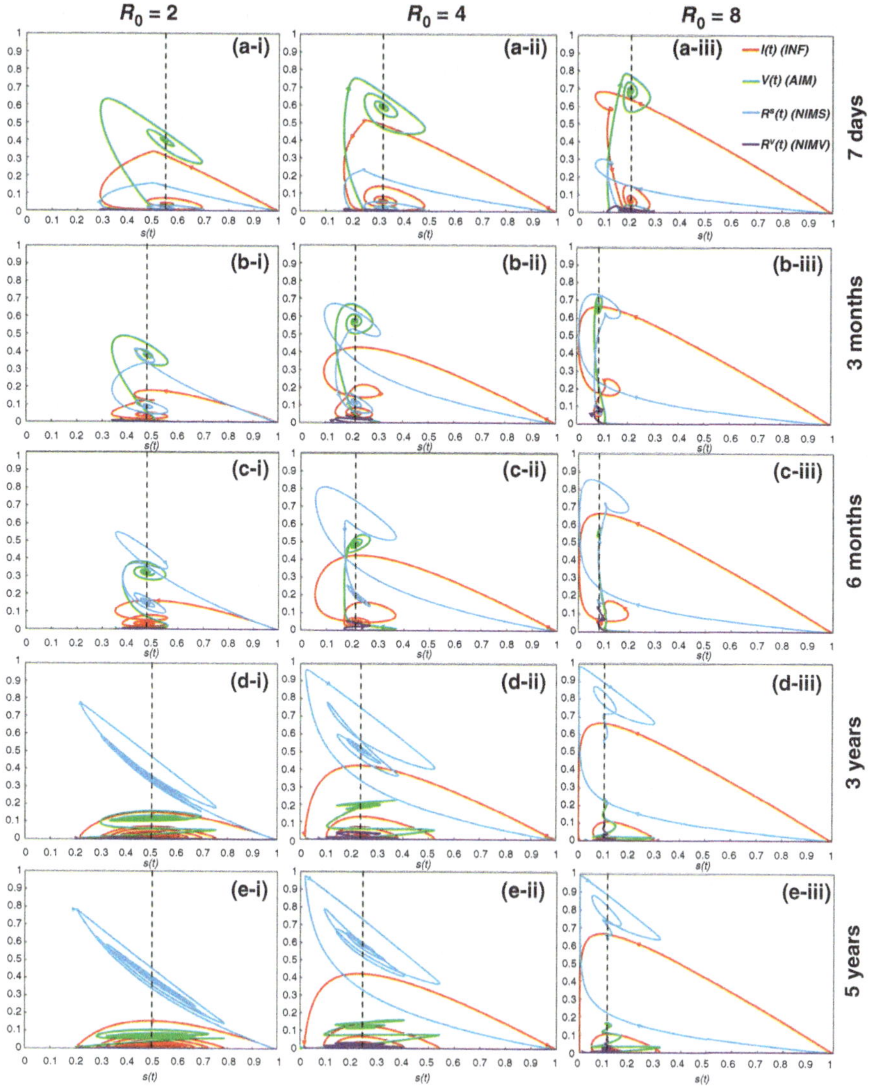

Fig. 7.4 Trajectories in phase space $[I(t), V(t), R^s(t), R^v(t)]$ vs. $S(t)$ for five different protection durability $(a - *)$ 7 days, $(b - *)$ 3 months, $(c - *)$ 6 months, $(d - *)$ 3 years, and $(e - *)$ 5 years, also for $(* - \text{i})$ $R_0 = 2$ $(\beta = 0.2)$, $(* - \text{ii})$ $R_0 = 4$ $(\beta = 0.4)$, and $(* - \text{i})$ $R_0 = 8$ $(\beta = 0.8)$. Here $\gamma = 0.1$, $\eta = 0.9$, and $C_r = 0.1$

immunity effectively works as long as its duration is long span, whereas the artificial immunity brought by a vaccination rather works well when the duration in short.

To elaborate this, Fig. 7.5 compares the cases of 7 days and 5 years when the set of η and C_r is varied. Regardless of the combination of efficiency and cost of vaccination, thus even if it coming to the situation where vaccination is quite

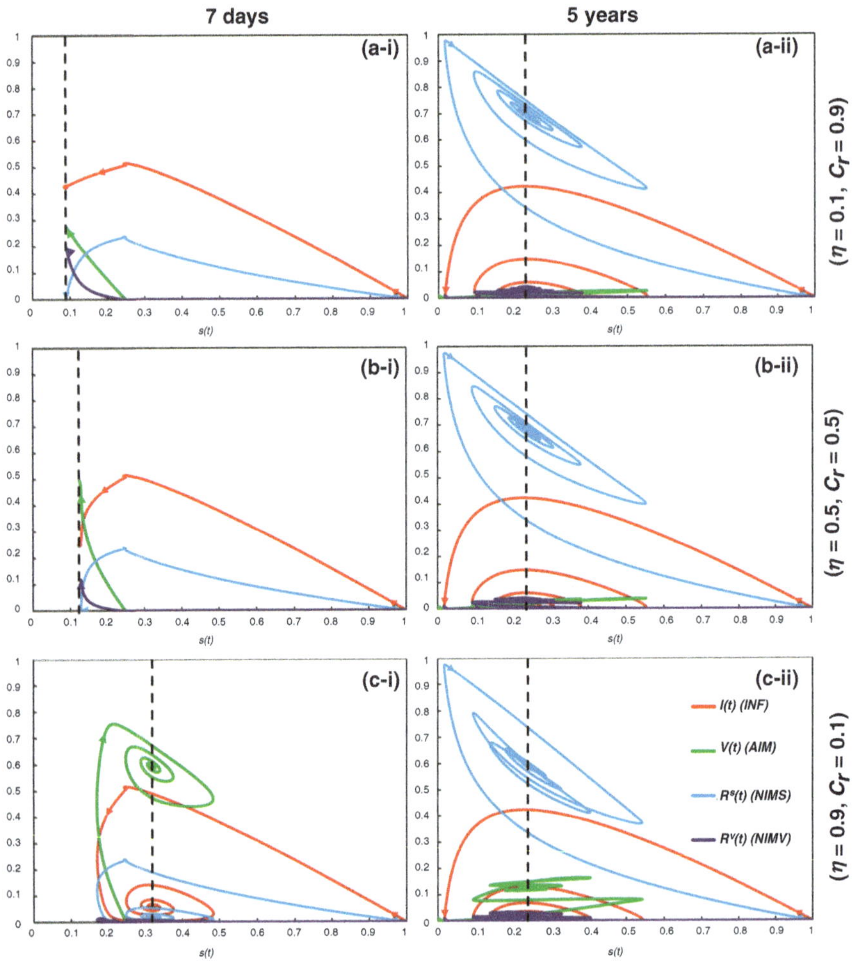

Fig. 7.5 Trajectories in phase space $[I(t), V(t), R^S(t), R^V(t)]$ vs. $S(t)$ for combining values of efficiency and relative cost (η, C_r) as (a $-$ $*$) (0.1,0.9), (b $-$ $*$) (0.5,0.5), and (c $-$ $*$) (0.9,0.1) with protection durability of (*-i) 7 days and (b-*) 5 years

advantageous with high efficiency and less cost; $(\eta, C_r) = (0.9, 0.1)$, or coming to the situation where vaccination becomes less disadvantageous; $(\eta, C_r) = (0.1, 0.9)$, more of less it is proved the tendency that the naturally-obtained immunity works with a long duration and the artificial one from a vaccination rather works with a short duration.

7.3.3 Phase Diagram Analysis

Figure 7.6 displays phase diagrams drawn on 2D plane of C_r (vaccination cost) vs η (vaccine efficiency) for (A) infected individuals; INF, (B) vaccinated (artificial immunity); AIM, (C) unvaccinated and infected & recovered (naturally-obtained immunity); NIMS, and (D) vaccinated but infected & recovered (naturally-obtained immunity); NIMV. In sub-panels: a-*, b-*, and c-*, we varied the transmission rate: $\beta = 0.2$, 0.4, and 0.8, while in sub-panels: *-i, *-ii, and *-iii, we varied duration of immunity: 7 days, 6 months, and 5 years.

First, we see that a cheaper and more reliable vaccine reduces the number of infectious individuals even though a meager protection duration presumed (deep-blue region in sub-panel a-* in panel A). Even with increase of β from 0.2 to 0.8 a longer duration shows none of infected individuals at all (typically in the case of 5 years; sub-panel c-* in panel A). However, there is remarkably high infected region when the protection by immunity becomes shorter (a-*). The longest duration; 5 years (c-*) ensures disease-free (panel A), since an individual is protected by either the NIMS or NIMV state ((c-*) in panels C and D). As a general tendency, comparing panel B with D, we can note that NIMV shows high when an individual fails to acquire artificial immunity through a vaccine because of a lower vaccination efficiency.

Observing panel B of Fig. 7.6 closely, we note that the higher AIM region colored with yellow-red appears (highlighted by black dotted triangles in a-i, ii, and iii; panel B). When presuming more severe disease situations; $\beta = 0.4$ and 0.8 (sub-panels a-ii and a-iii), such a high AIM region suggests that more vaccinators appears in lower cost and higher-efficiency region. It is quite conceivable. However, interestingly and surprisingly, in the case of $\beta = 0.2$ (sub-panel a-i), such a high AIM region rather appears in lower cost but medium efficiency area. Why lower cost and higher-efficiency region does not show higher AIM? Referring to previous study,[5] too high efficiency in relation with vaccination cost promotes the incentive of free riding amid individuals. Although the present model differs from the previous model in detail points, an analogous scenario might take place. In other words, in the region with much higher efficiency than the focal region (black dotted triangle in the case of $\beta = 0.2$; sub-panels a-i), a large number of free riders, denoted by S, come into the system. However, a smaller fraction of vaccinators could still support "herd immunity" sufficiently, thereby driving the system to perfect eradication of the disease (panel A). Thus, a large fraction of non-vaccinators can be described as being "successful free riders."

Now let us turn to discuss what happens when the duration of artificial immunity; vaccination denoted by AIM is different from that of naturally-obtained immunity; NIMS and NIMV. Figure 7.7 responses to this question. Figure 7.7 (a) displays INF (panels A, C, and E) and AIM (panels B, D, and F), while Fig. 7.7 (b) gives NIMS (panels A, C, and E) and NIMV (panels B, D, and F). We vary AIM (row direction) for 7 days (*-ii), 6 months (*-iii), and 5 years (*-iv), while NIMS and NIMV(column

[5]Kuga and Tanimoto (2018), p. 023407.

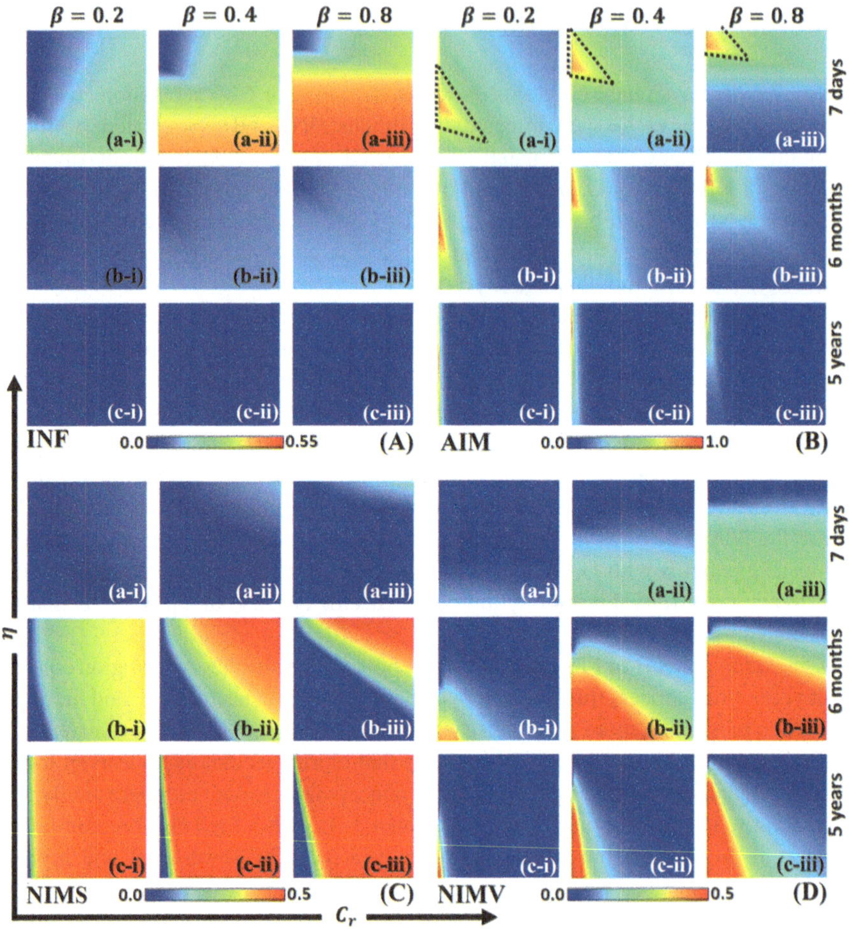

Fig. 7.6 Phase diagrams of INF, AIM, NIMS, and NIMV along vaccine efficiency η and relative cost of vaccination C_r in panels A – D, respectively. Each panel shows the infection rate and durability of protection period (β, duration) as (a – i) (0.2, 7 days), (a – ii) (0.4, 7 dsays), (a – iii) (0.8, 7 dsays), (b – i) (0.2, 6 months), (b – ii) (0.4, 6 months), (b – iii) (0.8, 6 months), (c – i) (0.2, 5 years), (c – ii) (0.4, 5 years), and (c – iii) (0.8, 5 years) with $\gamma = 0.1$

direction) are varied 6 months (a-*), 3 years (b-*), and 5 years (c-*). In addition, we deliver "no vaccine" case, displayed as sub-panels (*-i) by assuming none of artificial vaccination at all by letting both $x = 0.0$ and $V_0(t) = 0$. Thus, the results given in sub-panels (*-i) of panels B, D, and F in both Fig. 7.7 (a) and (b) are trivial because there is none of vaccinated person in the system.

Comparing with the "no vaccine" case, we are able to note that a vaccination bringing artificial immunity even if its duration is extremely short (7 days) is able to oppress a disease spreading, which can be more clearly observed in the region with a lower vaccination cost when the duration of naturally-obtained immunity is short (6 months).

Looking beyond different combinations of durability for AIM and NIMS (NIMV), note that natural immunity with longer duration (5 years) can eradicate

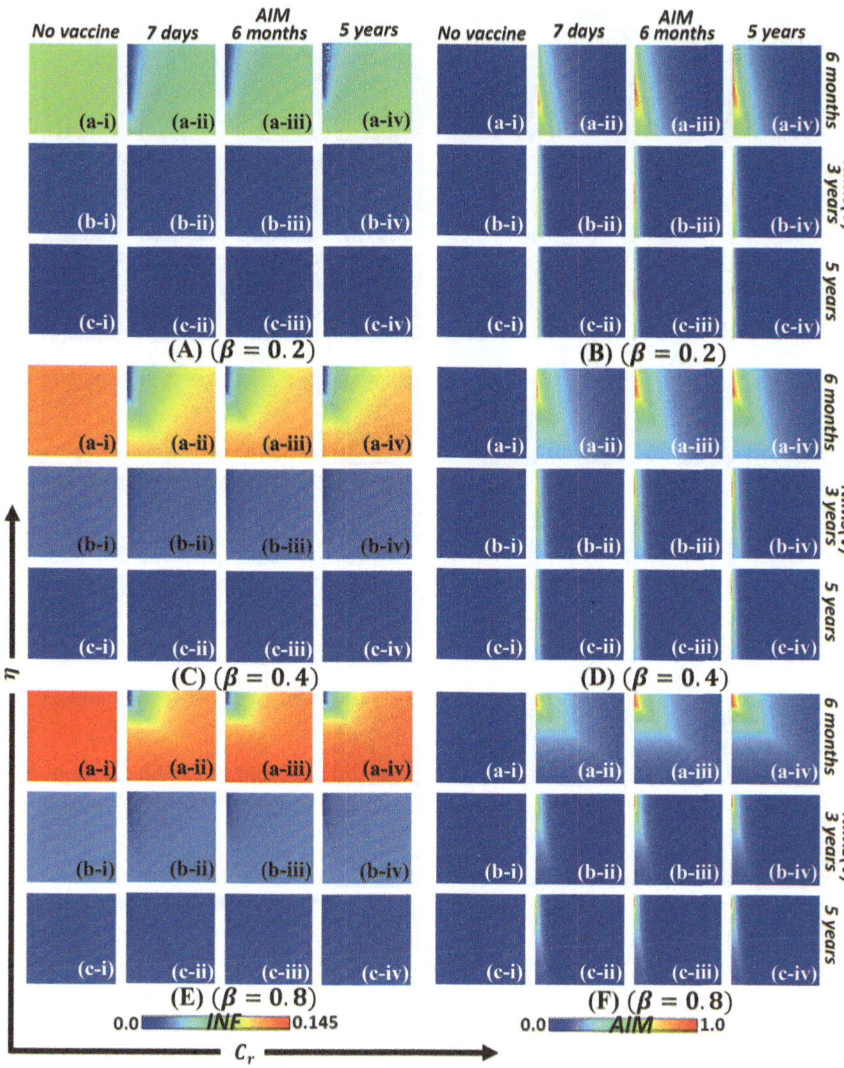

Fig. 7.7 (**a**) Phase diagrams of INF and AIM along vaccine efficiency η and relative cost of vaccination C_r. Panels A, C, and E show the heat maps of INF, while panels B, D, and F show those of AIM. Panels A and B, C and D, and E and F vary the disease transmittance; $\beta = 0.2$, 0.4, and 0.8. Sub − panels ∗ − i present the "no vaccination" case. In each panel, the durability of protection for AIM is (∗ − ii) 7 days, (∗ − iii) 6 months, and (∗ − iv) 5 years. The durability of protection for NIMS(V) is (a − ∗) 6 months, (b − ∗) 3 years, and (c − ∗) 5 years. (**b**) Phase diagrams of NIMS and NIMV along vaccine efficiency η and relative cost of vaccination C_r. Panels A, C, and E show the heat maps of NIMS, while panels B, D, and F show those of NIMV. Panels A and B, C and D, and E and F vary the disease transmittance; $\beta = 0.2$, 0.4, and 0.8. Sub-panels *-i present the "no vaccination" case. In each panel, the durability of protection for AIM is (*-ii) 7 days, (*-iii) 6 months, and (*-iv) 5 years. The durability of protection for NIMS(V) is (a-*) 6 months, (b-*) 3 years, and (c-*) 5 years

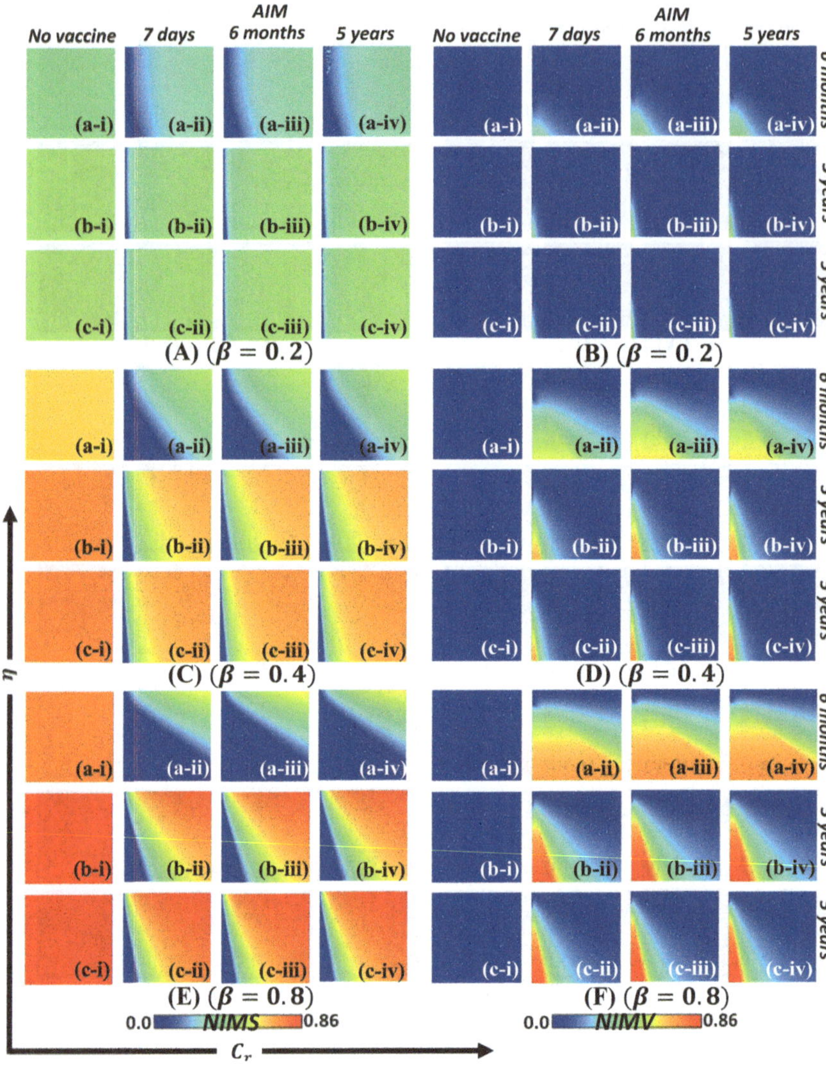

Fig. 7.7 (continued)

infection irrespective to whether artificial immunity is introduced or not. This sheds some light on the fact that a vaccination only contributes in a limited way for eradicating a disease if the naturally-obtained immunity through infection lasts for a longer time. The effect, or say coverage from the naturally-obtained immunity, results from a simple reality dominated by the physical mechanism of 'disease-spreading'. In contrast, the effect of vaccination or coverage of artificial immunity is strongly influenced by human decision in a social context; whether an individual favoring to take a pre-emptive vaccination or trying to free ride on others' contribution (or believing an epidemic would not be serious). As the public-health authority standpoint, the former effect seems more predictable than the latter one, because they might think that the prediction from an objective mechanism is more reliable than that from a subjective mechanism. But, ironically, they cannot help relying on vaccination as one of the most representative public-health policies to safe their own people from the potential thread of an epidemic.

To explore the effect of public subsidy covering a vaccine cost, we present phase diagrams of "forced vaccination," where the government interferes to realize a time-constant vaccination rate of $x = 0.5$. It would be meaningful if we compare this case with another situation presuming free vaccination ($C_r = 0.0$), where the government pays the vaccination cost for all individuals who want to be vaccinated (in this sense, this setting cannot be described as "forced vaccination"). We also show cases of voluntary vaccination with their own payment ($C_r = 0.1$ and $C_r = 0.9$) for comprehensive comparison. Figure 7.8 delivers the result. Unlike the previous heat maps, the result is drawn as 2D heat maps of transmission rate; β, vs vaccination efficiency;. η. Panels A-*, B-*, C-*, and D-* show the aforementioned settings. Meanwhile, panels *-1, *-2, *-3, and *-4, respectively, show the INF, AIM, NIMS, and NIMV, where sub-panels (i)–(iv) show the results from different combinations of durations for AIM and NIMS (NIMV), namely (AIM, NIMS(NIMV))= (7 days, 6 months), (5 years, 6 months), (7 days, 5 years). and (5 years, 5 years), respectively.

Throughout all cases concerning vaccination cost including the case of forced vaccine, disease may break out when presuming a short duration of naturally-obtained immunity; 6 months, but never happens when presuming a longer natural immunity; 5 years (*-1).

When presuming a higher transmittance and a lower efficiency of vaccination, and also presuming relatively low vaccination cost; 0.1, or forced vaccination as well as the cost-free case, quite a few people get naturally-obtained immunity through vaccination (observing NIMV). Nevertheless, vaccination does little to control the disease, which is proved by lower AIM in the focal region. In fact, the disease cannot be eradicated in such settings when presuming shorter NIMS (NIMV); 6 months. More interestingly, however, in the cases of both forced vaccination and cost-free vaccination, when presuming a higher efficiency and a relatively low transmittance are presumed, the effect from artificial immunity by a vaccine shot (AIM) works quite well, which significantly helps to eradicate the disease (INF). This finding may justify that the public-health authority should take an intervention through a forced-vaccination policy or a financially supporting policy only when a disease is quite strong (its transmission rate is high) and the vaccine is less reliable. It is because an individual would be less pushed to a voluntarily vaccination due to the vaccine effect being questionable. But if the public authority strongly propels such policies, by leading vast majority of population to forcefully obtaining the artificial immunity, the disease remission can be possible. The remaining question is whether the government has a sufficient financial background or not.

Now, to examine the interplay between various transmission rates; β, and recover rates; γ, while keeping the ratio of those two at constant ($R_0 = \beta/\gamma = 4$), we should come to Fig. 7.9, which gives 2D heat maps of C_r vs η. Panels A–D present INF, AIM, NIMS, and NIMV, respectively, where sub-panels *-i, *-ii, *-iii, and *-iv

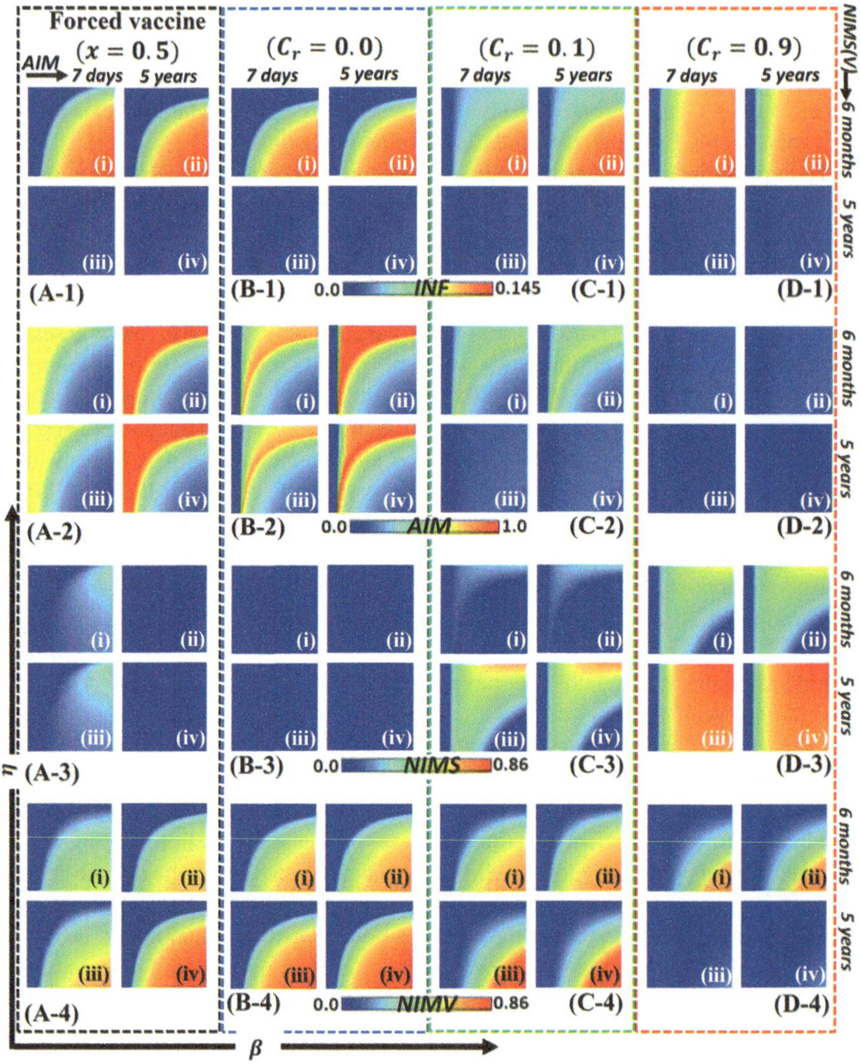

Fig. 7.8 Phase diagrams of (*-1) INF, (*-2) AIM, (*-3) NIMS, and (*-4) NIMV along vaccine efficiency η and infection rate β. Panels A $-$ $*$ display the phase diagrams of the forced $-$ vaccination case without considering the evolutionary process, where $x = 0.5$. Each panel shows the durability of protection for (AIM, NIMS(V)) as (i) (7 days, 6 months), (ii) (5 years, 6 months), (iii) (7 days, 5 years), and (iv) (5 years, 5 years)

show the results for $(\beta, \gamma) = (0.01, 0.0025)$, $(0.2, 0.05)$, $(0.4, 0.1)$, and $(0.8, 0.2)$, respectively, and sub-panels (a-*) and (b-*) show those for the 7 days and 5 years periods of protection durability (for AIM and NIMS (NIMV)). Note that whether a disease-free equilibrium or an endemic state is reached cannot be classified only by R_0 anymore, which is also affected by the durations of both the artificial and

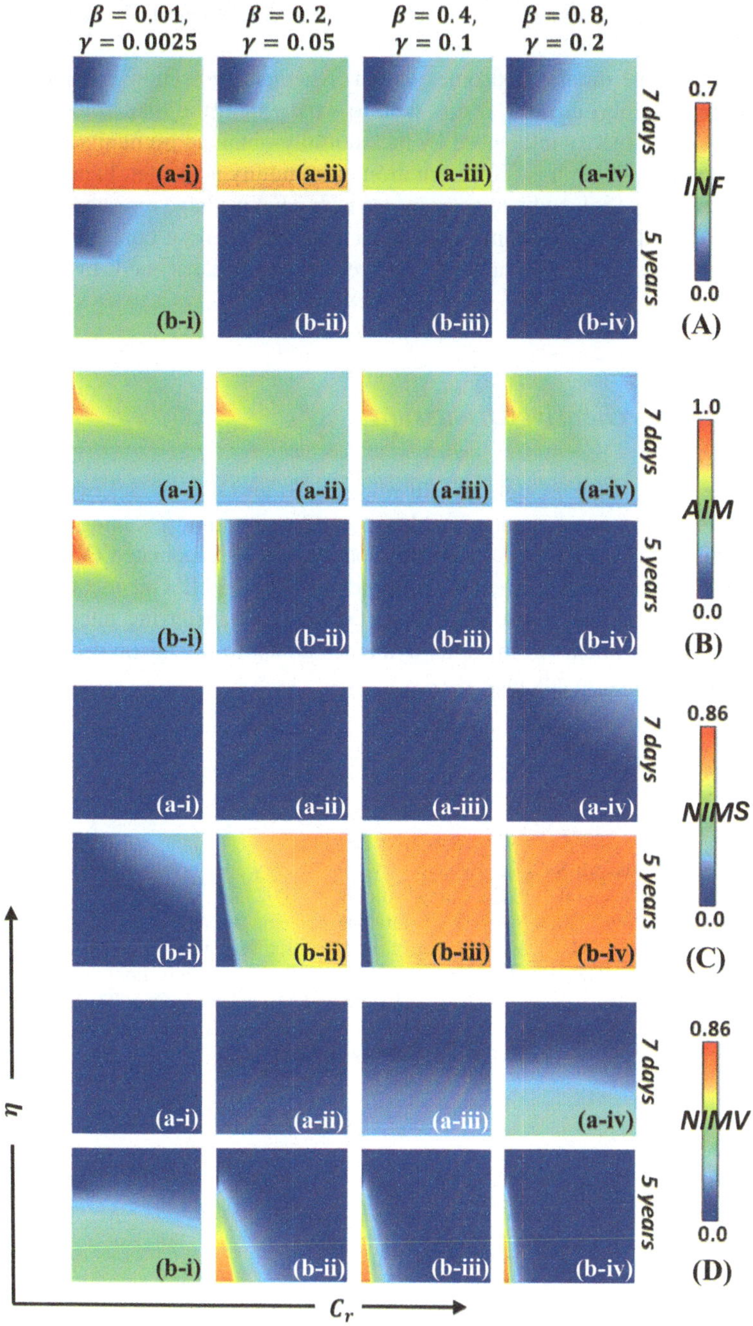

Fig. 7.9 Phase diagrams of (A) INF, (B) AIM, (C) NIMS, and (D) NIMV along vaccine efficiency η and relative cost of vaccination C_r. Each panel presents the combining values of (β, γ) as $(* - \text{i})$ $(0.01, 0.0025)$, $(* - \text{ii})$ $(0.2, 0.05)$, $(* - \text{iii})$ $(0.4, 0.1)$, and $(* - \text{iv})$ $(0.8, 0.2)$. Also, a-* and b-* display the duration of protection for AIM and NIMS(V) as 7 days and 5 years, respectively

naturally-obtained immune systems as well as the combination of vaccination cost and its efficiency.

As long as the duration becomes longer, whether eradication is achieved (i.e., disease-free equilibrium)—or if not, then to what extent the final epidemic size is increased—is primarily influenced by the naturally-obtained immunity system than the artificial one. In contrast, if the duration of immunity is shorter, then vaccination plays an important role, which is consistent with what we have repeatedly discussed do far. If the duration is longer, the vaccination somehow contributes to relax spreading a disease, especially in the region of cheaper and more efficient vaccination.

7.3.4 Comprehensive Discussion

Not relying on a PDE model but still referring to an ODE model, what we have discussed in the present chapter enables to emulate immunity waning effects. The kernel concept is introducing several compartments for vaccinated (V) and recovered (R) states to reproduce pseudo age-structures of V and R. One inconvenience of the present model vis-à-vis a PDE model is that the transferring rates connecting those sub-compartments to make its dynamics adjust to a realistic decaying curve should be identified in advance.

References

Bauch CT, Bhattacharyya S (2012) Evolutionary game theory and social learning can determine how vaccine scares unfold. PLOS Comput Biol 8(4):e1002452

Kabir KMA, Tanimoto J (2021) Analysis of individual strategies for artificial and natural immunity with imperfectness and durability of protection. J Theoret Biol 509:110531

Kuga K, Tanimoto J (2018) Which is more effective for suppressing an infectious disease: imperfect vaccination or defense against contagion? J Statistical Mech Theory Exp:023407

Zhang S, Guo H (2018) Global analysis of age-structured multi-stage epidemic models for infectious diseases. Appl Math Comput 337:214–233

Chapter 8
Pre-emptive Vaccination Versus Antiviral Treatment

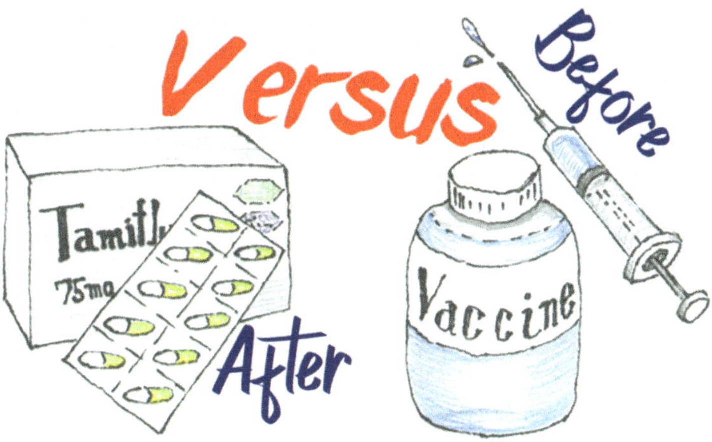

An either-or choice between pre-emptive vaccination or antiviral treatment for influenza raises an interesting question: which strategy should be taken to minimize the risk or maximize the expected benefit? In terms of information advantage, antiviral treatment, which is taken later, would be more advantageous than the prior-investment option that is pre-emptive vaccination. Is this always true, or does it depend on the situation? This chapter answers this question by establishing another vaccination game.

J. Tanimoto, *Sociophysics Approach to Epidemics*, Evolutionary Economics and Social Complexity Science 23, https://doi.org/10.1007/978-981-33-6481-3_8

8.1 Introduction and Background: Behavioral Incentives in a Vaccination-Dilemma Setting with an Optional Treatment

As we have discussed in the previous chapters, pre-emptive vaccination is an effective means of combating infectious diseases. Even when they confer only transient immunity and/ or an efficacy that is far from perfect (as is in the case of influenza), vaccines produce a positive economic effect by keeping the society healthier. Wide vaccination coverage in particular gives rise to herd immunity whereby the fraction of vaccinated individuals is high enough to offer a degree of protection, even to those who have not been vaccinated. Vaccination, however, comes at a cost that can be obvious, e.g., paying a doctor's bill for getting vaccinated, and entails some risk from a vaccine's side effects. To avoid this cost, some individuals resort to free riding in the hope that others will vaccinate and protect them from the disease; yet such behavior ultimately undermines herd immunity and increases the chances of an epidemic outbreak. This is why we have discussed the so-called vaccination game and its conceptual extension in the intervention game. Meanwhile we should note there is another important option for an individual besides taking a vaccine: antiviral treatment. Once people get infected, they resort to *ex-post* treatments such as the widespread use of Oseltamivir (Tamiflu) against influenza, which is also costly in monetary terms and accompanied by potentially serious side effects. They can, however, avoid taking a pre-emptive vaccination, i.e., the *ex-ante* provision, as long as they rely on an *ex-post* treatment. Such a treatment might be more convenient and its expected profit can be more beneficial than an *ex-ante* provision depending on the spread of disease.

To quantify such social-dilemma situations, we establish a new vaccination model as below.

8.2 Model Structure

We constructed an SVITR compartmental epidemic model to which, in addition to the usual susceptible (S), infectious (I), and removed (R) compartments, we add a compartment for agents under treatment (T) and another for agents who chose pre-emptive vaccination (V). Mathematical modeling of antiviral treatments has heretofore largely focused upon the effectiveness of drugs in containing epidemics and the implications of drug-resistant viral strains for the spread of disease.[1] Coupling the present SVITR epidemic model with game-theoretic concepts allows us to comprehensively analyze the incentives that drive individual human behaviors when

[1]There have been many previous studies on this point. For representative ones, see Regoes and Bonhoeffer (2006), pp. 389–391.

Alexander et al. (2007), pp. 1675–1684.

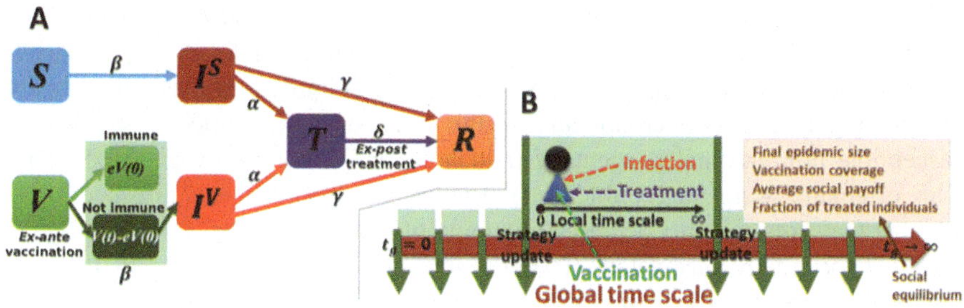

Fig. 8.1 Schematic-model diagram. (**a**) In our SITR/V epidemic model, susceptible (S) individuals get infected at a rate β, which applies even to the fraction of vaccinated individuals for whom vaccination failed to induce immunity. This fraction, as well as that of vaccinated and immune individuals, is determined by vaccine effectiveness e. Disease carriers are considered infectious (I), in which case they may receive an antiviral treatment with probability rate α and then recover at rate δ, or recover without treatment at rate $\gamma < \delta$. Once an individual is recovered, they are removed (R) from further consideration on the local timescale at which the epidemic progresses. (**b**) Aside from the disease progression on the local timescale, the evolutionary decision – making process takes place on a global timescale, t_g. Here, an individual decides whether to vaccinate or not at the onset of an epidemic season based on how they fared in the previous epidemic season relative to others. Those agents who fared well are unlikely to change their strategy with respect to vaccination, and vice versa for agents who fared poorly

facing dilemma situations. Specifically, agents in the model make *ex-ante* decisions on whether to vaccinate or not at the onset of the epidemic. These decisions are based on the total cost incurred during the previous epidemic season, which generally is some combination of vaccination, infection, and treatment costs. Additionally, agents who become infectious during the current epidemic season may receive *ex-post* treatment, but we assume that the decision is in the hands of a medical professional (i.e., a healthcare establishment), rather than the agent themselves (irrespective of their own decision-making process). The described setup (Fig. 8.1a) naturally leads to two separate timescales: *ex-ante* decisions on vaccination (i.e., strategy updating) take place on a "global" or "evolutionary" timescale, whereas an epidemic season and *ex-post* decisions on treatment unwind on a "local" timescale (Fig. 8.1b), to use the terminology of our previous chapters.

8.2.1 Formulation of the SVITR Model[2]

As Fig. 8.1 shows, the SVITR model most naturally fits with seasonal-influenza-like diseases for which vaccination confers only temporary immunity for a single season. The season is then controlled by a local timescale, while the global timescale is inter-seasonal. With minor adjustments, however, it is possible to think of the local

[2]Kabir et al. (2019), p. 062402.

timescale as that of a generation of agents who make vaccination decisions for a lifetime. The global timescale would then be intergenerational. From the perspective of mathematical epidemiology, the SVITR model belongs to a class of compartmental epidemic models comprising, in this case, six compartments: $S(t)$ denotes susceptible non-vaccinated agents, $I^S(t)$ denotes infectious non-vaccinated agents who originally were susceptible, $V(t)$ denotes vaccinated agents irrespective of whether they subsequently acquired immunity of not, $I^V(t)$ denotes infectious agents who were originally vaccinated, $T(t)$ denotes treated agents, and finally $R(t)$ denotes removed agents. The dynamics of these compartments are given by:

$$\frac{dS}{dt} = -\beta S(t)\left(I^S(t) + I^V(t)\right), \tag{8.1a}$$

$$\frac{dV}{dt} = -\beta(V(t) - eV(0))\left(I^S(t) + I^V(t)\right), \tag{8.1b}$$

$$\frac{dI^S}{dt} = \beta S(t)\left(I^S(t) + I^V(t)\right) - \alpha I^S(t) - \gamma I^S(t), \tag{8.1c}$$

$$\frac{dI^V}{dt} = \beta(V(t) - eV(0))\left(I^S(t) + I^V(t)\right) - \alpha I^V(t) - \gamma I^V(t), \tag{8.1d}$$

$$\frac{dT}{dt} = \alpha I^S(t) + \alpha I^V(t) - \delta T(t), \tag{8.1e}$$

$$\frac{dR}{dt} = \gamma I^S(t) + \gamma I^V(t) + \delta T(t). \tag{8.1f}$$

Here, $\beta > 0$ is transmission rate [day^{-1} person^{-1}], $\alpha > 0$ is the antiviral treatment-probability rate [day^{-1}], $\gamma > 0$ is recovery rate [day^{-1}], $\delta > \gamma$ is the accelerated-recovery rate due to treatment [day^{-1}], and $0 < e \leq 1$ is vaccine effectiveness, introduced in Sect. 3.2.1. We worked with a normalized population such that

$$S(t) + V(t) + I^S(t) + I^V(t) + T(t) + R(t) = 1, \tag{8.2}$$

which implies that each of the six state variables refers to a population fraction in a particular state. The initial condition in any season is $S(0) = 1 - x$ and $V(0) = x$, where x is the fraction of vaccinators (vaccination coverage), which is determined by the evolutionary consideration defined on the global timescale, as described below.

One quite important assumption embedded in the model dynamics is, as aforementioned, that the state transfer from either I^S or I^V to T is time-constant, rather than evolving like x. This implies that whether antiviral treatment is introduced or not is in the hands of a medical doctor not obeying their patient's evaluation.

8.2.2 Reproduction Number

As we discussed in Sect. 3.1.1, in a classic SIR model, the basic reproduction number R_0 is the average number of secondary infections caused by a single infectious agent in a completely susceptible population. It can be shown that $R_0 = \beta/\gamma$, and that if $R_0 > 1$, the infection spreads through the population, whereas in the opposite case, the infection dies out in the long run. In the SVITR model, however, instead of the whole population, only a fraction $S(0) + (1 - e)V(0)$ is susceptible due to vaccination. Furthermore, the clearance rate of infectious individuals is not just γ, but $\gamma + \alpha$ due to the admission to treatment. It is therefore more appropriate to look at the control-reproduction number[3] R_c defined as

$$R_c = \frac{\beta}{\gamma + \alpha}[S(0) + (1 - e)V(0)]. \tag{8.3}$$

For this control-reproduction number, it still holds that the infection spreads through the population if $R_c > 1$, but dies out otherwise.

8.2.3 Payoff Structure

An agent's fate during an epidemic season determines their costs and ultimately the total payoff. Only forgoing vaccination and staying healthy are costless. Getting infected costs C_i, getting vaccinated costs $C_v < C_i$, and similarly getting treated costs $C_t < C_i$. We simplified these considerations by setting $C_i = 1$ and then working with a relative cost of vaccination $C_V = C_v/C_i$ and a relative cost of treatment $C_T = C_t/C_i$.

To calculate the total payoff of an agent at the end of an epidemic season (i.e., when an equilibrium is reached on a local timescale), we reclassified agents into six classes: healthy vaccinated H_V°, healthy non-vaccinated H_V^n, infected vaccinated and treated $I_{VT}^{\circ\circ}$, infected vaccinated and non-treated $I_{VT}^{\circ n}$, infected non-vaccinated and treated $I_{VT}^{n\circ}$, and infected non-vaccinated and non-treated I_{VT}^{nn}. A superscript of an open circle means "taking the provision denoted by the subscript below it," while "n" indicates "not taking that subscript." The subscripts are either vaccination V or treatment T. Each class can be linked to the results of the SVITR model. To see how, we first note that when the SVITR model reaches equilibrium, i.e., $t \to \infty$, we have $S(\infty) + V(\infty) + R(\infty) = 1$, while $I^S(\infty) = I^V(\infty) = T(\infty) = 0$. By definition, we further have that $H_V^\circ = V(\infty)$ and $H_V^n = S(\infty)$. Each of the remaining four classes, $I_{VT}^{\circ n}$, I_{VT}^{nn}, $I_{VT}^{\circ\circ}$, and $I_{VT}^{n\circ}$, can be thought of as a contribution to $R(\infty)$. We can separate these contributions using Eqs. (8.1e) and (8.1f), from which it follows that R

[3] van den Driessche and Watmough (2008).

Table 8.1 Agent class (upper row) and the corresponding payoff (lower row) depending on how agents of each class fared during an epidemic season

Strategy/state	Healthy	Infected	Infected & Treated
Vaccinated	H_{V}° $-C_{\mathrm{V}}$	$I_{\mathrm{VT}}^{\circ\mathrm{n}}$ $-C_{\mathrm{V}}-1$	$I_{\mathrm{VT}}^{\circ\circ}$ $-(C_{\mathrm{V}}+C_{\mathrm{T}})-1$
Non-vaccinated	$H_{\mathrm{V}}^{\mathrm{n}}$ 0	$I_{\mathrm{VT}}^{\mathrm{nn}}$ -1	$I_{\mathrm{VT}}^{\mathrm{n}\circ}$ $-C_{\mathrm{T}}-1$

$(\infty) = \gamma \int I^V(t)dt + \gamma \int I^S(t)dt + \delta \int T(t)dt$ and $\delta \int T(t)dt = \alpha \int I^V(t)dt + \alpha \int I^S(t) dt + T(\infty)$, where the integration is performed from $t = 0$ to $t \to \infty$. Because at equilibrium, the epidemic season is finished and thus $T(\infty) = 0$, we finally obtain $R(\infty) = \gamma \int I^V(t)dt + \gamma \int I^S(t)dt + \alpha \int I^V(t)dt + \alpha \int I^S(t)dt$, showing that $I_{\mathrm{VT}}^{\circ\mathrm{n}} = \gamma \int I^V(t)dt$, $I_{\mathrm{VT}}^{\mathrm{nn}} = \gamma \int I^S(t)dt$, $I_{\mathrm{VT}}^{\circ\circ} = \alpha \int I^V(t)dt$, and $I_{\mathrm{VT}}^{\mathrm{n}\circ} = \alpha \int I^S(t)dt$.

Based on the payoff-structure specification so far, an agent's total payoff is determined by their class membership, as summarized in Table 8.1. In addition, it is possible to calculate the average social payoff, $\langle \pi \rangle$, the expected payoff for a vaccinator, $\langle \pi_{\mathrm{C}} \rangle$, and the expected payoff for a non-vaccinator, $\langle \pi_{\mathrm{D}} \rangle$, as follows:

$$\langle \pi \rangle = -C_{\mathrm{V}} \cdot H_{\mathrm{V}}^{\circ} - (C_{\mathrm{V}} + C_{\mathrm{T}} + 1) \cdot I_{\mathrm{VT}}^{\circ\circ} - (C_{\mathrm{V}} + 1) \cdot I_{\mathrm{VT}}^{\circ\mathrm{n}} - (C_{\mathrm{T}} + 1) \cdot I_{\mathrm{VT}}^{\mathrm{n}\circ} - I_{\mathrm{VT}}^{\mathrm{nn}}, \tag{8.4}$$

$$\langle \pi_{\mathrm{C}} \rangle = \frac{1}{x} \left[-C_{\mathrm{V}} \cdot H_{\mathrm{V}}^{\circ} - (C_{\mathrm{V}} + C_{\mathrm{T}} + 1) \cdot I_{\mathrm{VT}}^{\circ\circ} - (C_{\mathrm{V}} + 1) \cdot I_{\mathrm{VT}}^{\circ\mathrm{n}} \right], \tag{8.5}$$

$$\langle \pi_{\mathrm{D}} \rangle = \frac{1}{1-x} \left[-(C_{\mathrm{T}} + 1) \cdot I_{\mathrm{VT}}^{\mathrm{n}\circ} - I_{\mathrm{VT}}^{\mathrm{nn}} \right]. \tag{8.6}$$

8.2.4 Strategy Updating and Global Dynamics

At the beginning of an epidemic season, after the previous epidemic cycle on the local timescale has unwound and reached an equilibrium, agents decide whether to vaccinate or not, thus updating their strategy. This update is based on imitating what works better than the agent's own strategy.

Following to the previous chapters, in the present model, we apply individual-based risk assessment (IB-RA) and strategy-based risk assessment (SB-RA), as introduced in Sect. 3.2.2. We utilize the mean-field approximation (MFA) and thoroughly presume the noise parameter in Fermi function in each state probability function κ of 0.1.

8.2.4.1 Individual-Based Risk Assessment (IB-RA)

In the present model, there are eighteen state-transition-probability functions:

$$P(H_V^{\circ} \leftarrow H_V^{n}) = \frac{1}{1 + \exp[-(0 - (-C_V))/\kappa]}, \tag{8.7a}$$

$$P(H_V^{\circ} \leftarrow I_{VT}^{n\circ}) = \frac{1}{1 + \exp[-((-C_T - 1) - (-C_V))/\kappa]}, \tag{8.7b}$$

$$P(H_V^{\circ} \leftarrow I_{VT}^{nn}) = \frac{1}{1 + \exp[-(-1 - (-C_V))/\kappa]}, \tag{8.7c}$$

$$P(H_V^{n} \leftarrow H_V^{\circ}) = \frac{1}{1 + \exp[-(-C_V - 0)/\kappa]}, \tag{8.7d}$$

$$P(H_V^{n} \leftarrow I_{VT}^{\circ\circ}) = \frac{1}{1 + \exp[-(-(C_V + C_T + 1) - 0)/\kappa]}, \tag{8.7e}$$

$$P(H_V^{n} \leftarrow I_{VT}^{\circ n}) = \frac{1}{1 + \exp[-((-C_V - 1) - 0)/\kappa]}, \tag{8.7f}$$

$$P(I_{VT}^{n\circ} \leftarrow H_V^{\circ}) = \frac{1}{1 + \exp[-(-C_V - (-C_T - 1))/\kappa]}, \tag{8.7g}$$

$$P(I_{VT}^{nn} \leftarrow H_V^{\circ}) = \frac{1}{1 + \exp[-(-C_V - (-1))/\kappa]}, \tag{8.7h}$$

$$P(I_{VT}^{\circ\circ} \leftarrow H_V^{n}) = \frac{1}{1 + \exp[-(0 - (-C_V - C_T - 1))/\kappa]}, \tag{8.7i}$$

$$P(I_{VT}^{\circ n} \leftarrow H_V^{n}) = \frac{1}{1 + \exp[-(0 - (-C_V - 1))/\kappa]}, \tag{8.7j}$$

$$P(I_{VT}^{\circ\circ} \leftarrow I_{VT}^{n\circ}) = \frac{1}{1 + \exp[-((-C_T - 1) - (-C_V - C_T - 1))/\kappa]}, \tag{8.7k}$$

$$P(I_{VT}^{\circ\circ} \leftarrow I_{VT}^{nn}) = \frac{1}{1 + \exp[-(-1 - (-C_V - C_T - 1))/\kappa]}, \tag{8.7l}$$

$$P(I_{VT}^{\circ n} \leftarrow I_{VT}^{n\circ}) = \frac{1}{1 + \exp[-((-C_T - 1) - (-C_V - 1))/\kappa]}, \tag{8.7m}$$

$$P(I_{VT}^{\circ n} \leftarrow I_{VT}^{nn}) = \frac{1}{1 + \exp[-(-1 - (-C_V - 1))/\kappa]}, \tag{8.7n}$$

$$P(I_{VT}^{n\circ} \leftarrow I_{VT}^{\circ\circ}) = \frac{1}{1 + \exp[-((-C_V - C_T - 1) - (-C_T - 1))/\kappa]}, \tag{8.7o}$$

$$P(I_{VT}^{nn} \leftarrow I_{VT}^{\circ\circ}) = \frac{1}{1 + \exp[-((-C_V - C_T - 1) - (-1))/\kappa]}, \tag{8.7p}$$

$$P\left(I_{\mathrm{VT}}^{\mathrm{n}\circ} \leftarrow I_{\mathrm{VT}}^{\circ\mathrm{n}}\right) = \frac{1}{1 + \exp\left[-\left(\left(-C_{\mathrm{V}} - 1\right) - \left(-C_{\mathrm{T}} - 1\right)\right)/\kappa\right]}, \tag{8.7q}$$

$$P\left(I_{\mathrm{VT}}^{\mathrm{nn}} \leftarrow I_{\mathrm{VT}}^{\circ\mathrm{n}}\right) = \frac{1}{1 + \exp\left[-\left(\left(-C_{\mathrm{V}} - 1\right) - (-1)\right)/\kappa\right]}. \tag{8.7r}$$

At the end of each epidemic season, each individual is allowed to update his strategy depending upon the last season's payoff. The fraction of vaccinators, $x = H_{\mathrm{V}}^{\circ}(x) + I_{\mathrm{VT}}^{\circ\circ}(x) + I_{\mathrm{VT}}^{\circ\mathrm{n}}(x)$, increases whenever there is a net tendency among agents to imitate one of the vaccinator classes H_{V}°, $I_{\mathrm{VT}}^{\circ\mathrm{n}}$, and $I_{\mathrm{VT}}^{\circ\circ}$. Conversely, the variable x decreases whenever agents predominantly imitate one of the non-vaccinator classes, i.e., $H_{\mathrm{V}}^{\mathrm{n}}$, $I_{\mathrm{VT}}^{\mathrm{n}\circ}$, and $I_{\mathrm{VT}}^{\mathrm{nn}}$. In the case of individual-based risk assessment (IB-RA), the total number of possibilities is 18, because a member of each of the three vaccinator classes can imitate any of the three non-vaccinator classes and vice versa. Accordingly, we have

$$\begin{aligned}
\frac{\mathrm{d}x}{\mathrm{d}t} &= H_{\mathrm{V}}^{\mathrm{n}}(x)H_{\mathrm{V}}^{\circ}(x)\left[\mathrm{P}\left(H_{\mathrm{V}}^{\mathrm{n}} \leftarrow H_{\mathrm{V}}^{\circ}\right) - \mathrm{P}\left(H_{\mathrm{V}}^{\circ} \leftarrow H_{\mathrm{V}}^{\mathrm{n}}\right)\right] \\
&+ I_{\mathrm{VT}}^{\mathrm{n}\circ}(x)H_{\mathrm{V}}^{\circ}(x)\left[\mathrm{P}\left(I_{\mathrm{VT}}^{\mathrm{n}\circ} \leftarrow H_{\mathrm{V}}^{\circ}\right) - \mathrm{P}\left(H_{\mathrm{V}}^{\circ} \leftarrow I_{\mathrm{VT}}^{\mathrm{n}\circ}\right)\right] \\
&+ I_{\mathrm{VT}}^{\mathrm{nn}}(x)H_{\mathrm{V}}^{\circ}(x)\left[\mathrm{P}\left(I_{\mathrm{VT}}^{\mathrm{nn}} \leftarrow H_{\mathrm{V}}^{\circ}\right) - \mathrm{P}\left(H_{\mathrm{V}}^{\circ} \leftarrow I_{\mathrm{VT}}^{\mathrm{nn}}\right)\right] \\
&+ H_{\mathrm{V}}^{\mathrm{n}}(x)I_{\mathrm{VT}}^{\circ\circ}(x)\left[\mathrm{P}\left(H_{\mathrm{V}}^{\mathrm{n}} \leftarrow I_{\mathrm{VT}}^{\circ\circ}\right) - \mathrm{P}\left(I_{\mathrm{VT}}^{\circ\circ} \leftarrow H_{\mathrm{V}}^{\mathrm{n}}\right)\right] \\
&+ H_{\mathrm{V}}^{\mathrm{n}}(x)I_{\mathrm{VT}}^{\circ\mathrm{n}}(x)\left[\mathrm{P}\left(H_{\mathrm{V}}^{\mathrm{n}} \leftarrow I_{\mathrm{VT}}^{\circ\mathrm{n}}\right) - \mathrm{P}\left(H_{\mathrm{V}}^{\mathrm{n}} \leftarrow I_{\mathrm{VT}}^{\circ\mathrm{n}}\right)\right] \\
&+ I_{\mathrm{VT}}^{\mathrm{n}\circ}(x)I_{\mathrm{VT}}^{\circ\circ}(x)\left[\mathrm{P}\left(I_{\mathrm{VT}}^{\mathrm{n}\circ} \leftarrow I_{\mathrm{VT}}^{\circ\circ}\right) - \mathrm{P}\left(I_{\mathrm{VT}}^{\circ\circ} \leftarrow I_{\mathrm{VT}}^{\mathrm{n}\circ}\right)\right] \\
&+ I_{\mathrm{VT}}^{\mathrm{nn}}(x)I_{\mathrm{VT}}^{\circ\circ}(x)\left[\mathrm{P}\left(I_{\mathrm{VT}}^{\mathrm{nn}} \leftarrow I_{\mathrm{VT}}^{\circ\circ}\right) - \mathrm{P}\left(I_{\mathrm{VT}}^{\circ\circ} \leftarrow I_{\mathrm{VT}}^{\mathrm{nn}}\right)\right] \\
&+ I_{\mathrm{VT}}^{\mathrm{n}\circ}(x)I_{\mathrm{VT}}^{\circ\mathrm{n}}(x)\left[\mathrm{P}\left(I_{\mathrm{VT}}^{\mathrm{n}\circ} \leftarrow I_{\mathrm{VT}}^{\circ\mathrm{n}}\right) - \mathrm{P}\left(I_{\mathrm{VT}}^{\circ\mathrm{n}} \leftarrow I_{\mathrm{VT}}^{\mathrm{n}\circ}\right)\right] \\
&+ I_{\mathrm{VT}}^{\mathrm{nn}}(x)I_{\mathrm{VT}}^{\circ\mathrm{n}}(x)\left[\mathrm{P}\left(I_{\mathrm{VT}}^{\mathrm{nn}} \leftarrow I_{\mathrm{VT}}^{\circ\mathrm{n}}\right) - \mathrm{P}\left(I_{\mathrm{VT}}^{\circ\mathrm{n}} \leftarrow I_{\mathrm{VT}}^{\mathrm{nn}}\right)\right],
\end{aligned} \tag{8.8}$$

where the notation CLASS(x) explicitly specifies that the fraction of agents belonging to a certain class depends on x, whereas products of the form $\mathrm{CLASS}_1(x) \cdot \mathrm{CLASS}_2(x)$ are probabilities that any two individual members of the two classes enter into contact for imitation to be possible.

8.2.4.2 Strategy-Based Risk Assessment (SB-RA)

In the present model, SB-RA requires the following six cases:

$$P\left(H_{\mathrm{V}}^{\circ} \leftarrow \mathrm{D}\right) = \frac{1}{1 + \exp\left[-\left(\langle \pi_{\mathrm{D}} \rangle - (-C_{\mathrm{V}})\right)/\kappa\right]}, \tag{8.9a}$$

$$P\left(I_{\mathrm{VT}}^{\circ\circ} \leftarrow \mathrm{D}\right) = \frac{1}{1 + \exp\left[-(\langle \pi_{\mathrm{D}} \rangle - (-(C_{\mathrm{V}} + C_{\mathrm{T}} + 1)))/\kappa\right]}, \tag{8.9b}$$

$$P\left(I_{\mathrm{VT}}^{\circ\mathrm{n}} \leftarrow \mathrm{D}\right) = \frac{1}{1 + \exp\left[-(\langle \pi_{\mathrm{D}} \rangle - (-C_{\mathrm{V}} - 1))/\kappa\right]}, \tag{8.9c}$$

$$P\left(H_{\mathrm{V}}^{\mathrm{n}} \leftarrow \mathrm{C}\right) = \frac{1}{1 + \exp\left[-(\langle \pi_{\mathrm{C}} \rangle - 0\,)/\kappa\right]}, \tag{8.9d}$$

$$P\left(I_{\mathrm{VT}}^{\mathrm{n}\circ} \leftarrow \mathrm{C}\right) = \frac{1}{1 + \exp\left[-(\langle \pi_{\mathrm{C}} \rangle - (-C_{\mathrm{T}} - 1)\,)/\kappa\right]}, \tag{8.9e}$$

$$P\left(I_{\mathrm{VT}}^{\mathrm{nn}} \leftarrow \mathrm{C}\right) = \frac{1}{1 + \exp\left[-(\langle \pi_{\mathrm{C}} \rangle - (-1))/\kappa\right]}. \tag{8.9f}$$

In the case of society-based risk assessment (SB-RA), any vaccinator (resp., non-vaccinator) compares their performance with the average performance of non-vaccinators (resp., vaccinators) as a whole. This leaves us with only the following six possibilities:

$$\begin{aligned}\frac{\mathrm{d}x}{\mathrm{d}t} = &-H_{\mathrm{V}}^{\circ}(x)\mathrm{D}(x)\mathrm{P}\left(H_{\mathrm{V}}^{\circ} \leftarrow \mathrm{D}\right) - I_{\mathrm{VT}}^{\circ\circ}(x)\mathrm{D}(x)\mathrm{P}\left(I_{\mathrm{VT}}^{\circ\circ} \leftarrow \mathrm{D}\right)\\ &- I_{\mathrm{VT}}^{\circ\mathrm{n}}(x)\mathrm{D}(x)\mathrm{P}\left(I_{\mathrm{VT}}^{\circ\mathrm{n}} \leftarrow \mathrm{D}\right)\\ &+H_{\mathrm{V}}^{\mathrm{n}}(x)\mathrm{C}(x)\mathrm{P}\left(H_{\mathrm{V}}^{\mathrm{n}} \leftarrow \mathrm{C}\right) + I_{\mathrm{VT}}^{\mathrm{n}\circ}(x)\mathrm{C}(x)\mathrm{P}\left(I_{\mathrm{VT}}^{\mathrm{n}\circ} \leftarrow \mathrm{C}\right)\\ &+ I_{\mathrm{VT}}^{\mathrm{nn}}(x)\mathrm{C}(x)\mathrm{P}\left(I_{\mathrm{VT}}^{\mathrm{nn}} \leftarrow \mathrm{C}\right),\end{aligned} \tag{8.10}$$

where $\mathrm{C}(x) = x$ is the fraction of all vaccinators and $\mathrm{D}(x) = 1 - x$ is the fraction of all non-vaccinators. C and D stand for cooperators and defectors, respectively.

8.2.5 Utility of Treatment

Initially, we introduced the antiviral-treatment-probability rate, α, and the accelerated-recovery rate due to treatment, δ, as independent parameters. If, however, δ were much larger than the natural recovery rate γ, the preference for treatment should also be much higher than if δ were only marginally higher than γ. Put more explicitly, δ^{-1} is the average number of days for recovery under treatment; γ^{-1} is the average number of days to unaided (i.e., natural) recovery. If $\gamma^{-1} - \delta^{-1} \gg 0$, then the utility of treatment is very high, and treatment should be a highly sought-after option. Parameter α should therefore be a function of the difference $\gamma^{-1} - \delta^{-1}$. The exact form of this functional dependence is unknown, but as with payoff differences, decision making under a utility difference is often captured using the smoothed best response. In our case, this is the Fermi pairwise rule:

$$\alpha = \frac{\omega}{1 + \exp\left[-\frac{\gamma - \delta^{-1}}{\kappa}\right]}, \tag{8.11}$$

where ω is the maximum antiviral-treatment-probability rate achieved when treatment dramatically accelerates recovery. Of note is that in the limit $\kappa \to 0$, Eq. (8.11) turns into a threshold rule, with $\alpha = \omega$ if $\gamma^{-1} > \delta^{-1}$ and $\alpha = 0$ otherwise. Values $\kappa > 0$ smooth the threshold rule into a sigmoidal function recognizable as the smoothed best response, which allows the selection of an inferior (in terms of payoff or utility) option with non-zero probability. Parameter κ is therefore often called the strength of irrational selection. As mentioned above, we set $\kappa = 0.1$.

8.3 Result and Discussion

In this section, we explore the SVITR model, first on its own and then coupled with evolutionary dynamical equations. Quantities of interest include vaccination coverage (VC), treatment adoption in terms of the fraction of treated agents during an epidemic season, final epidemic size (FES), and average social payoff (ASP) as a measure of policy burden to society. Aside from the already mentioned $\kappa = 0.1$, the parameters have the following default values unless specified otherwise: $\beta = 2.5/3$, $\alpha = 0.1$, $\gamma = 1/3$ ($R_0 = \beta/\gamma = 2.5$), $\delta = 0.5$, and where applicable $\omega = 0.1$.

8.3.1 SVITR Dynamics

Setting $\alpha = \delta = 0$ and assuming $x = e = 0.5$ recover the SVITR model into a traditional SVIR model whose outputs serve as a benchmark (Fig. 8.2 (a)). Reintroducing the option of treatment with $\alpha = \delta = 0.1$ considerably reduces the peak fraction of infectious agents in comparison to this benchmark while slightly lengthening the epidemic's duration (Fig. 8.2 (a)). Increasing the recovery rate under treatment from $\delta = 0.1$ to $\delta = 0.6$ has no bearing upon the dynamics of the infectious agents (Fig. 8.2 (a)), but it does decrease the peak fraction of treated agents (not shown), which is important in practice not to overwhelm healthcare institutions and medical resources. Of note is that it makes sense to consider situations with $\delta < \gamma$, despite the greatly diminished utility of a treatment prolonging recovery, because the peak fraction of infectious agents still gets reduced simply by diverting some of them to treatment. Doubling the treatment-probability rate to $\alpha = 0.2$ suppresses the peak fraction of infectious agents approximately 4.5-fold (Fig. 8.2 (a)), suggesting that eventually there may be no epidemic at all. This indeed transpires at $\alpha = 0.3$, when $R_c \approx 0.99 < 1$ (not shown).

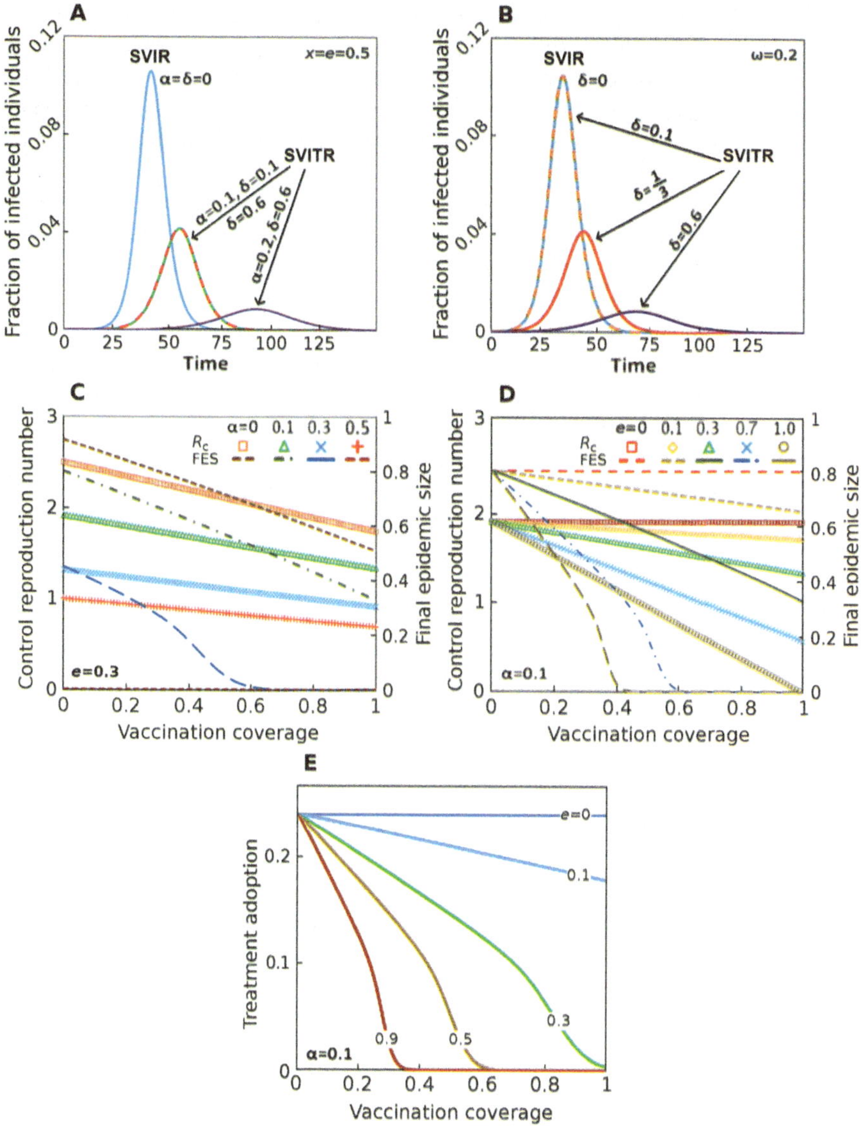

Fig. 8.2 SITR/V dynamics. (**a**, **b**), Progression of an epidemic season without and with Eq. (8.11) implemented, respectively. (**c**) and (**d**) control the reproduction number and final epidemic size as functions of the vaccination coverage for various treatment-probability rates and vaccine effectivenesses, respectively. (**e**), Treatment adoption as a function of vaccination coverage for various vaccine effectivenesses. Parameter values are $\beta = 2.5/3$ and $\gamma = 1/3$, with other values specified in the plots. See the accompanying description in the main text for details

When the utility of treatment, as prescribed by Eq. (8.11), is added to the mix, we see (in Fig. 8.2 (b)) that a sufficiently prolonged recovery (i.e., $\delta \ll \gamma$) leads to ignoring the treatment option (i.e., $\alpha = 0$), unaffected recovery (i.e., $\delta \approx \gamma$) leads to adoption of treatment at half the maximum probability rate (i.e., $\alpha = \omega/2$), and sufficiently accelerated recovery (i.e., $\delta \gg \gamma$) leads to maximum treatment adoption (i.e., $\alpha = \omega$).

In particular, the larger the treatment-probability rate, the lower the control-reproduction number and ultimately the FES (Fig. 8.2 (c)). This holds for any given VC, but the full synergistic effect of vaccination and treatment occurs when the control-reproduction number is pushed below unity, in which case the epidemic is avoided altogether (Fig. 8.2 (c)).

As with a larger treatment-probability rate, an increasing vaccine effectiveness also lowers the control-reproduction number and thus the FES although these positive effects are highly dependent upon the VC (Fig. 8.2 (d)).

Finally, the need for treatment diminishes with increasing VC, and may even be completely eliminated if the population attains herd immunity (Fig. 8.2 (e)). The latter, however, is only possible if vaccine effectiveness is sufficiently high (Fig. 8.2 (e)).

8.3.2 *Interplay Between Vaccination and Treatment Costs*

Figure 8.3 displays the VC drawn on a 2D plane of treatment cost vs vaccination for the two different strategy-updating rules. At a glance, we note that the general tendency observed in different update rules is consistent although a more detailed discussion is delivered in Sect. 8.3.3. Interestingly, when the vaccine effectiveness is low, people respond to vaccination in a binary manner that is determined by its cost, whereby everyone vaccinates if vaccine is cheap while no one vaccinates if it is expensive; and the cost of treatment has a relatively small effect upon the VC.

The people's response to vaccination cost becomes much more gradual as effectiveness increases, with the influence of treatment cost being much more pronounced at intermediate vaccine prices and effectivenesses. An improvement in the vaccine reliability causes an individual to deliberate on whether to take it or not, as it is not a simple question of whether a vaccine is cheap, but a more complex one influenced by the treatment cost as well as what probability of being infected in the next season. However, with a high vaccination effectiveness, the sensitivity to treatment cost becomes less significant again. This is because the question to an individual returns to whether or not a vaccination cost can be estimated below an acceptable level in relation with their expectation of being infected, which causes the manifest color gradation in the direction of vaccination cost, unlike the binary tendency observed when $e = 0.1$.

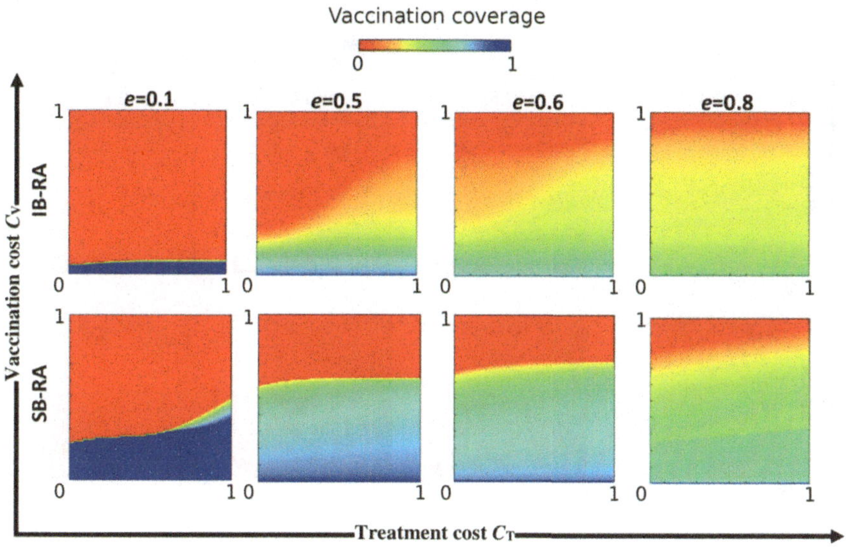

Fig. 8.3 Vaccination coverage is primarily controlled by vaccine cost and effectiveness. Shown is the vaccination coverage as a function of relative vaccination cost C_V and relative treatment cost C_T. The most expensive vaccines are rejected, even if they are very efficacious. Interestingly, cheaper vaccines may achieve less coverage with an increasing effectiveness, but this is because they better suppress outbreaks. Also interesting to note is that costlier treatments increase the vaccination coverage when the vaccine is rightly priced and sufficiently efficacious. The individual-based risk assessment (IB-RA; top row) generally leads to a somewhat lower vaccination coverage compared to the society-based risk assessment (SB-RA; bottom row). Parameters used are $\beta = 2.5/3$, $\alpha = 0.1$, $\gamma = 1/3$, and $\delta = 0.5$, while effectiveness improves from e=0.1 (leftmost column) to $e = 0.8$ (rightmost column)

Figures 8.4, 8.5, and 8.6 display treatment adoption, FES, and ASP drawn in the same manner as Fig. 8.3. Looking beyond VC reveals that treatment is secondary to vaccination in the sense that the vaccination cost, rather than the treatment cost, largely controls the adoption of treatment (Fig. 8.4). Simply put, the most expensive vaccines are avoided regardless of efficiency, and eventually the fraction of infected agents who receive treatment during the epidemic season is maximized (Fig. 8.5). For more reasonable cost-effectiveness combinations being assumed for vaccination, however, cheap and expensive treatments alike start to give way to vaccination (Fig. 8.4). A direct consequence of all this is that the primary determinants of the FES are vaccine characteristics. Specifically, the avoidance of expensive vaccines leads to a maximum FES that is kept in check only by the adoption of treatment (Fig. 8.5). The situation improves as vaccines become cheaper, with effectiveness being the primary determinant of how cheap is cheap enough (Fig. 8.5). The dependence of FES upon the cost of treatment is mostly weak. The burden to society in terms of the ASP is even less dependent upon the treatment cost, with this dependence being more pronounced when vaccines are expensive and almost

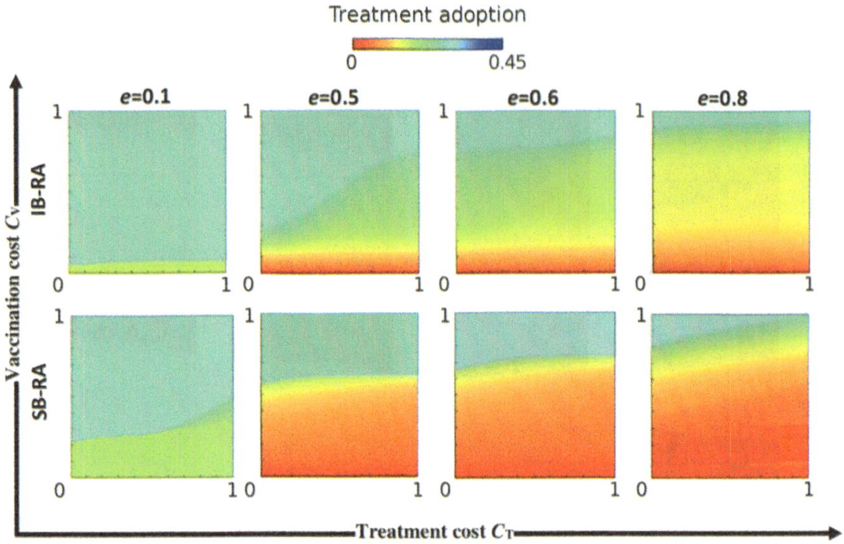

Fig. 8.4 Treatment is secondary to vaccination. Shown is the treatment adoption as a function of relative vaccination cost (C_V) and relative treatment cost (C_T). In the expensive-vaccine limit, 30% of infected individuals eventually get treated, regardless of the cost. As a vaccine becomes affordable, however (implying the right combination of cost and effectiveness), treatment is abandoned in favor of vaccination. These plots are therefore a mirror image of the plots in Fig. 8.3, and the individual-based risk assessment (IB-RA; top row) generally leads to a somewhat higher treatment adoption compared with society-based risk assessment (SB-RA; bottom row). The color bar is synchronized with Fig. 8.9. The parameters used are $\beta = 2.5/3$, $\alpha = 0.1$, $\gamma = 1/3$, and $\delta = 0.5$, while the effectiveness improves from $e = 0.1$ (leftmost column) to e=0.8 (rightmost column)

non-existent when vaccines are cheap (Fig. 8.6). An interesting consequence is that when expensive treatment incentivizes vaccination, it does so at relatively little cost to society.

8.3.3 Individual-Versus Society-Centered Decision Making

Judging from the phase diagrams in Figs. 8.3, 8.4, 8.5 and 8.6, a general impression is that society-based risk assessment (SB-RA) is advantageous over individual-based risk assessment (IB-RA), but the truth is more complicated. Looking closely into this issue, we find that when vaccines are cheap, SB-RA indeed supports a wider VC, and thus also a smaller FES (Fig. 8.7). However, as vaccines become more expensive, and especially if their effectiveness is high as well, it is IB-RA that supports a wider VC, and thus also a smaller FES (Fig. 8.7). The reason for this is that the SB-RA subdues contrarian decisions relative to the IB-RA. For example, in a

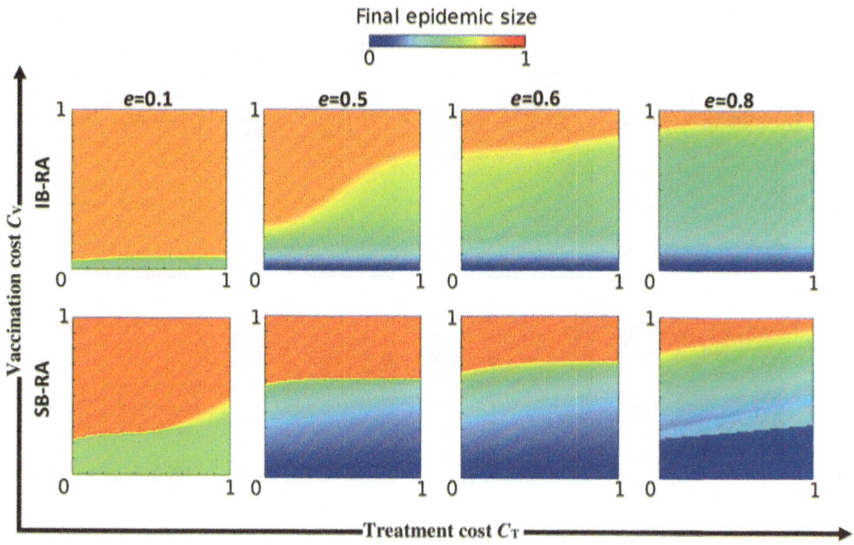

Fig. 8.5 Outbreak size mirrors the vaccination coverage. Put another way, the final epidemic size is largely controlled by the coverage, and thus vaccine cost and effectiveness. Shown is the final epidemic size as a function of relative vaccination cost, C_V, and relative treatment cost, C_T. A costlier treatment suppresses outbreaks by turning agents proactive instead of reactive, but this works only if the vaccine is rightly priced and sufficiently efficacious. The individual-based risk assessment (IB-RA; top row) generally leads to somewhat poorer outcomes compared with society-based risk assessment (SB-RA; bottom row). The parameters used are $\beta = 2.5/3$, $\alpha = 0.1$, $\gamma = 1/3$, and $\delta = 0.5$, while the effectiveness improves from $e = 0.1$ (leftmost column) to $e = 0.8$ (rightmost column)

population dominated by vaccinators (implying cheaper vaccines), a lone non-vaccinator is less likely to get infected and may fare above average by refusing to vaccinate, thus creating a strong incentive to imitate this behavior under IB-RA, ultimately causing a lower VC than under SB-RA. In a population dominated by non-vaccinators, it is a lone vaccinator who is likely to fare above average, thus reversing the outcome. The described distinction between IB-RA and SB-RA, as discussed later, contains important implications for the popularity of present-day anti-vaccination movements.

8.3.4 Interplay Between Vaccine and Treatment Characteristics

Aside from costs, it is worthwhile to emphasize how other parameters conspire to control outbreaks. Increasing the treatment-probability rate α first leads to a higher treatment adoption; but as treatment becomes more widely administered and the

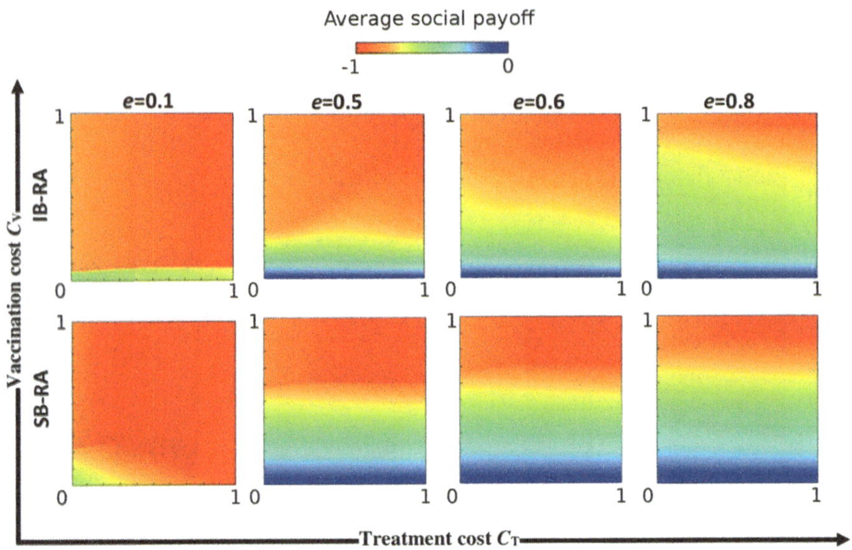

Fig. 8.6 Burden to society is weakly sensitive to the treatment cost. Shown is the average social payoff as a function of the relative vaccination cost (C_V) and relative treatment cost (C_T). Plots reveal a relatively small decrease in the average social payoff, even in the limit of expensive treatment. The individual-based risk assessment (IB-RA; top row) is generally somewhat more burdensome than the society-based risk assessment (SB-RA; bottom row). Parameters used are $\beta = 2.5/3$, $\alpha = 0.1$, $\gamma = 1/3$, and $\delta = 0.5$, while the effectiveness improves from $e = 0.1$ (leftmost column) to $e = 0.8$ (rightmost column)

epidemic is better controlled, further increasing α only decreases the treatment adoption until the epidemic is fully eradicated (left panels, Fig. 8.8). Interestingly, the results are insensitive to vaccine effectiveness if e is too low. Here, the precise meaning of "too low" depends upon the vaccination cost (left panels; Fig. 8.8). Moreover, if the utility of the drug is considered as prescribed by Eq. (8.11), treatment that prolongs recovery ($\delta \ll \gamma$) is ignored, whereas treatment that shortens recovery ($\delta \gg \gamma$) is adopted as much as is allowed by the vaccine effectiveness (right panels; Fig. 8.8).

The effects of the vaccine and treatment characteristics upon the FES only partly mirror those upon treatment adoption described above, and in fact reveal further complexities. In particular, increasing the treatment-probability rate α gradually reduces FES and even eradicates the disease when treatment is administered widely enough (left panels; Fig. 8.9). As with treatment adoption, an overly low vaccine effectiveness is inconsequential, but what constitutes "overly low" is decided by the vaccination cost (left panels; Fig. 8.9). With a highly effective vaccine available, we would expect an FES-reducing synergy between the vaccine and the treatment. It turns out, however, that the margin for such synergy is very small, and in most instances, treatment ends up interfering with the vaccine's ability to control FES (left panels; Fig. 8.9).

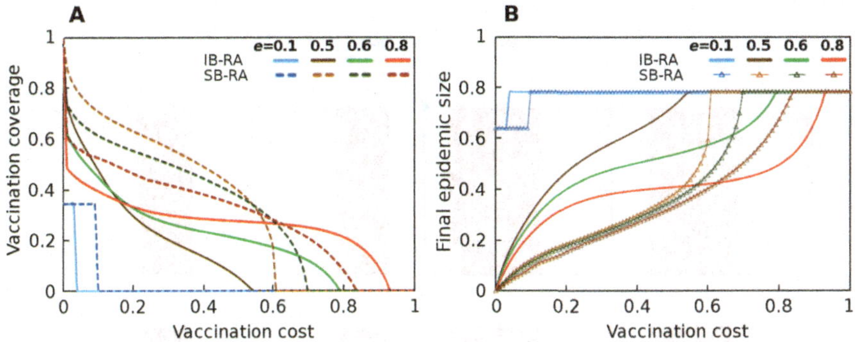

Fig. 8.7 Society-based risk assessment resists contrarian decisions. When vaccines are cheap and there are many vaccinators, a lone non-vaccinator is less likely to become infected, and may fare above average by taking a contrarian stance. In such circumstances, the individual-based risk assessment (IB-RA) provides a local signal to the surrounding vaccinators to imitate the successful non-vaccinator, with the overall result being a lower vaccination coverage and a larger final epidemic size than with society-based risk assessment (SB-RA). Reverse reasoning holds when vaccines are expensive and there are many non-vaccinators, in which case a lone vaccinator is more likely to avoid infection, and thus fare above average by taking a contrarian stance. The overall result is then a wider vaccination coverage and a smaller final infection size with IB-RA than SB-RA. (**a**) and (**b**) show the vaccination coverage and the final epidemic size, respectively, under IB-RA and SB-RA as a function of the relative vaccination cost C_V for a range of vaccine efficacies e. The parameters used are $\beta = 2.5/3$, $\alpha = 0.1$, $\gamma = 1/3$, and $C_T = 0.5$

Let us turn to the right panels, where the utility of the drug is accounted for via Eq. (8.11). Treatment that shortens recovery indeed somehow helps to decrease FES (right panels, Fig. 8.9). This interference of treatment with vaccination is more clearly illustrated in Fig. 8.10. The mechanism at work here is that treatment initially acts to reduce the FES, but as evolutionary time progresses, the restricted seasonal spread of the disease prompts some agents to stop vaccinating. The end result is that FES is larger than it would have been without the treatment option (see the caption of Fig. 8.10).

8.3.5 *Comprehensive Discussion*

We have largely focused upon the situation in which the treatment-probability rate is moderate, meaning that the availability of treatment considerably reduces the final epidemic size, but is insufficient to fully eliminate the epidemic. Put more technically, the control-reproduction number is considerably lower than the basic reproduction number, but still above unity. This situation is, in fact, most interesting from an epidemiological perspective, because excessive use of drugs hastens the evolution of resistance; furthermore, there may be technological limitations upon drug availability even when the price of the drug is not an issue. In this context, we found that treatment indeed takes a back seat to vaccination, because the cost of the latter, in

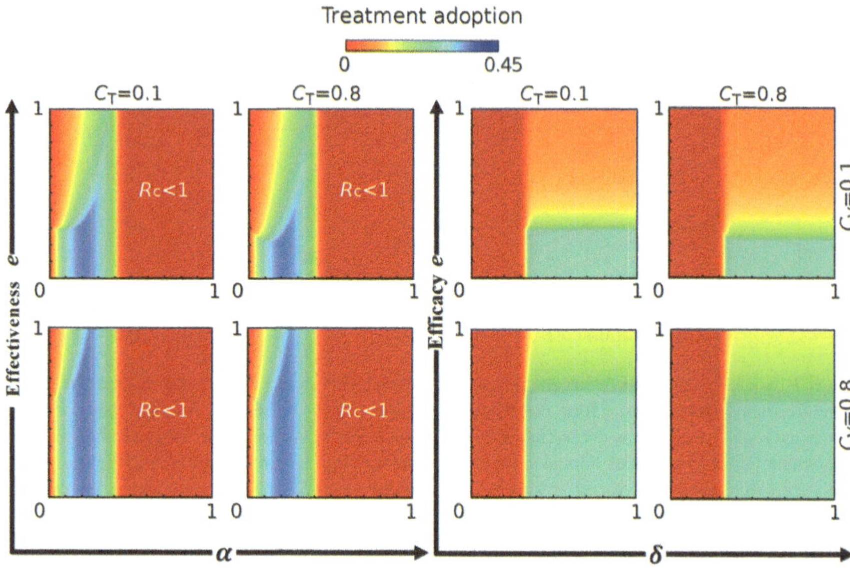

Fig. 8.8 How vaccine and treatment characteristics control the treatment adoption. When Eq. (8.11) is not implemented (i.e., α is independent of δ; left panels), increasing the treatment-probability rate α first leads to wider treatment adoption; however, as the epidemic is more effectively suppressed, treatment adoption begins to decrease. At sufficiently high values of parameter α ($\alpha > 0.4$), the control-reproduction number falls below unity and the epidemic is eradicated. Interestingly, the results are almost independent of the cost of treatment, yet vaccine characteristics matter. When effectiveness is low, parameter α alone controls treatment adoption, but with a sufficiently high effectiveness, fewer treatments are necessary. How much "high enough" depends upon the vaccination cost. When Eq. (8.11) is implemented (i.e., α is related to δ, right panels), treatments that prolong recovery are ignored, whereas treatments that shorten recovery are adopted equally, regardless of their cost. Instead, the vaccine effectiveness and vaccination cost are more important. The parameters used are $\beta = 2.5/3$ and $\gamma = 1/3$, while $\delta = 0.5$ in the left panels and $\omega = 0.1$ in the right panels

conjunction with effectiveness, primarily determines VC, and treatment adoption. The FES is consequently also much more sensitive to the vaccination cost and effectiveness than to the treatment cost. The most important effect of the treatment cost is that expensive treatment creates an incentive to resort to vaccination, especially for the right vaccine cost-effectiveness combination. Because the consequent increase in social burden is small, in situations with both quality vaccines and treatments, it makes sense to incentivize vaccination with higher treatment prices, especially if doing so can prolong the evolution of drug resistance. This is further justified by the narrow margin for truly synergistic effects of vaccines and treatment in suppressing the FES.

Present-day society is facing the emergence of mistrust toward vaccines, often centered around influential public figures who express skepticism against vaccination. It is interesting in this context that the individual-based risk assessment (IB-RA) is more prone to succumbing to such contrarian views than the society-

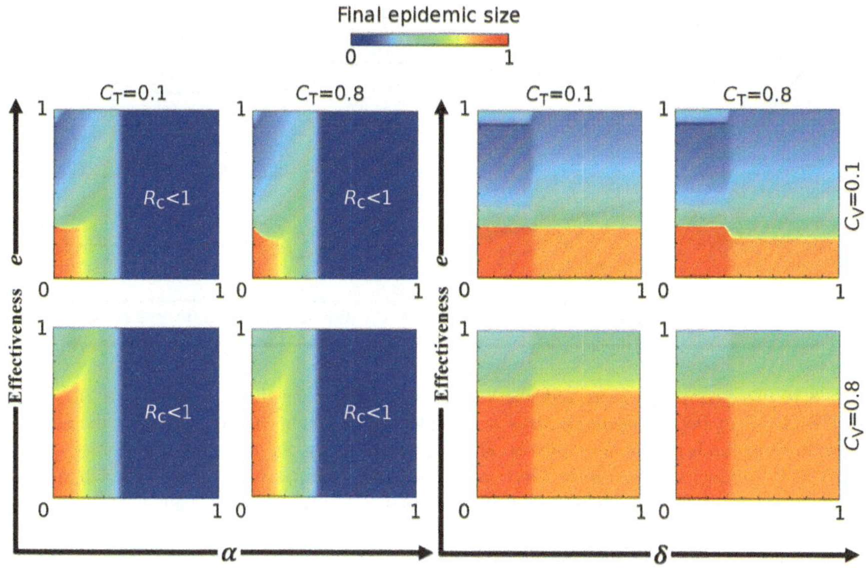

Fig. 8.9 How vaccine and treatment characteristics control the final epidemic size. When Eq. (8.11) is not implemented (i.e., α is independent of δ; left panels), increasing treatment probability α eventually leads to eradication of the disease ($\alpha > 0.4$). Vaccines with overly low effectiveness have no bearing upon the final epidemic size (FES), but what "overly low" means is decided by the vaccination cost. Interestingly, directing more agents to treatment interferes with the ability of a highly efficacious vaccine to control the epidemic, thus actually causing FES to increase alongside parameter α until treatment is administered often enough to overwhelm the disease. When Eq. (8.11) is implemented (i.e., α is related to δ, right panels), treatments that prolong recovery are ignored along with vaccines with overly low effectiveness, allowing the disease to spread freely. Again, what "overly low" means is decided by the vaccination cost. Treatments that shorten recovery are indeed adopted, but they reduce FES only when the vaccine effectiveness is low, and end up increasing FES when effectiveness is high. The parameters used are $\beta = 2.5/3$ and $\gamma = 1/3$, while $\delta = 0.5$ in the left panels and $\omega = 0.1$ in the right panels

based risk assessment (SB-RA). In particular, when vaccine coverage is high, individuals who refuse vaccination are protected by others, and thus fare very well by getting protection for free. Seeing no downside for non-vaccinators, IB-RA quickly leads to imitation of this behavior. SB-RA, however, implies a modicum of "collective memory," whereby the harm that may befall non-vaccinators is more difficult to ignore. The question, therefore, is how to reinforce this collective memory enough to guide health-related decisions that may save lives.

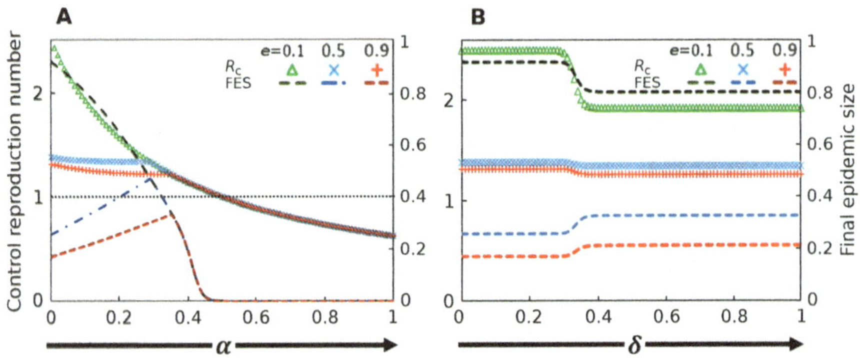

Fig. 8.10 Treatment interferes with the vaccine's control of the final epidemic size. (**a**), when Eq. (8.11) is not implemented (i.e., α is independent of δ), treatment helps to lower the control-reproduction number (R_c). In terms of FES, however, when the treatment-probability rate is moderate, a reduced disease prevalence prompts some vaccinators to forgo vaccination, thus actually increasing FES in comparison to the no-treatment setting. (**b**), the same mechanism is at work even when Eq. (8.11) is implemented (i.e., α is related to δ). Here, treatment adoption helps to reduce FES only in the case of a low-effectiveness vaccine. If the vaccine is effective enough, introducing treatment at a moderate probability rates worsens the disease prevalence. The parameters used are $\beta = 2.5/3$ and $\gamma = 1/3$, while the relative costs of vaccination and treatment are $C_V = C_T = 0.1$, respectively

References

Alexander ME, Bowman CS, Feng Z, Gardam N, Moghadas SM, Röst G, Wu J, Yan P (2007) Emergence of drug resistance: implications for antiviral control of pandemic influenza. Proc R Soc B 274:1675–1684

Kabir KMA, Jusup M, Tanimoto J (2019) Behavioral incentives in a vaccination-dilemma setting with optional treatment. Phys Rev E 100:062402

Regoes RR, Bonhoeffer S (2006) Emergence of drug-resistant influenza virus: Population dynamical considerations. Science 312:389–391

van den Driessche P, Watmough J (2008) Further notes on the basic reproduction number. In: Brauer F, van den Driessche P, Wu J (eds) Mathematical epidemiology. Lecture notes in mathematics, vol 1945. Springer, Berlin, Heidelberg

Chapter 9
Pre-emptive Vaccination Versus Late Vaccination

The conventional vaccination game always assumed pre-emptive vaccination, by which an individual decides whether or not to participate in a vaccination game prior to an epidemic season. This chapter develops a new vaccination-game framework where not only pre-emptive vaccination but also late vaccination is available, which is more realistic. The two-layer vaccinating opportunity allows the coevolution of individual behavior around vaccination to avoid infection. Based on the potential risk of infection that is influenced by how many people vaccinate while being infected, late vaccination allows an individual who is not taking a pre-emptive vaccination to participate in a literally "late-timed vaccination" at any arbitrary moment in an epidemic season. The theoretical model presented in this chapter is

J. Tanimoto, *Sociophysics Approach to Epidemics*, Evolutionary Economics and Social Complexity Science 23, https://doi.org/10.1007/978-981-33-6481-3_9

validated by the results obtained through multi-agent simulation (MAS). Furthermore, this chapter quantitatively demonstrates that the cost advantage of late vaccination over pre-emptive vaccination enables an individual to prefer late vaccination, and lower vaccine effectiveness compels an individual who previously intended to free ride to choose late vaccination.

9.1 Introduction and Background: Is Pre-Emptive or Late Vaccination More Beneficial?

As noted above, most previous works targeting a seasonal epidemic like influenza have assumed that pre-emptive vaccination is one of the most effective intervention provisions to subdue the spread of an epidemic. It is very true, however, as we have elucidated for the last several chapters, that pre-emptive vaccination inevitably evokes the so-called vaccination dilemma, which hampers surging vaccination coverage because there is strong incentive to individuals to free ride on herd immunity. By considering pre-emptive vaccination, the vaccination opportunity is limited to the beginning of each epidemic season. This may be justified in models mainly concerned with investigating the number of pre-emptively vaccinated people that is sufficient at a social-equilibrium point, but the following question has been ignored in the previous models: what extent of infection would influence people to participate in vaccination after the outbreak of an epidemic? In fact, such people have been regarded as non-vaccinators or free riders in previous studies, thus causing their overestimation during the outbreak of an epidemic, since only pre-emptive vaccination is allowed. In real situations, some of them can participate in "late vaccination" during a season. Some recently conducted studies[1] have tried to model such delayed and dynamic vaccination behavior, but none has successfully integrated the dynamics of late vaccinators' behavior with that of the pre-emptive vaccinators.

Hence, this chapter attempts to reproduce the situation in which both pre-emptive and late vaccination are allowed for an individual as possible provisions to avoid infection, and to evaluate its impact upon the final epidemic size (FES). Below, we utilize the term "late vaccination" to distinguish such dynamic vaccination from pre-emptive vaccination.

---

[1]Representative examples include
Liqun and Yanfeng (2019), p. 122032.
Alvarez-Zuzek et al. (2019), p. 012302.

9.2 Model Structure

The present model obeys the standard vaccination game where the local timescale controls how an epidemic (implicitly seasonal influenza) spreads, while the global timescale stipulates how an individual's strategy—either pre-emptively vaccinating or not—evolves over repeated seasons. However, a novel feature is that there is an option available to individuals who did not take pre-emptive vaccination at the beginning of a season to take late vaccination at an intermediate time during the season. In the following theoretical model, we presume a well-mixed and infinite population and apply the mean-field approximation (MFA); thus, the sum of all compartments remains unity, as the previous chapters presumed.

9.2.1 Formulation of the Dynamics of the Epidemic and Human Behavior

Every individual decides their vaccination strategy (i.e., whether to participate in vaccination pre-emptively at the beginning of each epidemic season). Using this model, vaccination can protect people from infection with a probability e, $(0 \leq e \leq 1)$, which denotes the vaccine effectiveness. Thus, the present model obeys the effectiveness model. If a vaccine is effective, an individual becomes immunized against disease (expressed as the IM state).

After this decision-making stage, the time evolution of an epidemic season occurs in a local time-step, which corresponds to the conventional SIR model. We establish Eqs. (9.1) to (9.12) to analyze the SIR dynamics of the pre-emptively vaccinated group (denoted PV) and the non-vaccinated group (denoted LV) at a global time-step g and local time-step t (see Fig. 9.1). In both groups, susceptible people are exposed to the risk of infection with a transmission rate β [day^{-1} person^{-1}] multiplied by the fraction of infected people, while infected people can recover from infection at a recovery rate of γ [day^{-1}].

As the disease spreads, susceptible non-vaccinators or successful free riders (SFR) have the possibility to take the vaccine at a rate x given by Eq. (9.10). This formula controls the individual's behavior toward late vaccination. Namely, in this equation, δ denotes the speed parameter that controls the frequency of late vaccination; C_{LV} denotes the relative cost of vaccination to the disease cost (defined as unity), which an individual must pay to participate in late vaccination, where $0 \leq C_{\mathrm{LV}} \leq 1$. The term δ / C_{LV} denotes the ease of participation in late vaccination due to social factors mainly caused by vaccination cost. I_{total} and V_{total} denote the momentary fractions of total infected and total vaccinated people, respectively. The ratio of I_{total} to V_{total} denotes the infection-risk level of a society at a given moment, enhancing the late vaccination of people during an epidemic season. The smallness parameter ϵ (=0.1) prevents the divergence of x when $C_{\mathrm{LV}} V_{\mathrm{total}}$ becomes zero. Eqs. (9.11) and (9.12) are applied to determine the fraction of SFR. SVLV and

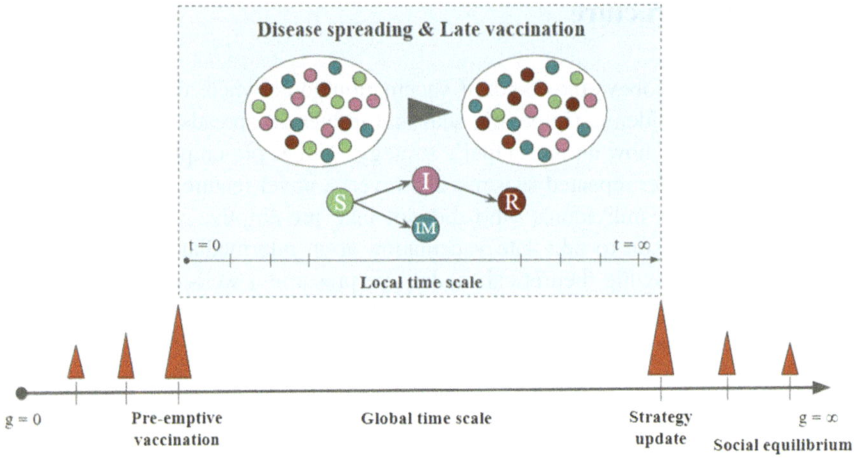

Fig. 9.1 Global timescale and local timescale. The global time-step starts with the pre-emptive vaccination of randomly selected people. In the following results, 10% of the entire population is selected as the initial vaccinators. Subsequently, disease spreading and late vaccination occur at the local timescale, which continues until infected people recover. Moreover, people determine their strategy for the next epidemic season based upon the results for the previous season. The set of the strategy update and epidemic season is repeated until the strategy fraction converges

Table 9.1 Fractions of six types of individual

Strategy/State	Healthy		Infected	
PV	SPV		FPV	
LV	Non-vaccinated	Vaccinated	Non-vaccinated	Vaccinated
	SFR	SVLV	FFR	FVLV

FVLV denote the abbreviations for "successfully" vaccinated late vaccinators and "failed" late vaccinators, respectively, which are also summarized in Table 9.1:

$$\frac{dS_{\mathrm{PV}}(g,t)}{dt} = -\beta S_{\mathrm{PV}}(g,t) I_{\mathrm{total}}(g,t), \tag{9.1}$$

$$\frac{dI_{\mathrm{PV}}(g,t)}{dt} = \beta S_{\mathrm{PV}}(g,t) I_{\mathrm{total}}(g,t) - \gamma I_{\mathrm{PV}}(g,t), \tag{9.2}$$

$$\frac{dIM_{\mathrm{PV}}(g,t)}{dt} = 0, \tag{9.3}$$

$$\frac{dS_{\mathrm{LV}}(g,t)}{dt} = -\beta S_{\mathrm{LV}}(g,t) I_{\mathrm{total}}(g,t) - \mathrm{exSFR}(g,t), \tag{9.4}$$

$$\frac{dI_{\mathrm{LV}}(g,t)}{dt} = \beta S_{\mathrm{LV}}(g,t) I_{\mathrm{total}}(g,t) - \gamma I_{\mathrm{LV}}(g,t), \tag{9.5}$$

$$\frac{dIM_{\mathrm{LV}}(g,t)}{dt} = \mathrm{exSFR}(g,t), \tag{9.6}$$

$$\frac{dR(g,t)}{dt} = \gamma I_{\mathrm{total}}(g,t), \tag{9.7}$$

$$I_{\mathrm{total}}(g,t) = I_{\mathrm{PV}}(g,t) + I_{\mathrm{LV}}(g,t), \tag{9.8}$$

$$V_{\mathrm{total}}(g,t) = \mathrm{PV}(g) + \mathrm{SVLV}(g,t) + \mathrm{FVLV}(g,t), \tag{9.9}$$

$$x = \frac{\delta I_{\mathrm{total}}}{C_{\mathrm{LV}} V_{\mathrm{total}} + \varepsilon}, \tag{9.10}$$

$$\frac{d\mathrm{SVLV}(g,t)}{dt} = x\mathrm{SFR}(g,t) - \beta I_{\mathrm{total}}(g,t)(\mathrm{SVLV}(g,t) - \mathrm{IM}_{\mathrm{LV}}(g,t)), \tag{9.11}$$

$$\frac{d\mathrm{FVLV}(g,t)}{dt} = \beta I_{\mathrm{total}}(g,t)(\mathrm{SVLV}(g,t) - IM_{\mathrm{LV}}(g,t)), \tag{9.12}$$

where t and g, respectively, indicate the time of the variable defined at the local and global timescales. It is quite important that we must carefully set the initial conditions at each season; these are given as:

$$S_{\mathrm{PV}}(g,0) = (1-e)\mathrm{PV}(g), \tag{9.13a}$$

$$I_{\mathrm{PV}}(g,0) = I_0\mathrm{PV}, \tag{9.13b}$$

$$\mathrm{IM}_{\mathrm{PV}}(g,0) = e\mathrm{PV}(g), \tag{9.13c}$$

$$S_{\mathrm{LV}}(g,0) = 1 - \mathrm{PV}(g,0) - I_{\mathrm{LV}}(g,0) = (1-I_0)(1-\mathrm{PV}), \tag{9.13d}$$

$$I_{\mathrm{LV}}(g,0) = I_0(1-\mathrm{PV}), \tag{9.13e}$$

$$IM_{\mathrm{LV}}(g,0) = 0, \tag{9.13f}$$

$$R(g,0) = 0, \tag{9.13g}$$

$$I_{\mathrm{total}}(g,0) = I_0, \tag{9.13h}$$

$$V_{\mathrm{total}}(g,0) = \mathrm{PV}(g,0), \tag{9.13i}$$

$$\mathrm{SVLV}(g,0) = 0, \tag{9.13j}$$

$$\mathrm{FVLV}(g,0) = 0. \tag{9.13k}$$

Additionally, we must set I_0 to a sufficiently small value. Throughout this chapter, we set $I_0 = 0.0005$.

Note that the local variable x indicates the fraction of late-vaccinated individuals. Another important variable relating to the fraction of vaccinees is *PV*(g), meaning the fraction of pre-emptively vaccinated individuals in season, with g, which is defined as globally evolved, i.e., variable in repeating seasons, of which dynamics is given in Eq. (9.15) as letter.

9.2.2 Payoff Structure

The epidemic stage starts after the pre-emptive vaccination term is finished in every season, and it continues until all infected people recover. Those who decide to participate in pre-emptive vaccination must pay a cost, C_{PV} ($0 \leq C_{\mathrm{PV}} \leq 1$), to obtain it. If the vaccine can successfully immunize them against contagion, then they do not need to pay an additional cost. Meanwhile, some people who are unfortunately infected due to the imperfect pre-emptive vaccination pay a cost of infection, C_i, that is defined as unity ($= 1$) without loss of generality as well.

Moreover, non-vaccinated people simply pay the infection cost C_i if they get infected, whereas those non-vaccinated who successfully avoided infection without taking a vaccine pay nothing. Additionally, as explained in Sect. 9.2.1, non-vaccinators who decide to participate in late vaccination must pay for its cost, C_{LV} ($0 \leq C_{\mathrm{LV}} \leq 1$), just like pre-emptive vaccinators. Some of them are still at risk of infection due to imperfect vaccination and may need to pay an additional cost C_i. Table 9.2 presents all possible payoffs for different combinations of strategy and health state at the end of an epidemic season.

9.2.3 Strategy Updating and Global Dynamics

The present model applies individual-based risk assessment (IB-RA), as was used in the previous chapters. Thus, every individual randomly picks another individual and determines whether they adopt the selected individual's strategy by evaluating their payoff difference with the Fermi function. We set $\kappa = 0.1$ for the Fermi function. The total number of the transition probabilities obeying IB-RA is sixteen, resulting from $4 \times 2 + 2 \times 4$. Let us elaborate upon this. Referring to Table 9.1, the late-vaccinated individuals in a current season (labeled SFR, SVLV, FFR, and FVLV) may move to another choice in the next season. That is to say, that they may take a pre-emptive vaccination (labeled PV), under which category there are two possible reference classes: SPV and FPV. This is the origin of the 4×2 term. Likewise, pre-emptively vaccinated individuals, labeled SPV and FPV, may decide not to pre-emptively vaccinate (labeled LV), under which category there are four possible reference classes: SFR, SVLV, FFR, and FVLV. This is the origin of the 2×4 term. Let us enumerate these sixteen transition probabilities below:

Table 9.2 Payoff structure

Strategy/State	Healthy		Infected	
PV	$-C_{\mathrm{PV}}$		$-C_{\mathrm{PV}} - 1$	
LV	Non-vaccinated	Vaccinated	Non-vaccinated	Vaccinated
	0	$-C_{LV}$	-1	$-C_{LV} - 1$

$$P(\mathrm{SFR} \leftarrow \mathrm{SPV}) = \frac{1}{1 + \exp\left[-(0 - (-C_{\mathrm{PV}}))/\kappa\right]}, \tag{9.14a}$$

$$P(\mathrm{SFR} \leftarrow \mathrm{FPV}) = \frac{1}{1 + \exp\left[-(0 - (-C_{\mathrm{PV}} - 1))/\kappa\right]}, \tag{9.14b}$$

$$P(\mathrm{SVLV} \leftarrow \mathrm{SPV}) = \frac{1}{1 + \exp\left[-(-C_{\mathrm{LV}} - (-C_{\mathrm{PV}}))/\kappa\right]}, \tag{9.14c}$$

$$P(\mathrm{SVLV} \leftarrow \mathrm{FPV}) = \frac{1}{1 + \exp\left[-(-C_{\mathrm{LV}} - (-C_{\mathrm{PV}} - 1))/\kappa\right]}, \tag{9.14d}$$

$$P(\mathrm{FFR} \leftarrow \mathrm{SPV}) = \frac{1}{1 + \exp\left[-(-1 - (-C_{\mathrm{PV}}))/\kappa\right]}, \tag{9.14e}$$

$$P(\mathrm{FFR} \leftarrow \mathrm{FPV}) = \frac{1}{1 + \exp\left[-(-1 - (-C_{\mathrm{PV}} - 1))/\kappa\right]}, \tag{9.14f}$$

$$P(\mathrm{FVLV} \leftarrow \mathrm{SPV}) = \frac{1}{1 + \exp\left[-(-C_{\mathrm{LV}} - 1 - (-C_{\mathrm{PV}}))/\kappa\right]}, \tag{9.14g}$$

$$P(\mathrm{FVLV} \leftarrow \mathrm{FPV}) = \frac{1}{1 + \exp\left[-(-C_{\mathrm{LV}} - 1 - (-C_{\mathrm{PV}} - 1))/\kappa\right]}, \tag{9.14h}$$

$$P(\mathrm{SPV} \leftarrow \mathrm{SFR}) = \frac{1}{1 + \exp\left[-(-C_{\mathrm{PV}} - (0))/\kappa\right]}, \tag{9.14i}$$

$$P(\mathrm{SPV} \leftarrow \mathrm{SVLV}) = \frac{1}{1 + \exp\left[-(-C_{\mathrm{PV}} - (-C_{\mathrm{LV}}))/\kappa\right]}, \tag{9.14j}$$

$$P(\mathrm{SPV} \leftarrow \mathrm{FFR}) = \frac{1}{1 + \exp\left[-(-C_{PV} - (-1))/\kappa\right]}, \tag{9.14k}$$

$$P(\mathrm{SPV} \leftarrow \mathrm{FVLV}) = \frac{1}{1 + \exp\left[-(-C_{\mathrm{PV}} - (-C_{\mathrm{LV}} - 1))/\kappa\right]}, \tag{9.14l}$$

$$P(\mathrm{FPV} \leftarrow \mathrm{SFR}) = \frac{1}{1 + \exp\left[-(-C_{PV} - 1 - (0))/\kappa\right]}, \tag{9.14m}$$

$$P(\mathrm{FPV} \leftarrow \mathrm{SVLV}) = \frac{1}{1 + \exp\left[-(-C_{\mathrm{PV}} - 1 - (-C_{LV}))/\kappa\right]}, \tag{9.14n}$$

$$P(\mathrm{FPV} \leftarrow \mathrm{FFR}) = \frac{1}{1 + \exp\left[-(-C_{\mathrm{PV}} - 1 - (-1))/\kappa\right]}, \tag{9.14o}$$

$$P(\mathrm{FPV} \leftarrow \mathrm{FVLV}) = \frac{1}{1 + \exp\left[-(-C_{\mathrm{PV}} - 1 - (-C_{\mathrm{LV}} - 1))/\kappa\right]}. \tag{9.14p}$$

The evolutionary process of the fraction of pre-emptively vaccinated people, PV (g), can be described using the following dynamical equation:

$$
\begin{aligned}
\frac{d\mathrm{PV}}{dt} = {} & \mathrm{SFR}(g,\infty)\mathrm{SPV}(g,\infty)P(\mathrm{SFR} \leftarrow \mathrm{SPV}) \\
& + \mathrm{SFR}(g,\infty)\mathrm{FPV}(g,\infty)P(\mathrm{SFR} \leftarrow \mathrm{FPV}) \\
& + \mathrm{SVLV}(g,\infty)\mathrm{SPV}(g,\infty)P(\mathrm{SVLV} \leftarrow \mathrm{SPV}) \\
& + \mathrm{SVLV}(g,\infty)\mathrm{FPV}(g,\infty)P(\mathrm{SVLV} \leftarrow \mathrm{FPV}) \\
& + \mathrm{FFR}(g,\infty)\mathrm{SPV}(g,\infty)P(\mathrm{FFR} \leftarrow \mathrm{SPV}) \\
& + \mathrm{FFR}(g,\infty)\mathrm{FPV}(g,\infty)P(\mathrm{FFR} \leftarrow \mathrm{FPV}) \\
& + \mathrm{FVLV}(g,\infty)\mathrm{SPV}(g,\infty)P(\mathrm{FVLV} \leftarrow \mathrm{SPV}) \\
& + \mathrm{FVLV}(g,\infty)\mathrm{FPV}(g,\infty)P(\mathrm{FVLV} \leftarrow \mathrm{FPV}) \\
& - \mathrm{SPV}(g,\infty)\mathrm{SFR}(g,\infty)P(\mathrm{SPV} \leftarrow \mathrm{SFR}) \\
& - \mathrm{SPV}(g,\infty)\mathrm{SVLV}(g,\infty)P(\mathrm{SPV} \leftarrow \mathrm{SVLV}) \\
& - \mathrm{SPV}(g,\infty)\mathrm{FFR}(g,\infty)P(\mathrm{SPV} \leftarrow \mathrm{FFR}) \\
& - \mathrm{SPV}(g,\infty)\mathrm{FVLV}(g,\infty)P(\mathrm{SPV} \leftarrow \mathrm{FVLV}) \\
& - \mathrm{FPV}(g,\infty)\mathrm{SFR}(g,\infty)P(\mathrm{FPV} \leftarrow \mathrm{SFR}) \\
& - \mathrm{FPV}(g,\infty)\mathrm{SVLV}(g,\infty)P(\mathrm{FPV} \leftarrow \mathrm{SVLV}) \\
& - \mathrm{FPV}(g,\infty)\mathrm{FFR}(g,\infty)P(\mathrm{FPV} \leftarrow \mathrm{FFR}) \\
& - \mathrm{FPV}(g,\infty)\mathrm{FVLV}(g,\infty)P(\mathrm{FPV} \leftarrow \mathrm{FVLV}).
\end{aligned} \tag{9.15}
$$

9.3 Result and Discussion

Figure 9.2 depicts the phase diagrams for varying pre-emptive (X-axis) and late (Y-axis) vaccination costs of FES, the total vaccinator fraction (F_{V}), the fraction of pre-emptively vaccinated people (F_{PV}), the fraction of late-vaccination-strategy holders who actually participated in vaccination (F_{VLV}), and the average social payoff, ASP, for vaccine effectiveness of $e = 0.1$, 0.5, 0.8, and 1.0. Generally, when $C_{\mathrm{LV}} < C_{\mathrm{PV}}$, people prefer late vaccination to pre-emptive vaccination simply because the more cost-effective strategy for obtaining immunity to contagion (that is late vaccination) is rationally selected. Meanwhile, it interestingly turns out that a lower FES appears when presuming $C_{\mathrm{LV}} > C_{\mathrm{PV}}$ than when presuming $C_{\mathrm{LV}} < C_{\mathrm{PV}}$. This is because the cost benefit of late vaccination drives more individuals to take late vaccination than pre-emptive vaccination, and late vaccination's contribution to suppressing the spread of a disease in society inevitably appears late in the season by letting some of late vaccinators (with fraction of e) immunized vis-à-vis pre-emptive vaccination contributing that works at the beginning of a season. Hence, except for

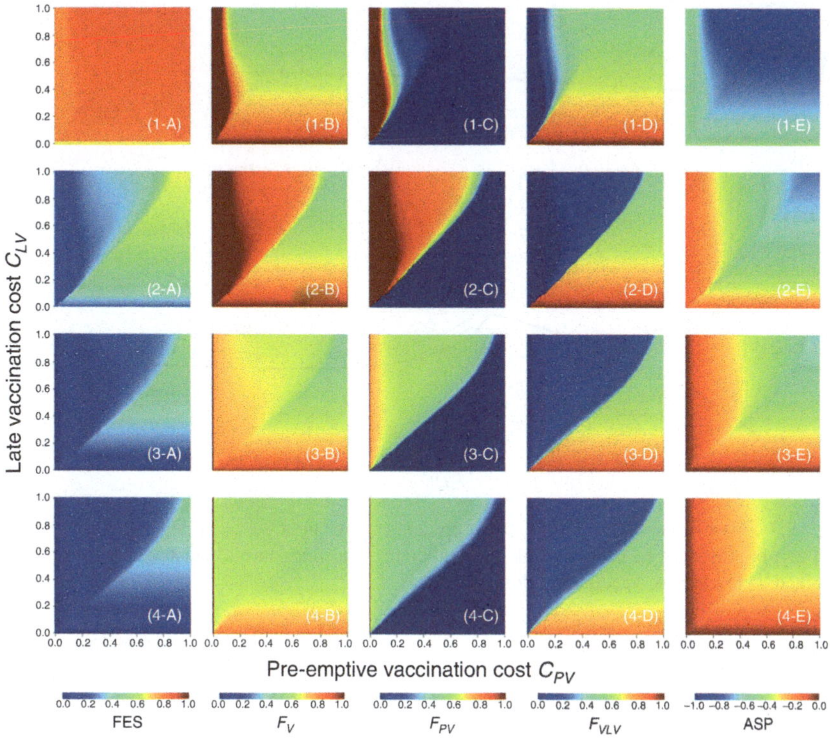

Fig. 9.2 FES (labeled (*-A)), total vaccinator fraction (labeled (*-B)), fraction of pre-emptive vaccinators (labeled (*-C)), fraction of late vaccinators who commit vaccination (labeled (*-D)), and social-average payoff (labeled (*-E)) for vaccine effectivenesses of $e = 0.1$ (labeled (1-*)), $e = 0.5$ (labeled with (2-*)), $e = 0.8$ (labeled with (3-*)), and $e = 1.0$ (labeled with (4-*))

the particular case given below, pre-emptive vaccination should be more addressed than late vaccination from a governmental public-health perspective.

Additionally, when late vaccination is available with an extremely low cost below a certain threshold value, it evidently displays a significantly lower FES than that observed when presuming $C_{LV} > C_{PV}$. This is because the effect of late vaccination can be strongly enhanced from the beginning of a season as long as a small value of C_{LV} is possible. The dynamics depicted by Eq. (9.10) consequently makes most people participate in late vaccination from an earlier time in the season.

Figure 9.3 illustrates the phase transition observed in FES with increasing late-vaccination cost for various vaccine effectiveness. In each case except $e = 0.1$ (the lowest-vaccine-efficacy case), each line evidently shows a phase transition. Below that critical point, late vaccination is dominant due to low cost, while beyond the critical point, pre-emptive vaccination dominates to reduce the FES that once increased along with C_{LV}.

Figure 9.4 shows the time evolution of the fractions of four health states (susceptible (S), infected (I), recovered (R), and immunized (IM)), as well as the fraction

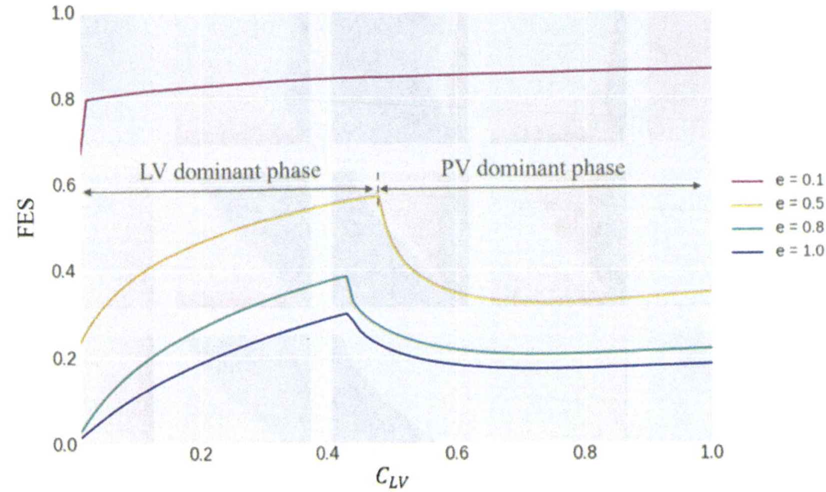

Fig. 9.3 Phase transition of the final epidemic size at $C_{\mathrm{PV}} = 0.5$. The phase transition can be observed in the midst of C_{LV}, separating the lines into a late-vaccination-dominant area and a pre-emptive-vaccination-dominant area

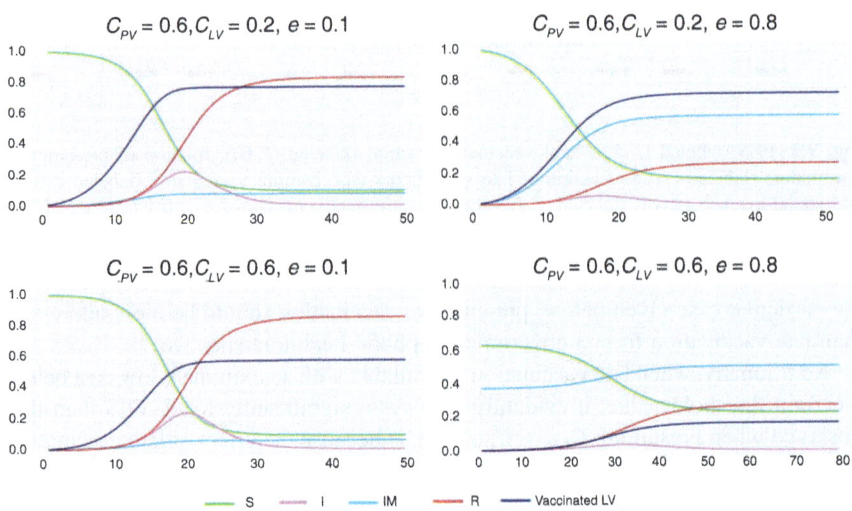

Fig. 9.4 Time evolution of the compartment fractions and the fraction of late vaccinators in the final epidemic season before reaching social equilibrium. The X-axis indicates the elapsed days

of late vaccinators (denoted Vaccinated, LV), who actually participate in late vaccination in the final epidemic season before attaining social equilibrium. A comparison between the left and right panels demonstrates the vaccine effectiveness.

If a low vaccine effectiveness is imposed, it seems less clear how the introduction of late vaccination with an inexpensive cost compared to that of pre-emptive vaccination works. This is because, although a slightly higher rate of late vaccinees is observed in the top-left panel than in the case with a higher C_{LV} (bottom-left panel), the levels of infected individuals in both cases seem almost comparable. Under such circumstances, vaccinated people, regardless of whether they use pre-emptive or late vaccination, are unavoidably infected.

However, in the case of attractive late vaccination having a high reliability with a low cost (right panels), the introduction of late vaccination can work clearly, where the lower $-C_{LV}$ case induces more late vaccinations than the higher- C_{LV} case. In both cases, infections can be successfully suppressed vis-à-vis the lower-effectiveness case (left panels).

Now, let us explore the context in which an individual chooses late vaccination as their strategy. Individuals who choose late vaccination or, inversely speaking, who do not choose pre-emptive vaccination have a so-called information advantage, because they postpone the decision on whether to the expend the cost for vaccination until after the start of a season. More importantly, they are able to secure the possibility of doing nothing, i.e., avoiding vaccination but expecting not to be infected until the end of a season. This implies that they may be able to become "SFR," taking neither pre-emptive nor late vaccination but avoiding infection due to herd immunity. However, they always experience a higher risk of infection than that to a vaccinator when exposed, which creates a social dilemma: "expecting your fortune or buying certainty through insurance." This dilemma is unequivocally influenced by the cost and effectiveness of vaccination. To quantify such things, let us see Fig. 9.5.

The upper row of Fig. 9.5 depicts the fraction of free riders (F_{FR}), which is the sum of SFR and FFR, as defined in Table 9.1. The middle row shows the ratio of free riders to late vaccinators, as quantified by $\frac{F_{FR}}{F_{LV}} = \frac{F_{FR}}{\mathrm{SFR+SVLV+FFR+FVLV}}$. Let us call this ratio the "expectation to free ride," denoted by E_{FR}. This ratio demonstrates people's willingness not to do anything in order to free ride on herd immunity. The bottom row illustrates the ratio of SFR out of all free riders, i.e., $\frac{\mathrm{SFR}}{F_{FR}}$.

It is noteworthy that the higher $-E_{FR}$ region in the red-highlighted triangle (indicating $C_{LV} > C_{PV}$), which is illustrated in yellow-red to red, significantly spreads with the increase of the vaccine effectiveness up to $e = 0.5$. Why do people's free riding expectations erupt? One possibility is that the numerator of E_{FR}, i.e., F_{FR}, increases. We note that F_{FR} in the region obviously decreases by considering the top row. Thus, this possibility is fully denied. The only remaining cause is that the denominator of E_{FR} decreases, which can be paraphrased by saying that the number of pre-emptive vaccinators increases. It is quite rational not to aim at a free ride, but rather to depend upon self-vaccination when an individual is exposed to an environment of relatively low vaccine effectiveness, since self-vaccination might be rather more reliable than other people's vaccination effect. In other words, this would be less likely to bring about herd immunity. Moreover, if vaccinating, it is more rational to choose pre-emptive vaccination than late vaccination because the former outperforms the latter in forming herd immunity and there is cost merit ($C_{LV} > C_{PV}$). To this end, people become very responsive to participating in

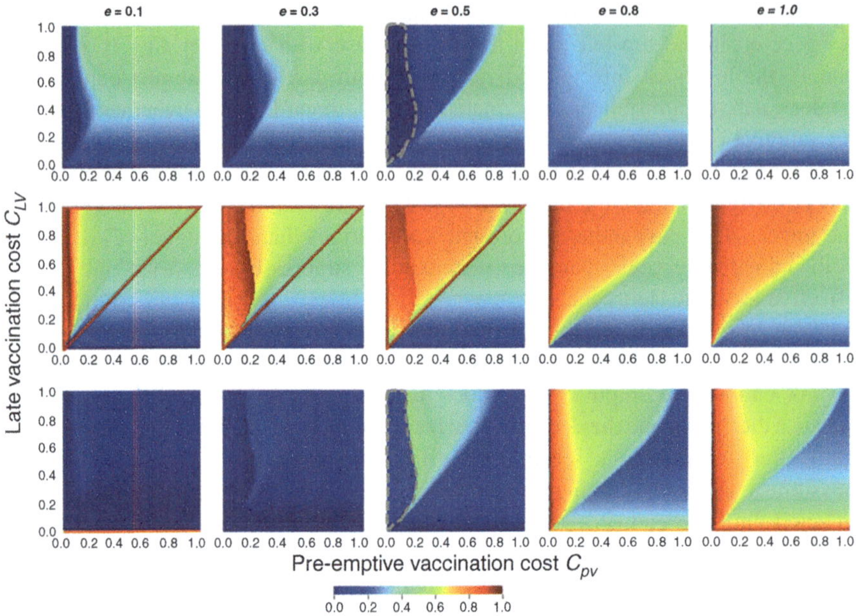

Fig. 9.5 Fraction of free riders (upper row), ratio of free riders in the LV group (middle row), and ratio of successful free riders out of all attempted free riders (lower row). Each column presumes each effectiveness

pre-emptive vaccination (decreasing the fraction of non-pre-emptively vaccinated individuals) with the increase in effectiveness. Another interesting question to consider is the following: what proportion of individuals would be SFR amid the people trying to participate in vaccination? The bottom row considers this question. For $e = 0.5$, reasonable fractions of free riders can be successful (colored with green to yellow) in the triangular region, except for the region indicated by a gray dotted line. There, no free riders exist at all (the same is true for the gray dotted line at the top panel with $e = 0.5$). This is because all of them are pre-emptively vaccinated due to a quite low cost when compared with late vaccination.

As a provisional summary, we would like to confirm the assertion that, in the region of lower vaccination reliability, an increase in effectiveness decreases the option for late vaccination, whereas it increases participation in pre-emptive vaccination. However, when a higher vaccination reliability is ensured, an increase in effectiveness does not increase the number of pre-emptive vaccinators, but rather discourages vaccination on the whole, even when the cost merit of pre-emptive vaccination is ensured (compare the region of $C_{LV} > C_{PV}$ in Fig. 9.2 (3-C) and (4-C) with that in (2-C)).

The reason why people sensitively shift from late vaccination to pre-emptive vaccination under such a low-effectiveness condition is presented below. In a society with a high risk of infection, vaccination generally works as a better option than betting on the chance of a free ride, because the low efficacy of the vaccine hampers the establishment of herd immunity. Meanwhile, to prevent infection, pre-emptive vaccination is always more effective than late vaccination. As a general tendency, pre-emptive vaccination is much more enhanced with the increase of effectiveness as long as its effectiveness is not very high. By contrast, when the vaccine efficacy becomes high, people tend to move away from pre-emptive vaccination since they are attracted by a greater expectation to free ride (E_{FR}). In fact, even in the region with a cost premium for pre-emptive vaccination ($C_{\text{LV}} > C_{\text{PV}}$), a large ratio of SFR compared with the lower-effectiveness cases is found (see $e = 0.8$ and 1 in the bottom row with the lower-effectiveness cases).

Finally, to validate the present theoretical model discussed above, we conducted a series of stochastic agent-based simulations, i.e., MAS, according to what we introduced in Sect. 3.3. We presumed a complete graph as the underling spatial structure on which the total population is set to 10^4 for comparison with the theoretical result, Fig. 9.2. Based on the MFA, the simulation results demonstrated a perfectly well-mixed population when compared with the theoretical results. The effective transmission rate, β_e, introduced in Sect. 3.3.2 is set to 0.000086, which is determined to be the minimal value satisfying the threshold value of FES ($= 0.9$) without introducing any vaccination behavior. The other parameters are fixed to the same value as in the analytical simulation. Figure 9.6 depicts the results obtained by taking an ensemble average of 100 independent episodes. All of the corresponding figures generally agree well with the analytical results presented in Fig. 9.2.

This chapter introduced a new theoretical model for vaccination behavior in the situation where individuals can participate in pre-emptive vaccination based on conventional models of the vaccination game, or participate in late vaccination as the second option at any arbitrary time in an epidemic season. The developed mathematical framework for such "late vaccination" presumes that an individual would participate in vaccination if the risk in infection is increasing.

The kernel to depict the dynamics of late vaccination was given in Eq. (9.10). We did not follow the so-called behavior model that we discussed in Sect. 7.1, but rather established the original Eq. (9.10). Let us compare Eqs. (7.2) and (9.10). Note that the defined variable x in Eq. (7.2) does not indicate the vaccination fraction but the vaccination rate, which is consistent with the definition of x in Eq. (9.10). The behavior model takes account of the time-derivative of x, but Eq. (9.10) does not. Eq. (9.10) rather presumes that the property x, meaning the late-vaccination fraction, immediately balances the other field variables, including the fractions of infected and vaccinated individuals, unlike the behavior model presumes. We should note that this substantial difference in how the local dynamics is modeled is quite important.

By "local," we are referring to the local timescale describing both dynamic processes of epidemic spread and human decision making about whether to commit to late vaccination; this is as opposed to the global timescale, which describes another dynamics of decision on pre-emptive vaccination. In the present framework, local and global timescales allow an individual to use the time-asymmetric strategies of pre-emptive vaccination and either late vaccination or doing nothing. Late vaccination is asymmetrically advantageous over pre-emptive vaccination because an individual does not pay a vaccine cost at the beginning of an epidemic season,

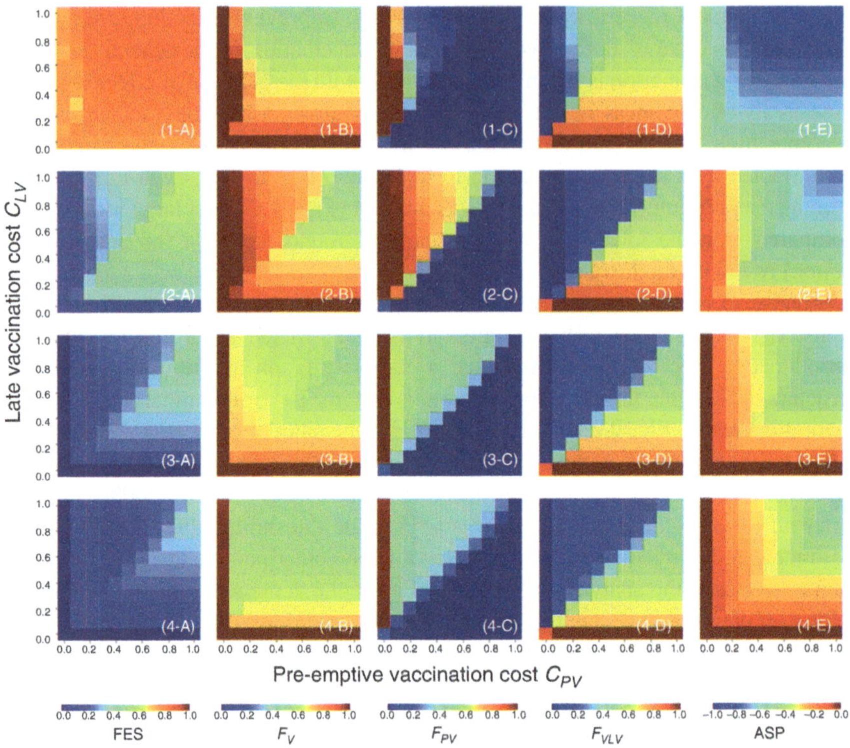

Fig. 9.6 Final epidemic size (labeled (*-A)), fraction of total vaccinators (labeled (*-B)), fraction of pre-emptive vaccinators (labeled (*-C)), fraction of late vaccinators who commit to vaccination (labeled (*-D)), and social-average payoff (labeled (*-E)) for vaccine effectivenesses of $e = 0.1$ (labeled (1-*)), $e = 0.5$ (labeled (2-*)), $e = 0.8$ (labeled (3-*)), and $e = 1.0$ (labeled (4-*)), as obtained from multi-agent simulation

even if one consequently vaccinates late. Such asymmetry inevitably brings about a social dilemma. Everyone favors not taking the pre-emptive option, even though pre-emptive vaccination more effectively subdues the disease than late vaccination on the grounds of late vaccination's information advantage. From a public-health standpoint, solving this dilemma to encourage people to cooperate by pre-emptive vaccination is absolutely important. Thus, what the present model can elucidate may be quite meaningful. For instance, to dilute the dilemma, the authority is able to offer a certain premium only to a pre-emptive vaccine backed by the subsidy, so as to lead more people to pre-emptive vaccination, which can be quantified by a series of 2D heat maps of C_{PV}–C_{LV}, as we discussed in this chapter.

The numerical results demonstrated in this chapter clarify that a low late-vaccination cost enables people to avoid pre-emptive vaccination, and that this results in a highly infected society unless the cost becomes much cheaper than that of pre-emptive vaccination. What we found interesting is that lower vaccine efficacies ironically tend to dilute the social dilemma above. Less reliable vaccines make people aware that they must protect themselves by not relying on others' contributions. As vaccine effectiveness increases, however, people tend toward not wanting to pay, but to be protected by the contributions of others. From a social-design viewpoint, the introduction of a cost premium for only pre-emptive vaccination may be meaningful for solving the social dilemma.

References

Alvarez-Zuzek LG, Di Muro MA, Havlin S, Braunstein LA (2019) Dynamic vaccination in partially overlapped multiplex network. Phys Rev E 99:012302

Liqun L, Yanfeng O (2019) Dynamic vaccination game in a heterogeneous mixing population. Physica A 533:122032

Chapter 10
Influenza Vaccine Uptake

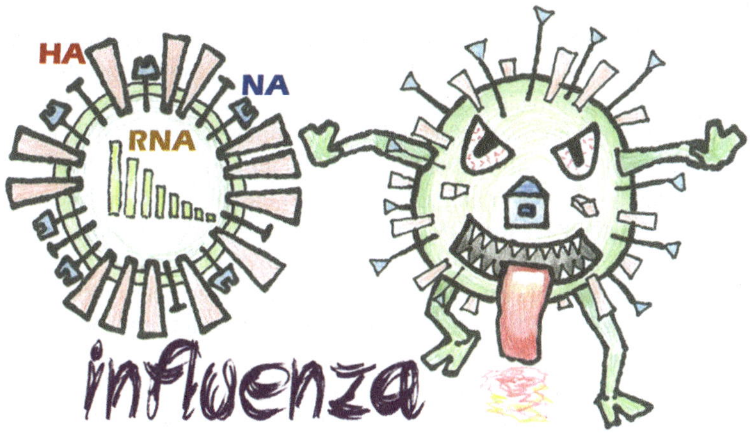

There are two main types of influenza viruses, A and B, with several subtypes that are commonly found to circulate among humans. The trivalent (TIV) flu vaccine targets two strains of influenza A and one strain of influenza B. The quadrivalent (QIV) vaccine targets one extra B-virus strain, ensuring better protection against influenza; however, the use of the QIV vaccine can be costly; hence, it imposes an extra financial burden upon individuals as well as society as a whole. Because there are multiple alternative vaccines in addition to multiple strains, an individual's choice must take account of a complex situation, including a strategy aimed at free-riding on herd immunity. This scenario might create a social dilemma in choosing vaccine types at the individual level. Hence, this chapter prescribes a means of dealing with this problem based on the vaccination-game framework.

J. Tanimoto, *Sociophysics Approach to Epidemics*, Evolutionary Economics and Social Complexity Science 23, https://doi.org/10.1007/978-981-33-6481-3_10

10.1 Introduction and Background: Multiple Strains and Multiple Vaccines

There are four main types of influenza viruses: A, B, C, and D[1], of which influenza A and B circulate most commonly among humans and cause seasonal epidemics worldwide; however, circulation patterns and strain prevalence can be region-specific and time-specific. According to a report published by the Centers for Disease Control and Prevention (CDC)[2], the United States 2017–18 influenza season (October 1, 2017–May 19, 2018) faced severity of both influenza viruses; influenza A viruses prevailed at the beginning of the season, but influenza B predominated in the latter part. As reported by previous studies, influenza A viruses have two strain subtypes (H1N1 and H3N2), while influenza B viruses have the subtypes (B/Victoria and B/Yamagata), which have been commonly found to circulate[3].

Pre-emptive vaccination is one of the most effective protections for combating seasonal influenza. The traditional trivalent (TIV) vaccine targets three strain subtypes of influenza viruses (two strains of influenza A and one of the B-virus strains); on the other hand, the quadrivalent (QIV) vaccine contains one extra strain of influenza B virus, offering improved protection against influenza B virus infections. Although each year, public-health authorities recommend which strain should be contained in vaccines for the upcoming season, due to the mismatch between the recommended B-virus strain for the TIV vaccine and the circulating virus strain, influenza morbidity may increase. Inclusion of an extra B strain in the QIV vaccine offers better efficacy than the TIV vaccine. Because it provides better protection, some developed countries have been turning toward QIV vaccines, although TIV vaccines are still recommended. Nevertheless, many other countries are still heavily relying upon the TIV vaccine. Moreover, the use of the QIV vaccine is inevitably more costly as compared with TIV, which causes people to hesitate in choosing a vaccination. Even in developed countries such as Japan, a vaccine shot against seasonal influenza for a typical adult is not covered by public medical-treatment insurance. Thus, even if the expected cost of vaccination using QIV vis-à-vis the disease cost is lower, an individual who is myopic and less cautious about long-term benefit would be exposed to a typical social-dilemma situation.

In this chapter we follow the vaccination-game framework we have studied. Thus, at the end of each season, individuals assess their perceived payoffs based on the previous season's experience in terms of vaccine efficacy, cost, etc., and

---

[1]WHO: In press. World Health Organization. Factsheet Influenza (seasonal). Available from: https// www.who.int/en/news-room/fact-sheets/detail/influenza-(seasonal) (Accessed August 22, 2019).

[2]Garten, R., Blanton, L., Elal, A.I. et al; Update: Influenza Activity in the United States During the 2017–18 Season and Composition of the 2018–19 Influenza Vaccine, *Morbidity and Mortality Weekly Report* **67**, 634–642, 2018.

[3]Yang, J.R., Huang, Y.P., Chang, F.Y., Hsu, L.C., Lin, Y.C., Huang, H.Y., Wu, F.T., Wu, H.S., Liu, M.T.; Phylogenetic and Evolutionary History of Influenza B Viruses, which Caused a Large Epidemic in 2011-2012, Taiwan, *PLoS One* **7**, e47179, 2012.

decide whether to vaccinate and which vaccine to use for the next season. This strategic adaptation after each season is based upon imitating others' strategy with better payoffs. In a nutshell, this chapter presumes strategy-based risk assessment (SB-RA). The group who choose QIV vaccine are expected to get better protection against the disease than that taking the TIV vaccine, although this may depend upon the circulation of influenza B virus. However, the cost of the TIV vaccine should be no more than the cost of the QIV vaccine. Taking all of these factors into account, we combine the disease-spreading dynamics of influenza with the vaccination decision-making process through an evolutionary framework.

10.2 Model Structure[4]

We entangle the simultaneous spread of two influenza viruses (A and B) and the evolution of the vaccination (QIV, TIV, or none) decision by constructing a repetitive sequence of a two-stage process in an infinite and well-mixed population, as we have presumed in previous chapters. Figure 10.1 illustrates the whole dynamical setup. The central inset in Fig. 10.1, labeled with "disease spreading," indicates that what happens in a single season obeys the dynamics described in Sect. 10.2.1 as below, which are controlled by the local timescale. Other events described in Fig. 10.1 take place at the end of each season. Thus, as a whole, the social equilibrium of the final epidemic size (FES) and other properties are explored in repeating seasons controlled by the global timescale. This fully follows the standard template of the vaccination game that we discussed in the previous chapters.

10.2.1 Dynamics of Epidemic Spread

Following our previous vaccination-game models, the first stage of the present dynamical system is the disease-spreading process based on a susceptible–infected–recovered (SIR)-like model coupled with vaccination. Initially, we divide the whole population into three categories: susceptible (unvaccinated) denoted by S, QIV vaccinees denoted by V_Q, and TIV vaccinees denoted by V_T. It is not usually recommended to take both vaccines (TIV and QIV) in a single season; therefore, we disregard all individuals taking both vaccines. We use the effectiveness model introduced in Sect. 3.2.1 to represent the vaccine efficacy. As influenza vaccines are not 100% perfect, we consider imperfect vaccinations in our disease modeling, where both vaccines are presumed to provide the same level of effectiveness against

[4]Arefin, M.R., Tanaka, M., Kabir, K.M.A., Tanimoto, J.; Interplay between cost and effectiveness in influenza vaccination uptake: vaccination game approach, *Proceedings of the Royal Society A* **475**, 20190608, 2019.

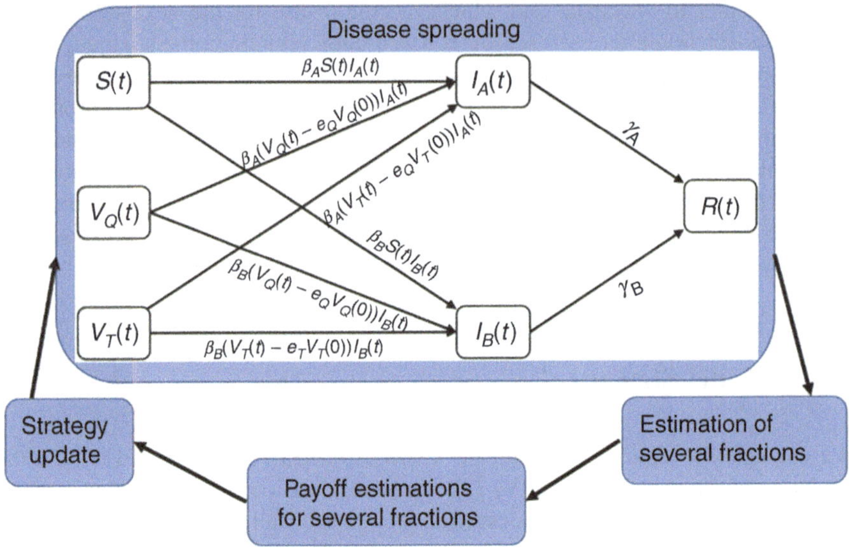

Fig. 10.1 The layout of the whole dynamical setup. The vaccine effectiveness of the TIV vaccine against the influenza B virus is assumed to be e_T, which is lower than that of the QIV vaccine. However, both vaccines are assumed to have the same effectiveness (e_Q) against the influenza A virus. Once the disease-spreading process ends, we estimate several fractions such as vaccinated and healthy, vaccinated but infected, infected with influenza A (B), etc., and evaluate their payoffs. These fractions then update their strategies for the next season. This process is repeated until we reach a steady state. The arrows depict the sequence of the evolutionary process

the A virus but different efficacies against the B virus. The inclusion of these parameters allows us to consider some individuals from vaccinated groups who fail to get immunity from vaccination (QIV or TIV) and still face the risk of infection like susceptible individuals. The unvaccinated or vaccinated people who fail to get immunity may become infected with either influenza virus (A or B); then, these infected people transfer to the recovery state after suffering from illness. Therefore, at time t, an individual can be in any of the six compartments: $S(t)$, $V_Q(t)$, $V_T(t)$, infected with influenza A ($I_A(t)$), infected with influenza B ($I_B(t)$), or in the recovered state, $R(t)$. We presume e_Q $(0 \leq e_Q \leq 1)$ as the vaccine efficacy of QIV against both influenza A and B viruses. As the TIV (QIV) vaccine contains two strains of influenza A virus and one (two) strain(s) of influenza B virus, we assume that both vaccines bestow same level of effectiveness (e_Q $(0 \leq e_Q \leq 1)$) against influenza A, but TIV vaccine provides relatively lower effectiveness, e_T $(0 \leq e_T \leq e_Q)$, against influenza B. If $V_Q(0)$ is the fraction of vaccinees at the beginning of an epidemic season, then a proportion e_Q of $V_Q(0)$ (i.e., $e_Q V_Q(0)$) get perfect immunity from QIV vaccine and the remaining fraction, $V_Q(t) - e_Q V_Q(0)$, become vulnerable to disease, where $V_Q(t)$ is the fraction of vaccinators in the V_Q compartment at time t. Similar logic has been used for TIV vaccinees. The transmission and recovery rates for influenza A and B viruses are assumed to be β_A [day^{-1} person^{-1}], γ_A [day^{-1}], and

β_B, γ_B, respectively. It is worth mentioning that the compartment variables represent several fractions of the total population; therefore, we are not considering a population of any particular size. Although coinfection and superinfection with two influenza viruses are possible in some cases[5], the incidence rate is not so significant. Therefore, our model disregards the incidence of coinfection and superinfection with two influenza viruses (A and B) and presumes long-term cross-immunity between them. Although we assume long-term cross-immunity between the two viruses, short-term immunity is possible in reality. The incidence of short-term cross-immunity between two viruses may allow some individuals recovered from one virus to become infected with the other one after a short time interval[6]. To this end, the model dynamics is given by:

$$\frac{dS(t)}{dt} = -\beta_A S(t) I_A(t) - \beta_B S(t) I_B(t), \tag{10.1-1}$$

$$\frac{dV_Q(t)}{dt} = -\beta_A (V_Q(t) - e_Q V_Q(0)) I_A(t) - \beta_B (V_Q(t) - e_Q V_Q(0)) I_B(t), \tag{10.1-2}$$

$$\frac{dV_T(t)}{dt} = -\beta_A (V_T(t) - e_Q V_T(0)) I_A(t) - \beta_B (V_T(t) - e_T V_T(0)) I_B(t), \tag{10.1-3}$$

$$\begin{aligned}\frac{dI_A(t)}{dt} &= \beta_A S(t) I_A(t) + \beta_A (V_Q(t) - e_Q V_Q(0)) I_A(t) \\ &\quad + \beta_A (V_T(t) - e_Q V_T(0)) I_A(t) - \gamma_A I_A(t),\end{aligned} \tag{10.1-4}$$

$$\begin{aligned}\frac{dI_B(t)}{dt} &= \beta_B S(t) I_B(t) + \beta_B (V_Q(t) - e_Q V_Q(0)) I_B(t) \\ &\quad + \beta_B (V_T(t) - e_T V_T(0)) I_B(t) - \gamma_B I_B(t),\end{aligned} \tag{10.1-5}$$

$$\frac{dR(t)}{dt} = \gamma_A I_A(t) + \gamma_B I_B(t). \tag{10.1-6}$$

We have imposed the following constraint:

$$S(t) + V_Q(t) + V_T(t) + I_A(t) + I_B(t) + R(t) = 1. \tag{10.2}$$

Let us presume that $x, y \in [0, 1]$ are the fractions of QIV and TIV vaccinees at the beginning of a season. Thus, we presume the initial condition at a certain season to be $V_Q(0) = x$, $V_T(0) = y$, $S(0) = 1 - x - y - \epsilon_A - \epsilon_B$, where $\epsilon_A(\rightarrow 0^+)$ and $\epsilon_B(\rightarrow 0^+)$ are initial infections with influenza A and B, respectively.

---

[5]Tramuto, F., Maida, C.M., Magliozzo, F., Amodio, E., Vitale, F.; Occurrence of a case of influenza A (H1N1) pdm09 and B co-infection during the epidemic season 2012-2013, *Infection, Genetics and Evolution* **23**, 95–98, 2014.

[6]Zarnitsyna, V.I., Bulusheva, I., Handel, A., Longini, I.M., Halloran, M. E., Antia, R.; Intermediate levels of vaccination coverage may minimize seasonal-influenza outbreaks, *PLoS One* **13 (6)**, e0199674, 2018.

After each epidemic season, we estimate several groups by numerically calculating fluxes from one state to another. We mainly estimate nine fractions of individuals, which are (i) QIV vaccinees and healthy-HV_Q, (ii) ((iii)) QIV vaccinees, but infected with influenza A (B) virus-V_QI_A (V_QI_B), (iv) TIV vaccinees and healthy-HV_T, (v) ((vi)) TIV vaccinees but infected with influenza A (B) virus-V_TI_A (V_TI_B), (vii) successful free riders—*SFR* (who are unvaccinated but did not suffer from any influenza virus), and (viii) ((ix)) failed free riders—FFR_A (FFR_B): unvaccinated individuals who are infected by either influenza virus. Individuals who remain in the S, V_Q, and V_T states at equilibrium ($t \to \infty$)—that is, $S(\infty)$, $V_Q(\infty)$, and $V_T(\infty)$—are termed successful free riders, healthy QIV vaccinees, and healthy TIV vaccinees, respectively. Suppose that $\varphi_{P \to Q}$ is the total proportion transferred from state P to state Q during an epidemic season. Thus, QIV vaccinees who become infected with influenza A (B) virus can be evaluated from the flux, $\varphi_{V_Q \to I_A}$ ($\varphi_{V_Q \to I_B}$). In a similar manner, we can estimate this for TIV vaccinees. The failed free riders can be estimated by the fluxes $\varphi_{S \to I_A}$and $\varphi_{S \to I_B}$.

10.2.2 Payoff Structure

Once an epidemic season ends, several groups estimate their payoffs prior to the onset of the next epidemic season. Suppose that the costs of vaccination and infection are, respectively, C_V and C_I. The vaccination cost includes the vaccine price with possible side effects due to vaccination. Since the infection cost is higher than the vaccination cost, we define the relative cost of vaccination as $C_r = C_V/C_I$, $(0 \leq C_r \leq 1)$. Without loss of generality, we can choose $C_I = 1$. Then, the relative costs for the QIV and TIV vaccines can be, respectively, assumed to be C_Q and C_T. Since the cost for TIV is assumed to be no more than that for QIV, we presume that $0 \leq C_Q \leq 1$and $0 \leq C_T \leq C_Q$. Therefore, QIV (TIV) vaccinees who remain healthy during the epidemic can have a payoff, $-C_Q$ $(-C_T)$; however, QIV (TIV) vaccinees who become infected with either influenza virus (A or B) have a payoff$-C_Q - 1$ $(-C_T - 1)$. In addition, payoffs for successful and failed free riders are set to 0 and -1, respectively. All fractions, along with their payoffs, are summarized in Table 10-1.

Using Table 10.1, we estimate the average social payoff ($\langle \pi \rangle$), and the average payoffs for QIV vaccinators ($\langle \pi_{V_Q} \rangle$), TIV vaccinators ($\langle \pi_{V_T} \rangle$), and non-vaccinators ($\langle \pi_{NV} \rangle$) as follows:

$$\begin{aligned}\langle \pi \rangle = & -C_Q HV_Q + (-C_Q - 1)(V_QI_A + V_QI_B) - C_T HV_T + (-C_T - 1) \\ & \times (V_TI_A + V_TI_B) - (FFR_A + FFR_B),\end{aligned}$$

(10.3-1)

Table 10.1 Several fractions of individuals (at equilibrium) with their payoffs (within brackets)

Strategy/State	Healthy	Infected with Influenza A	Infected with Influenza B
QIV vaccinees (V_Q)	$HV_Q\,(-C_Q)$	$V_QI_A\,(-C_Q-1)$	$V_QI_B\,(-C_Q-1)$
TIV vaccinees (V_T)	$HV_T\,(-C_T)$	$V_TI_A\,(-C_T-1)$	$V_TI_B\,(-C_T-1)$
non-vaccinated (NV)	$SFR\,(0)$	$FFR_A(-1)$	$FFR_B(-1)$

$$\langle \pi_{V_Q} \rangle = (-C_Q HV_Q + (-C_Q - 1)(V_Q I_A + V_Q I_B)\,)/x, \tag{10.3-2}$$

$$\langle \pi_{V_T} \rangle = (-C_T HV_T + (-C_T - 1)(V_T I_A + V_T I_B))/y, \tag{10.3-3}$$

$$\langle \pi_{NV} \rangle = -(FFR_A + FFR_B)/(1 - x - y). \tag{10.3-4}$$

10.2.3 Strategy Updating and Global Dynamics

The present model presumes strategy-based risk assessment (SB-RA), as introduced in Sect. 3.2.2. According to the concept described in Eq. (3.27), there are eighteen state-transition-probability functions (=9 × (3 − 1), where 9 comes from the number of possible states listed in Table 10.1), as listed below:

$$P(HV_Q \leftarrow V_T) = \frac{1}{1 + \exp\left[-(\langle \pi_{V_T} \rangle - (-C_Q))/k\right]}, \tag{10.4-1}$$

$$P(V_Q I_A \leftarrow V_T) = \frac{1}{1 + \exp\left[-(\langle \pi_{V_T} \rangle - (-C_Q - 1))/k\right]}, \tag{10.4-2}$$

$$P(V_Q I_B \leftarrow V_T) = \frac{1}{1 + \exp\left[-(\langle \pi_{V_T} \rangle - (-C_Q - 1))/k\right]}, \tag{10.4-3}$$

$$P(HV_Q \leftarrow NV) = \frac{1}{1 + \exp\left[-(\langle \pi_{NV} \rangle - (-C_Q))/k\right]}, \tag{10.4-4}$$

$$P(V_Q I_A \leftarrow NV) = \frac{1}{1 + \exp\left[-(\langle \pi_{NV} \rangle - (-C_Q - 1))/k\right]}, \tag{10.4-5}$$

$$P(V_Q I_B \leftarrow NV) = \frac{1}{1 + \exp\left[-(\langle \pi_{NV} \rangle - (-C_Q - 1))/k\right]}, \tag{10.4-6}$$

$$P(HV_T \leftarrow V_Q) = \frac{1}{1 + \exp\left[-(\langle \pi_{V_Q} \rangle - (-C_T))/k\right]}, \tag{10.4-7}$$

$$P(V_T I_A \leftarrow V_Q) = \frac{1}{1 + \exp\left[-(\langle \pi_{V_Q} \rangle - (-C_T - 1))/k\right]}, \tag{10.4-8}$$

$$P(V_T I_B \leftarrow V_Q) = \frac{1}{1 + \exp\left[-(\langle \pi_{V_Q} \rangle - (-C_T - 1))/k\right]}, \quad (10.4\text{-}9)$$

$$P(HV_T \leftarrow NV) = \frac{1}{1 + \exp\left[-(\langle \pi_{NV} \rangle - (-C_T))/k\right]}, \quad (10.4\text{-}10)$$

$$P(V_T I_A \leftarrow NV) = \frac{1}{1 + \exp\left[-(\langle \pi_{NV} \rangle - (-C_T - 1))/k\right]}, \quad (10.4\text{-}11)$$

$$P(V_T I_B \leftarrow NV) = \frac{1}{1 + \exp\left[-(\langle \pi_{NV} \rangle - (-C_T - 1))/k\right]}, \quad (10.4\text{-}12)$$

$$P(SFR \leftarrow V_T) = \frac{1}{1 + \exp\left[-(\langle \pi_{V_T} \rangle - (0))/k\right]}, \quad (10.4\text{-}13)$$

$$P(FFR_A \leftarrow V_T) = \frac{1}{1 + \exp\left[-(\langle \pi_{V_T} \rangle - (-1))/k\right]}, \quad (10.4\text{-}14)$$

$$P(FFR_B \leftarrow V_T) = \frac{1}{1 + \exp\left[-(\langle \pi_{V_T} \rangle - (-1))/k\right]}, \quad (10.4\text{-}15)$$

$$P(SFR \leftarrow V_Q) = \frac{1}{1 + \exp\left[-(\langle \pi_{V_Q} \rangle - (0))/k\right]}, \quad (10.4\text{-}16)$$

$$P(FFR_A \leftarrow V_Q) = \frac{1}{1 + \exp\left[-(\langle \pi_{V_Q} \rangle - (-1))/k\right]}, \quad (10.4\text{-}17)$$

$$P(FFR_B \leftarrow V_Q) = \frac{1}{1 + \exp\left[-(\langle \pi_{V_Q} \rangle - (-1))/k\right]}. \quad (10.4\text{-}18)$$

At the end of each epidemic season, each individual is allowed to update their strategy depending upon last season's payoff. As mentioned before, individuals update their vaccination strategies prior to the onset of the next epidemic season. The evolution of vaccination strategies (QIV or TIV or unvaccinated) is estimated using the following evolutionary equations, which have been derived by extending the master equation of the mean-field framework in a well-mixed population as:

$$\begin{aligned}\frac{dx}{dt} = & -HV_Q.V_T.P(HV_Q \leftarrow V_T) - V_Q I_A.V_T.P(V_Q I_A \leftarrow V_T) \\ & - V_Q I_B.V_T.P(V_Q I_B \leftarrow V_T) - \\ & HV_Q.NV.P(HV_Q \leftarrow NV) - V_Q I_A.NV.P(V_Q I_A \leftarrow NV) \\ & - V_Q I_B.NV.P(V_Q I_B \leftarrow NV) + \\ & HV_T.V_Q.P(HV_T \leftarrow V_Q) + V_T I_A.V_Q.P(V_T I_A \leftarrow V_Q) + V_T I_B.V_T.P(V_T I_B \leftarrow V_Q) + \\ & SFR.V_Q.P(SFR \leftarrow V_Q) + FFR_A.V_Q.P(FFR_A \leftarrow V_Q) \\ & + FFR_B.V_Q.P(FFR_B \leftarrow V_Q),\end{aligned} \quad (10.5\text{-}1)$$

$$\frac{dy}{dt} = -HV_T.V_Q.P(HV_T \leftarrow V_Q) - V_TI_A.V_Q.P(V_TI_A \leftarrow V_Q)$$
$$- V_TI_B.V_Q.P(V_TI_B \leftarrow V_Q)-$$
$$HV_T.NV.P(HV_T \leftarrow NV) - V_TI_A.NV.P(V_TI_A \leftarrow NV)$$
$$- V_TI_B.NV.P(V_TI_B \leftarrow NV)+$$
$$HV_Q.V_T.P(HV_Q \leftarrow V_T) + V_QI_A.V_T.P(V_QI_A \leftarrow V_T) + V_QI_B.V_T.P(V_QI_B \leftarrow V_T)+$$
$$SFR.V_T.P(SFR \leftarrow V_T) + FFR_A.V_T.P(FFR_A \leftarrow V_T)$$
$$+ FFR_B.V_T.P(FFR_B \leftarrow V_T). \qquad (10.5\text{-}2)$$

Time variable t in Eq. (10.1-5) is defined as the global timescale, which differs from that appearing in Eq. (10.1-1), which is defined as the local timescale.

10.3 Result and Discussion

This section presents the results of the whole dynamical process. We vary several parameters including transmission rates, vaccine effectiveness, and vaccination cost to illustrate different scenarios. The symptoms of typical influenza sickness are resolved after a period of 3 to 7 days, so the average durability in the infected stage is 5 days. Therefore, we presume $\gamma_A = \gamma_B = 1/5 = 0.2$ throughout our numerical experiments.

10.3.1 Dynamics in a Single Season

Let us first briefly discuss the relative dynamics of infections due to influenza A and B viruses for different transmission rates and degrees of initial infections in a single season without considering the game-theory aspect. Figure 10.2 presents the result. The vaccination coverage for this case is chosen as $V_Q(0) = V_T(0) = 0.333$, with $e_Q = 0.6$, $e_T = 0.4$. As $e_T \leq e_Q$, the influenza B virus seems to dominate in the case of equal transmission rates ($\beta_A = \beta_B$) (Fig. 10.2 (a)-(b)). However, the virus with a higher transmission rate is found to dominate regardless of the initial infection (Fig. 10.2 (d)); hence, the sensitivity of the initial infections seems insignificant compared to the transmission rates, i.e., the transmission rates have a greater impact upon virus dominance.

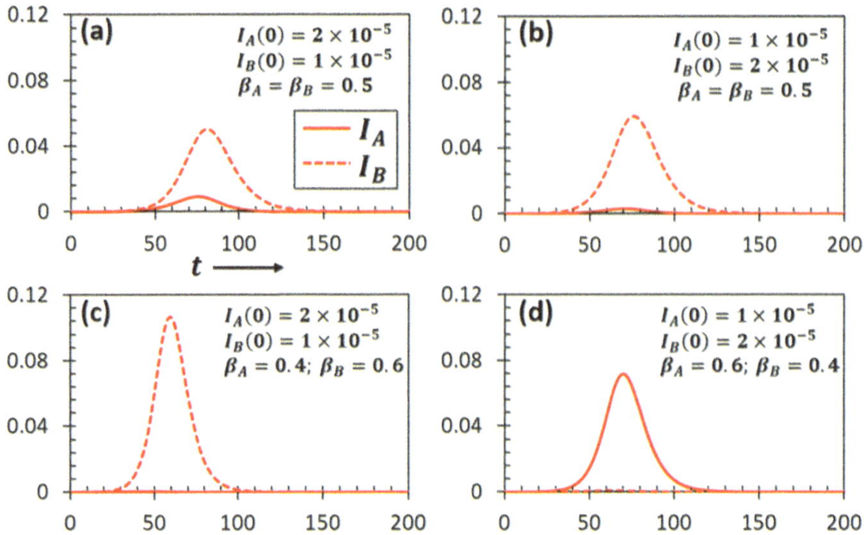

Fig. 10.2 Time series of the fractions of infection by two influenza viruses (A and B) for different initial conditions and transmission rates without considering a game-theoretic approach. The vaccination coverage for each vaccine is assumed to be approximately 33% with vaccine effectivenesses $e_Q = 0.6$ and $e_T = 0.4$. The recovery rate for each flu virus is chosen to be 0.2. Under the current setup, when $\beta_A = \beta_B$ (panels (a,b)), the B virus is found to dominate because $e_T \leq e_Q$, whereas the flu virus with a higher transmission rate (panels (c,d)) dominates the other one. The transmission rate seems more influential in virus dominance than the degree of initial dominance of infection by each virus

10.3.2 Evolutionary Outcome of Vaccination Coverage

We estimate the coevolution of both types of vaccination at the end of each season. Due to the complexity of Eq. (10.1-5), it is difficult to theoretically derive all possible equilibria; however, it is still possible to derive them numerically by tuning different parameters. Considering equal transmission rates for both viruses, we estimate six possible combinations of evolutionary outcomes (Fig. 10.3), which are: $(x, y) \equiv (1, 0)$; $(x, y) \equiv (0, 0)$; $(x, y) \equiv (x^*, y^*)$; $(x, y) \equiv (x^*, 0)$; $(x, y) \equiv (0, 1)$; $(x, y) \equiv (0, y^*)$;where $0 < x^*, y^* < 1$. Figure 10.3 suggests that there is no bi-stability arising from Eq. (10.5). One possible reason is that the present model, as a non-linear dynamical system, instinctively has strong sinks or steeply curved orbits, being insensible to initial conditions. Let us note that we choose the initial values of x and y in such a way that $x + y \leq 1$. Moreover, when we vary the initial values of x, the initial value of y is set as 0.33 and vice versa.

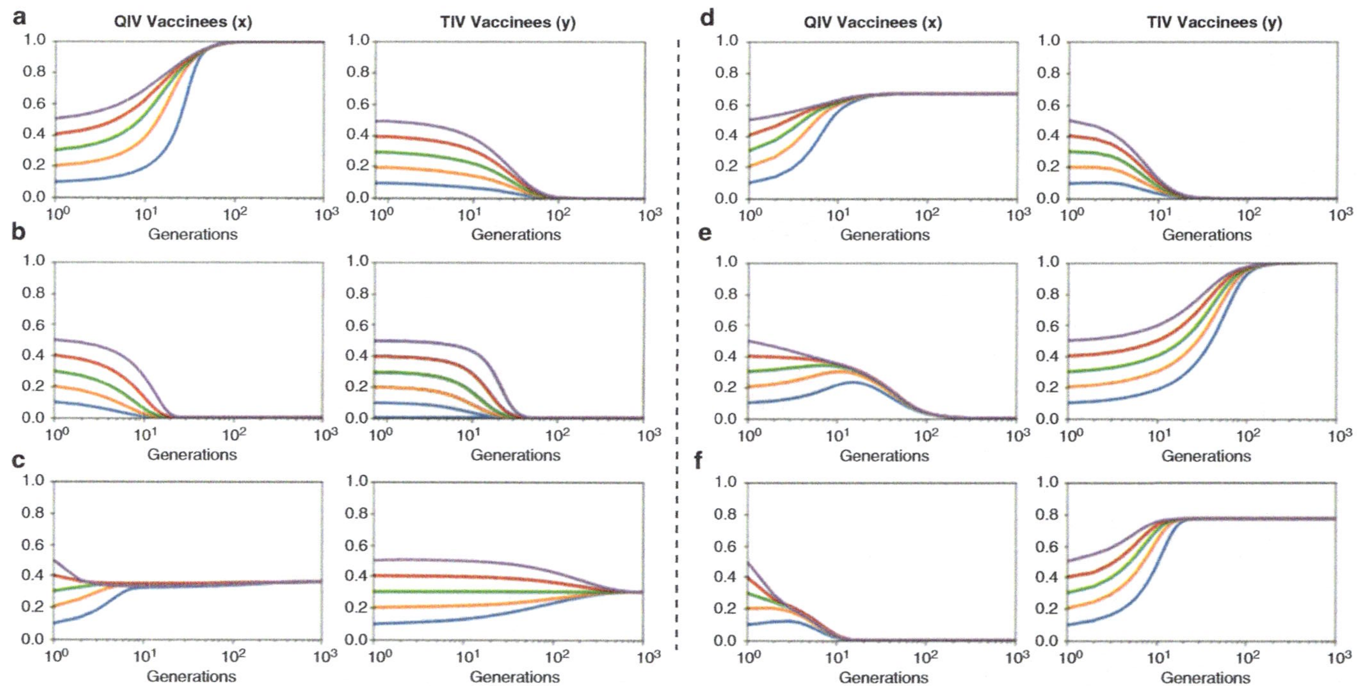

Fig. 10.3 Six possible combinations of QIV and TIV vaccination coverages at equilibrium attained from the evolutionary Eq. (10-5). The evolutionary dynamics of the vaccination coverage depends upon the choice of parameters. Clearly there is no bi-stability at equilibrium. There are six possible mixtures of dynamics: (a) $x = 1, y = 0$, when $C_Q = C_T = 0.4$, $e_Q = 0.4, e_T = 0.3$. (b) $x = 0, y = 0$ when $C_Q = 0.6, C_T = 0.4, e_Q = 0.4, e_T = 0.3$. (c) $x = x^*, y = y^*$, (coexistence phase of QIV and TIV; meaning the internal equilibria for both x and y) when $C_Q = 0.6, C_T = 0.4, e_Q = 0.8, e_T = 0.4$. (d) $x = x^*$(internal equilibrium), $y = 0$, when $C_Q = C_T = 0.4, e_Q = 0.7, e_T = 0.4$. (e) $x = 0, y = 1$ when $C_Q = 0.6, C_T = 0.4, e_Q = 0.6, e_T = 0.4$. (f) $x = 0, y = y^*$ (internal equilibrium), when $C_Q = 0.6$, $C_T = 0.4, e_Q = 0.8, e_T = 0.6$. All cases presume $\beta_A = \beta_B = 0.5, \gamma_A = \gamma_B = 0.2$. The initial value of y is kept as 0.33, while we vary the initial values of x and vice versa

10.3.3 Phase Diagrams

Case I ($\beta_A = \beta_B$)

Figure 10.4 portrays several heat maps depicting the FES for influenza A (B) virus—FES_A (FES_B), the vaccination coverage for the QIV (TIV) vaccine, and the average social payoff (ASP) (by referring to Eq. (10.3-1)) at social equilibrium as a function of (C_Q, C_T), presuming an equal transmission rate for both viruses (for the mean-field approach, we set $\beta_A = \beta_B = 0.5$ and $I_A(0) = 0.00001$, $I_B(0) = 0.00002$). It is worth mentioning that all heat maps in our analyses presume that the transition from blue to red moves from a good/better state to a bad/worse state of society in terms of infection, vaccination coverage, and ASP, and vice versa. Clearly, the infection due to the influenza B virus predominates the other, which is akin to what we observed in Figs. 102 (a) and (b) for the case of equal transmission rates. As long as $C_T < C_Q$ and C_T is below a certain threshold, a broader regime in the TIV

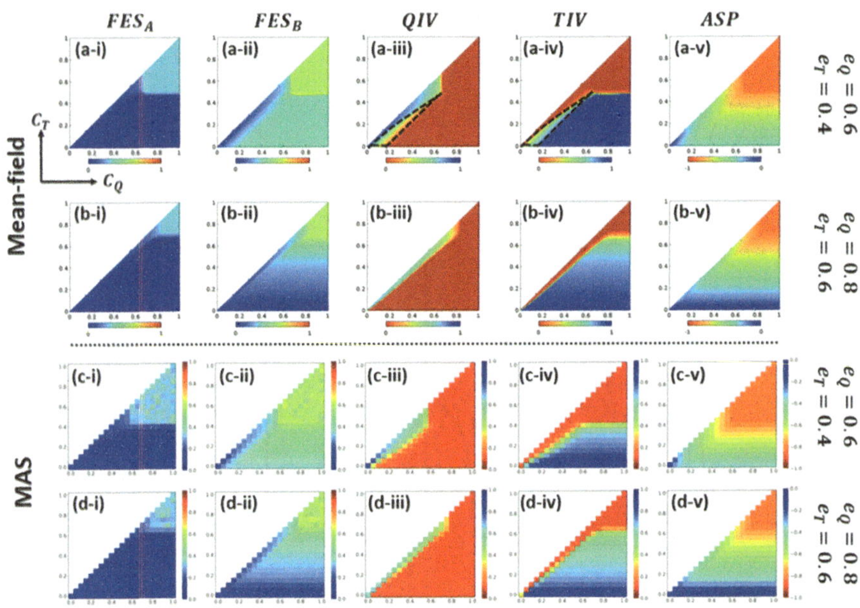

Fig. 10.4 C_Q versus C_T ($C_T \leq C_Q$) 2D heat maps with different effectiveness levels presuming mean-field (presuming $\beta_A = \beta_B = 0.5$, $I_A(0) = 0.00001$, $I_B(0) = 0.00002$) and multi-agent simulation (MAS) approaches in a well-mixed population. Chronologically, each column represents the final-epidemic size for influenza A (FES_A), the final-epidemic size for influenza B (FES_B), the fraction of QIV vaccinees, the fraction of TIV vaccinees, and the average social payoff. Clearly the infection due to influenza B dominates for the case $\beta_A = \beta_B$. The fraction of TIV vaccinees is seen to prevail for certain regime as long as C_T is below a certain threshold level; however, QIV vaccinees predominate whenever C_Q and C_T are comparable. Results from the MAS approach show an overall similar tendency to that of the mean-field approach. The MAS approach presumes 10,000 agents with $I_A(0) = 2$ agents, $I_B(0) = 4$ agents, $\beta_A = \beta_B = 5.19957 \times 10^{-5}$, and $\gamma_A = \gamma_B = 0.2$, and takes an ensemble average of 100 realizations

heat maps is colored in blue (panels (a–iv) and (b–iv)); the counterpart of this region in the QIV heat maps is colored in red, meaning that, because of the lower cost, most individuals are interested in taking the TIV vaccine until it bestows a considerable efficacy against virus B (that is, of course, below the efficacy of the QIV vaccine). Despite a major portion of the population taking the TIV vaccine, it cannot fully suppress the infection due to influenza B that can be clearly perceived from the FES_B heat maps, which accordingly bestow a lower average payoff to the society (see panel (a–v)). However, the increase of the effectiveness of the TIV vaccine against both viruses (if $e_Q = 0.8$, $e_T = 0.6$) would provide a better situation for society, as can be observed from the heat maps in the second row. Remarkably, the vaccination coverage for the QIV vaccine seems to be dominant whenever both vaccination costs are comparable. Moreover, we can perceive the coexistence of both vaccinees in the borderline region between the blue and red colors (yellowish regime enclosed by dotted lines in (a–iii) and (a–iv)).

The results obtained from the mean-field framework have been justified by a sequence of numerical simulations based on the multi-agent simulation (MAS) approach as introduced in Sect. 3.3, presuming a complete graph as an underlying network (since we assume a well-mixed population) with population size $N = 10{,}000$. The transmission-rate parameters are chosen as $\beta_A = \beta_B = 5.19957 \times 10^{-5}$. These are the effective transmission rates explained in Sect. 3.3.2, which are estimated as the minimal transmission rates that surpass the preset threshold FES of 0.9 without any vaccination. The corresponding initial infections are $I_A(0) = 2$ and $I_B(0) = 4$ agents. Note that we estimate the ensemble average for each 100 realizations. Generally, results obtained from the mean-field framework and the MAS approach show the same tendency overall, albeit having subtle discrepancies in terms of color scaling that come from the fact that the MAS approach assumes a finite population.

Now, let us focus upon analyzing similar phase diagrams by varying the vaccine efficacies (effectiveness). To this end, we generate several heat maps as above as a function of (e_Q, e_T), $0 \le e_T \le e_Q \le 1$ for the cases of similar and differing cost levels (Fig. 10.5). Obviously, similar costs for both vaccines would encourage people to choose the QIV vaccine instead of the TIV because the former can offer better protection against both viruses. This situation illustrates why the sensitivity comes only along the direction of e_Q (upper panels in Fig. 10.5). A lower degree of vaccine effectiveness would not entice individuals to take the vaccine (red region in Fig. 10.5 (a–iii)); however, if e_Q goes beyond a threshold level, the fraction of QIV vaccinees mounts to the highest level (blue region in Fig. 10.5 (a–iii)). A further increase of e_Q seems to supress both infections significantly, allowing some people to avoid vaccination by free riding upon herd immunity. This is why, after passing a transient regime with maximum vaccination coverage, the fraction of vaccinees decreases monotonically even with the increase of e_Q (yellowish region in Fig. 10.5 (a–iii)). That can be said to be proof of a social dilemma working behind the dynamics of our model, which is more carefully discussed in the following sub-section.

Nevertheless, if the cost for QIV is higher than that for TIV (lower panels in Fig. 10.5), individuals' vaccination choices differ depending upon the degrees of e_Q

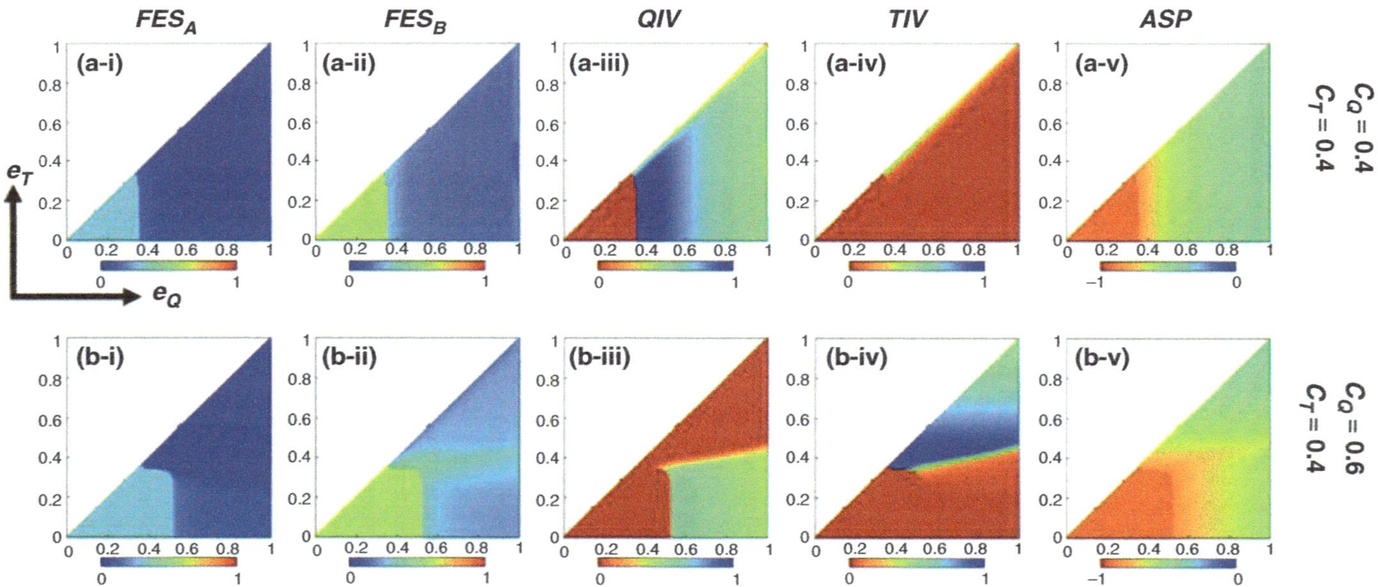

Fig. 10.5 e_Q versus e_T ($e_T \leq e_Q$) 2D-heat maps with similar (upper panels) and different (lower panels) cost levels, presuming a mean-field framework with parameter choices $\beta_A = \beta_B = 0.5$, $\gamma_A = \gamma_B = 0.2$ and initial conditions $I_A(0) = 0.00001$, $I_B(0) = 0.00002$. Similar costs encourage people (with a higher degree) to commit to the QIV vaccine, which consequently shows sensitivity in the direction of e_Q. However, in the case of different costs ($C_Q > C_T$), the QIV vaccine is favored over the TIV vaccine when $e_Q \geq 0.5$ (approximately); on the other hand, TIV is favored over QIV when $e_T \geq 0.4$ (approximately). The average social payoff for a similar cost seems to be better than that for different costs. Moreover, infection due to influenza A is lower than that due to influenza B

and e_T. More specifically, under the current settings, the QIV vaccine is favored over the TIV vaccine if e_Q is above approximately 50% and e_T is below approximately 40%; on the other hand, if $e_T$is above approximately 40%, then the TIV vaccine is preferred over the QIV vaccine, regardless of e_Q (of course, with the condition $e_T \leq e_Q$). Notably, in this case, the fraction of QIV vaccinees never reaches a maximum as the corresponding cost is higher; by contrast, the TIV vaccinees are seen to reach their highest level for a mid-range of e_T, although the number starts to decrease with the further increase of e_T that arises with the prevalence of free riders. Remarkably, the average payoff for society in the case of different costs seems lower than that for the equal cost (see heat maps for ASP in Fig. 10.5), as the former case imposes a higher financial burden upon society.

Case II ($\beta_A > \beta_B$)

The case $\beta_A > \beta_B$ ($\beta_A = 0.6, \beta_B = 0.4$) leads to the infection dominance of influenza A over B. See Fig. 106 (a) and (b), especially the top red triangular region. This situation can be controlled by the TIV vaccine alone, as it targets two strains of the A virus. Individuals then mostly prefer TIV vaccine over QIV vaccine, as the price of TIV is lower than that of QIV (see Fig. 10.6 (c) and (d)). Consequently, we perceive the sensitivity of choosing TIV only along the direction of C_T.

Case III ($\beta_A < \beta_B$)

Let us term to the right panels in Fig. 10.6. The predominance of influenza B over A creates a dilemma for choosing vaccine types. Individuals prefer the QIV vaccine over TIV whenever both costs are comparable or the cost difference between QIV and TIV is not very high (the region enclosed by black dotted lines in Fig. 10.6 (g)); however, the TIV vaccine seems mostly favorable in the case of a higher vaccination cost for QIV or a higher cost difference between QIV and TIV vaccine (region enclosed by red dotted lines in Fig. 10.6 (h)). Panels (g) and (h) in Fig. 10.6 also depict the coexistence of both types of vaccinees in the yellowish transient regions. Furthermore, comparing the light blue region (enclosed with black dotted lines) with the greenish region (enclosed by red dotted lines) in Fig. 10.6 (f), we observe that the QIV vaccine-dominant region imposes a greater disease-suppression effect than the TIV vaccine.

β_A versus β_B phase plane

We intend to observe the overall impact of the transmission rates upon disease propagation and vaccine uptake. To this end, we draw heat maps for total infection (FES) due to both viruses, the basic reproduction number (R_0), and the vaccination coverage for both vaccines (QIV, TIV) as functions of (β_A, β_B). Figure 10.7 displays the result. The vaccine-dependent basic reproduction number R_0[7] is defined as $R_0 = \max\{R_0^A, R_0^B\}$, where R_0^A and R_0^B are the vaccine-dependent basic reproduction number for the influenza A and B viruses, respectively. We derive R_0^A and R_0^B as

[7]van den Driessche, P.; Reproduction numbers of infectious disease models, *Infectious Disease Modelling* **2** **(3)**, 288–303, 2017

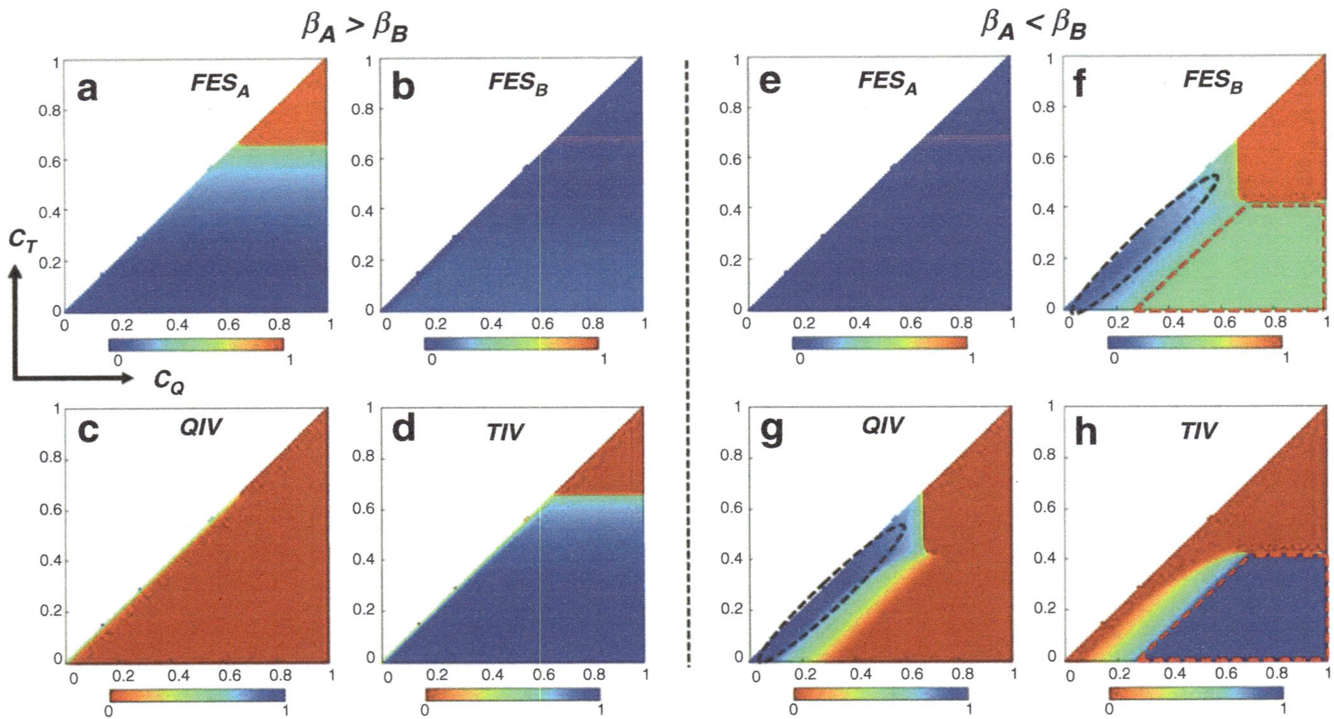

Fig. 10.6 The heat maps (a–d) show the fractions of infected individuals with influenza A and B (FES_A, FES_B) and the fractions of QIV and TIV vaccinees as functions of C_Q and C_T ($C_T \le C_Q$) when the transmission rate for virus A is higher than that for virus B ($\beta_A > \beta_B$; $\beta_A = 0.6$, $\beta_B = 0.4$). Panels (e--h) display the same objects for the case where $\beta_A < \beta_B$; $\beta_A = 0.4$, $\beta_B = 0.6$. We choose $e_Q = 0.6$ and $e_T = 0.4$ for both cases. We can perceive the infection dominance of viruses A and B according to the conditions $\beta_A > \beta_B$ and $\beta_A < \beta_B$. With the current settings, $\beta_A > \beta_B$ implies the total predominance of TIV vaccinees (panels (e, f)). However, if $\beta_A < \beta_B$, then QIV vaccinees prevail whenever the cost difference is not very high and both costs are below 0.6 (approximately); however, the prevalence of TIV vaccinees surges with the increase of C_Q and the cost difference. The region where QIV dominates has a greater impact on disease suppression than that where TIV dominates (panels (f,g,h))

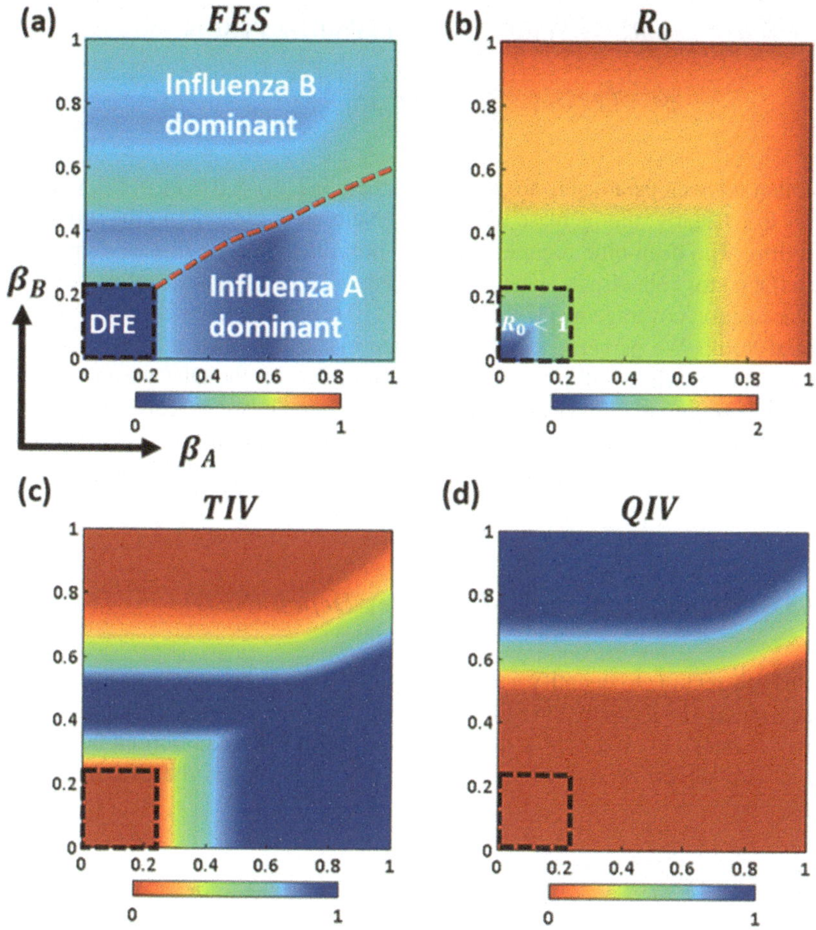

Fig. 10.7 The effect of transmission rates (β_A, β_B) upon the spread of disease (panel (a)), the vaccine-dependent basic reproduction number-R_0(panel (b)), and the vaccination coverage of both vaccines (panels (c,d)). The deep-blue region inside the box in panel (a) represents a disease-free equilibrium (DFE) for both viruses, where R_0 is below one (panel (b)), and accordingly there is no vaccination coverage of any type inside the box. Outside of that box, influenza A (B) becomes prevalent with the increase of $\beta_A(\beta_B)$. The red dotted line depicts the threshold level below (above) which influenza A (B) is dominant. TIV vaccinees predominate when β_A gets larger and QIV vaccinees predominate when β_B increases. The relevant parameters are set as $\gamma_A = \gamma_B = 0.2$; $e_Q = 0.6$, $e_T = 0.4$; $C_Q = 0.4$, $C_T = 0.2$

$$R_0^A = \frac{\beta_A}{\gamma_A}(S(0) + (1 - e_Q)(V_Q(0) + V_T(0))),$$

$$R_0^B = \frac{\beta_B}{\gamma_B}(S(0) + (1 - e_Q)V_Q(0) + (1 - e_T)V_T(0)).$$

Other relevant parameters are set as $\gamma_A = \gamma_B = 0.2$, $e_Q = 0.6$, $e_T = 0.4$, $C_Q = 0.4$, $C_T = 0.2$. Clearly, the increase of β_A (β_B) results in an upsurge in the influenza A (B) virus. The deep-blue regime inside the box illustrates the disease-free equilibrium (DFE) for both viruses, where $R_0 < 1$ (Fig. 10.7 (b)). Consequently, we observe no vaccination coverage inside that box (Fig. 10 (c) and (d)), which might be quite a natural result. We perceive a red dotted line in the FES heat map (Fig. 10.7 (a)) below (above) which influenza A (B) predominates. As the QIV vaccine provides better protection against virus B than the TIV vaccine, we perceive the predominance of QIV vaccinees in the case of higher β_B; contrarily, the TIV vaccine is favored over QIV for higher β_A, because its price is lower than that of QIV.

10.3.4 Analysis of Social-Efficiency Deficit (SED)

In this sub-section, to quantify a social dilemma working behind the dynamics of the present study, let us apply the SED, which was theoretically formulated in Sect. 2.3. Now, we intend to explore how much the payoff attained at equilibrium falls short of the desired payoff or social-optimum (SO) payoff, which is the SED that can be defined as

$$\text{SED}, \delta = ASP^{SO} - ASP^{EQM}, \tag{10.6}$$

where ASP^{EQM} and ASP^{SO}denote the ASP at equilibrium and social optimum, respectively. We use this parameter to explain the degree of the dilemma associated each vaccine choice.

As there are two vaccines available, individuals face a dual-dilemma situation: whether to choose the QIV vaccine or the TIV vaccine, or otherwise choose none, depending upon their assessment of the cost and effectiveness. These factors allow us to evaluate the SED corresponding to both vaccines to ponder the impact of both types of vaccinations upon society. The procedure given below can be followed in calculating SED for both vaccines. Figure 10.8 visually explains how SED can be identified. Suppose that we wish to derive SED as a function of (C_Q, C_T). Note that it is also possible to express SED as a function of (e_Q, e_T).

Step I) We estimate ASP at equilibrium (EQM) as a function of (C_Q, C_T), for instance, by considering the heat map for ASP given in Fig. 10.4 (a–v).

Step II) Each point on this heat map corresponds to a pair, $\left(C_Q^*, C_T^*\right)$. Also, there is a corresponding pair (x^*, y^*), i.e., the vaccination coverage for each vaccine, for

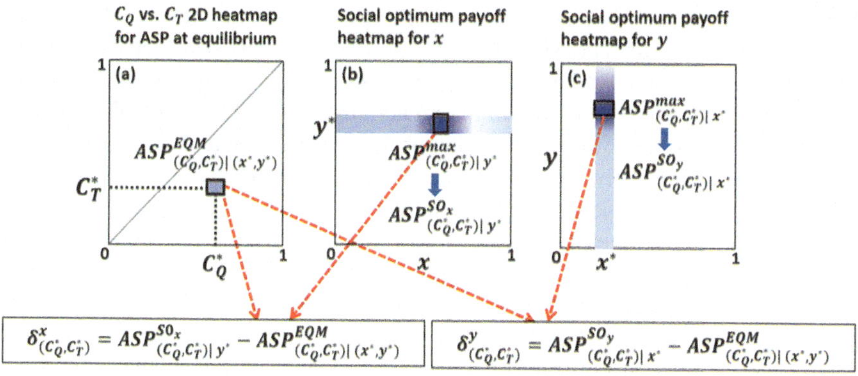

Fig. 10.8 A schematic of SED derivation. At first, we choose a point (C_Q^*,C_T^*) on the heat map (C_Q vs. C_T) illustrating the average social payoff (ASP) at equilibrium (EQM) (panel (a)), for which there is a corresponding point, (x^*, y^*), i.e., a corresponding pair of vaccinees (QIV, TIV). With fixed (C_Q^*,C_T^*) and y^*, we vary x ($0 \leq x \leq 1$) and estimate the maximum average payoff, which we call the social-optimum (SO) payoff for x at (C_Q^*,C_T^*) and y^*, i.e., $ASP^{SO_x}_{(C_Q^*,C_T^*|y^*)}$ (panel (b)). Afterwards, we derive SED for x at (C_Q^*, C_T^*) by subtracting ASP at equilibrium from the payoff at social optimum. We follow the same procedure to derive SED for y using panels (a) and (c)

each pair $\left(C_Q^*, C_T^*\right)$. Now, fixing $y^*(x^*)$ and varying $x(y)$ from 0 to 1, we estimate the maximum average payoff using Eq. (10.3-1), which we call the socially optimum payoff for $x(y)$ at the point $\left(C_Q^*, C_T^*\right)$ (Fig. 10.8 (a)-(c)).

Step III) Next, we derive SED for $x(y)$ at the point $\left(C_Q^*, C_T^*\right)$ by simply subtracting the ASP attained in step I from the SO payoff attained in step II and denote it by $\delta^x_{\left(C_Q^*,C_T^*\right)}\left(\delta^y_{\left(C_Q^*,C_T^*\right)}\right)$; that is,

$$\delta^x_{\left(C_Q^*,C_T^*\right)} = ASP^{SO_x}_{\left(C_Q^*,C_T^*\right)|\,y^*} - ASP^{EQM}_{\left(C_Q^*,C_T^*\right)|\,(x^*,y^*)},$$

$$\delta^y_{\left(C_Q^*,C_T^*\right)} = ASP^{SO_y}_{\left(C_Q^*,C_T^*\right)|\,y^*} - ASP^{EQM}_{\left(C_Q^*,C_T^*\right)|\,(x^*,y^*)}.$$

Now let us explain the heat maps for SED that are generated by following the above procedure. As an archetypical example, we choose the case illustrated in the upper panel of Fig. 10-4, where we assumed $\beta_A = \beta_B = 0.5$ and $e_Q = 0.6$, $e_T = 0.4$. In that case, we observed that most people are inclined to choose a TIV vaccine whenever the cost difference between C_Q and C_T becomes higher; however, individuals prefer QIV if the cost difference becomes lower. It is conceivable that the overall desired payoff or SO payoff for QIV vaccinees (x) would be relatively higher than that for TIV vaccinees (y), as presented in panels (a,b) of Fig. 10.9. However, in this case, the vaccination coverage for QIV seems much lower than that for TIV (see

Fig. 10.4 (a–iii) and (a–iv)), indicating that the dilemma of choosing the QIV vaccine is greater than that for choosing TIV. The term "dilemma" here comes from the literal definition of SED, whereby the realized situation is inferior to the ideal one. Obviously, this comes about due to the higher cost associated of the QIV vaccine. This factor results in a lower contribution of the QIV vaccine to ASP (see Fig. 10.4 (a–v)). Hence, we perceive a relatively higher gap between the SO payoff for x and ASP (at equilibrium); hence, more areas of the heat map for SED_x seem dark compared to its counterpart (SED_y). The whiteout region in the heat maps (Figs. 10.9 (c) and (d)) depicts the scenario having no SED, that is, having no gap between the SO payoff and the ASP, implying no dilemma at all. There are two whiteout regions in the SED_y-heat map (Fig. 10.9 (d)), which either come from 100%-TIV-vaccination coverage (region enclosed by blue dotted lines) resulting from the lower C_T and higher C_Q (blue region in Fig. 10.4 (a–iii)), or from the scenario whereby everyone becomes a free-rider (region enclosed by red dotted lines in Figs. 10-9 (c) and (d)). The latter situation can arise for higher vaccinations cost or for very low vaccine effectiveness. In this case, avoiding any type of vaccination is the best strategy (desired strategy), resulting in a D (unvaccinated)-dominant equilibrium (see Figs. 10.4 (a–iii) and (a–iv)). Hence, the payoff at this equilibrium also agrees with the SO payoff, so no SED nor dilemmas arises in this case. This scenario can be classified as a D-dominant trivial game based on our discussion in Chap. 2, where choosing defection (i.e., choosing none of the pre-emptive vaccines) is a social optimum, which seems ironic. This is because the vaccines are so meager that they are not able to lead a better social state than doing nothing. By contrast, the former case can be said to be a C-dominant trivial game, in which all individuals take vaccination. It is, however, not QIV but TIV that is inferior to QIV.

10.3.5 Comprehensive Discussion

Generally, it has been thought that people's vaccination choice is influenced by costs as well as vaccine efficacies. However, the present results show that individuals' vaccination options (QIV or TIV) are more affected by the cost difference between vaccines as long as the vaccine with the lower price is able to bestow a considerable level of efficiency, especially against virus B. Another interesting result concerns the variation of transmission rates for both viruses and its effect upon disease propagation and vaccination choice. A higher transmission rate of influenza A leads to a maximum coverage for the TIV vaccine, since it possesses the same degree of effectiveness against influenza A as that of the QIV vaccine but with a lower cost; by contrast, the higher transmission rate of influenza B results in an almost-complete dominance of QIV vaccinees, which is quite conceivable. These results may be meaningful for building an effective policy on the part of the public-health authority responsible for controlling seasonal-influenza pandemics.

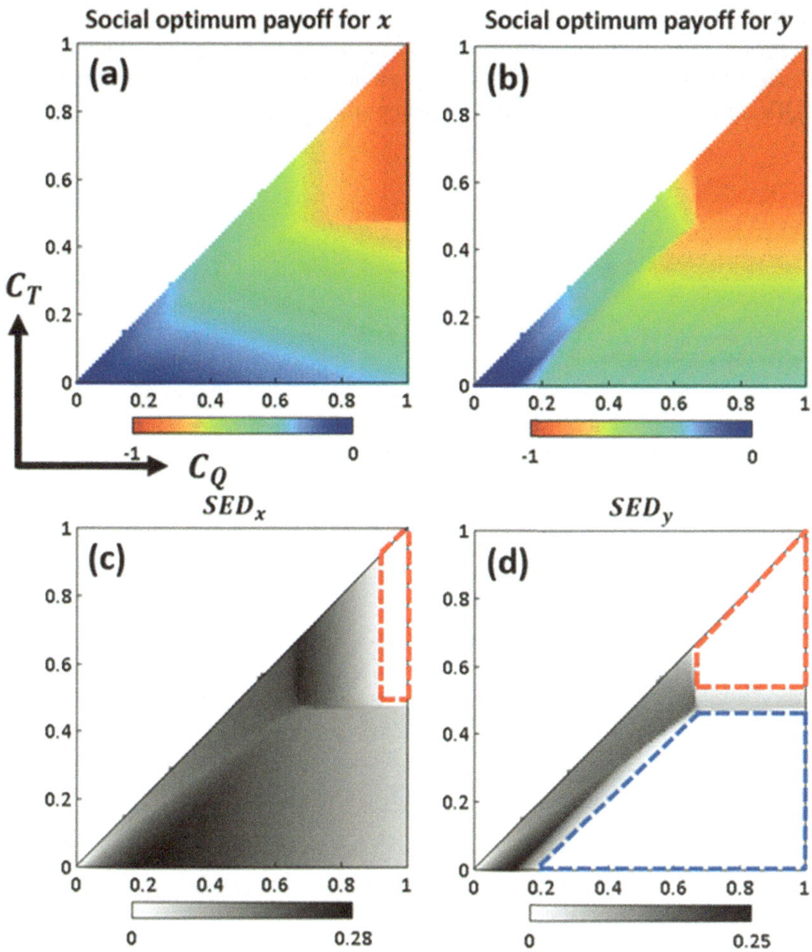

Fig. 10.9 Representation of the social-optimum payoff (panels (a,b)) and SED (panels (c,d)) for x (fraction of QIV vaccinees) and y (fraction of TIV vaccinees) as a function of (C_Q, C_T). SED heat maps are obtained by subtracting the ASP at equilibrium (Fig. 10.4 (a–v)) from the social-optimum (SO) payoffs for x and y (panels (a, b)). The relevant parameters are chosen as $e_Q = 0.6$, $e_T = 0.4$, $\beta_A = \beta_B = 0.5$, and $\gamma_A = \gamma_B = 0.2$. Referring to Fig. 10.4 (a–i)–(a–iv), these parameter choices depict the prevalence of the influenza B virus and a lower vaccination coverage for the QIV vaccine (especially when the cost difference is high). The whiteout regions in panels (c,d) depict there being no gap between SO payoffs (for x or y) and ASP at equilibrium. More areas in the SED heat map for x appear to have darker regimes than in the case for y, which indicates a higher gap between the SO payoffs for x and ASP at equilibrium

Our analysis also considers the extent to which the overall payoff or the ASP attained by voluntary pre-emptive vaccination (at equilibrium) falls short of the desired payoff or social-optimum payoff; we have called this the SED—the payoff gap between social optimum and equilibrium. This parameter also helps us to elucidate the degree of dilemma associated with each vaccination choice. It appears that, for the case of equal transmission rates, people face a greater dilemma when choosing QIV vaccine than TIV, which arises because of the higher vaccination cost for QIV, although it confers relatively better protection.

Chapter 11
Optimal Design of a Vaccination-Subsidy Policy

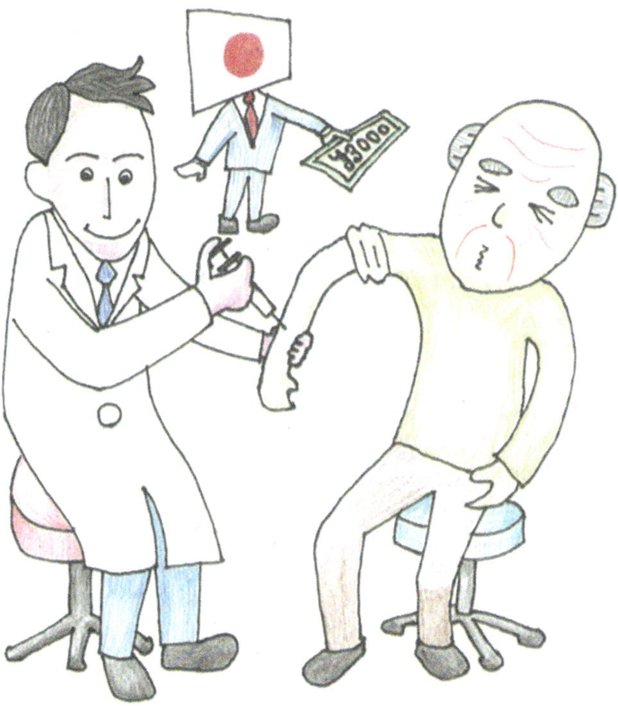

As we have discussed, pre-emptive vaccination is one of the most effective intervention measures for preventing the spread of a seasonal disease like influenza. This chapter explores a substantially important question of what kind of subsidy policy can minimize the holistic social cost, as quantified by the average social payoff (ASP). Roughly speaking, the concept of subsidizing vaccination is twofold: free-ticket and discount policies and funded under the same amount of the government

J. Tanimoto, *Sociophysics Approach to Epidemics*, Evolutionary Economics and Social Complexity Science 23, https://doi.org/10.1007/978-981-33-6481-3_11

budget. We explore whether or not another policy leading to a social optimal would be possible. In this chapter, we take a MAS approach, presuming a complex social network on which an epidemic spreads.

11.1 Introduction and Background: Free Ticket, Discount Ticket, or a Combination of the Two—Which Subsidy Policy Is Socially Optimal to Suppress the Spread of Disease?

Pre-emptive vaccination, if available, is regarded as the most effective preventative measure for reducing the global infection rate as well as the mortality associating with infectious diseases. Vaccination can be treated as voluntary as well as compulsory. While voluntary vaccination may fail to achieve herd immunity, mandatory vaccination may result in infringement of civil rights. However, it has been proved that vaccination campaigns that do not rely on individual financial coverage are quite effective as a provision for combating worldwide epidemics. In fact, global vaccination programs have reduced the prevalence of measles, pertussis, and polio to a great extent over the last several decades. By contrast, as long as a voluntary vaccination campaign is imposed, such a phenomenon may lead to the emergence of free riders who avoid vaccination, hoping to benefit from others' vaccination and to circumvent vaccine-related side effects. Therefore, deciding whether to vaccinate inevitably contains a social dilemma—"the vaccination dilemma" or "paradox of epidemiology"—which has been studied in previous chapters.

Achieving herd immunity can be difficult under a voluntary vaccination policy due to the prevalence of vaccine hesitancies that arise from vaccination cost, vaccine-related side effects, superstitions, etc., as mentioned above. To resolve such issues, governments or authorities often offer subsidies to inspire vaccination. For instance, in Japan, individuals over a certain age and other critical groups are given a subsidy to vaccinate against seasonal influenza, whereas other general people are not covered by public-health insurance when taking influenza vaccines. Although subsidies impose an extra burden on the economy, authorities in countries with universal medical-insurance systems like that found in Japan may be able to avoid rather higher medical-treatment social costs due to many infected people requiring care. Such external incentives as financial support from subsidies can alter individuals' vaccination decisions. Therefore, the choice of a subsidy must be carefully considered. To achieve a better outcome from subsidies, better policymaking is required; that is, subsidies should be optimally allocated such that herd immunity is established with a minimal social cost. Therefore, examining various provisions of subsidies and their comparative outcomes may help management to enact an efficient policymaking.

Several works have been contributed to address the vaccination-subsidy policy. Zhang et al.[1] established an evolutionary-vaccination-game model on a networked population to study the effect of two different subsidy policies: (i) dispensing free tickets to a certain fraction of individuals and (ii) offering a certain amount of subsidy (hereafter called a discount policy) to everyone. Another work along this vein was conducted by Zhang et al.[2], where the authors compared the discount-ticket policy with the free-ticket policy in which recipients of free tickets are randomly selected. Similar works exploring the interplay between subsidy policies and human behavioral responses can be found in Ding et al.[3] and Zhang et al.[4] Tang et al.[5] presented an interesting subsidy model whereby incentives are provided from an internal public fund (instead of a central-government intervention) contributed by individuals in a locality.

In the following discussion in this chapter, we imagine the vaccination-game model that we have discussed so far, whereby an agent either intends to voluntarily vaccinate (and is called a "cooperator"), or not (and be called a "defector"). Vaccination is introduced as a pre-emptive measure about which an agent draws a decision before entering the epidemic season. Because of the MAS approach this chapter takes to consider a more realistic situation than a well-mixed and infinite population, the Barabási–Albert scale-free graph[6] of average degree $\langle k \rangle = 8$ is assumed to be the underlying social network on which the epidemic spreads.

One subsidy policy is to offer a fixed discount to all cooperative agents (agents with the intention to vaccinate), irrespective of their degrees; this we call the "flat-discount" policy (see Fig. 11.1 (a)). Another protocol is to distribute free tickets to cooperative hub agents, which can be termed a "free-ticket" policy (Fig. 11.1 (b)). Taking inspiration from these two policies, we here design another subsidy policy within the framework of the vaccination game on a scale-free network, which we call the "degree-dependent subsidy," whereby the extent of a discount will be given according to the degree of cooperative agents (Fig. 11.1 (c)), i.e., according to the extent of the connection of cooperative agents. In other words, an agent with a higher degree will be given a higher incentive or discount, and vice versa. We will compare

[1]Zhang, H.F., Wu, Z.X., Xu, X.K., Small, M., Wang, L., Wang, B.H.; Impacts of subsidy policies on vaccination decisions in contact networks, *Physical Review E* **88**, 012813, 2013.

[2]Zhang, H.F., Wu, Z.X., Tang, M., Lai, Y.C.; Effects of behavioral response and vaccination policy on epidemic spreading - An approach based on evolutionary-game dynamics, *Scientific Reports* **4**, 5666, 2014.

[3]Ding, H., Xu, J.H., Wang, Z., Ren, Y.Z., Cui, G.H.; Subsidy strategy based on history information can stimulate voluntary vaccination behaviors on seasonal diseases, *Physica A* **503**, 390–399, 2018.

[4]Zhang, H.F., Shu, P.P., Wang, Z., Tang, M., Small, M.; Preferential imitation can invalidate targeted subsidy policies on seasonal-influenza diseases, *Applied Mathematics & Computation* **294**, 332–342, 2017.

[5]Tang, G.M., Cai, C.R., Wu, Z.X.; Evolutionary vaccination dynamics with internal support mechanisms, *Physica A* **473**, 135–143, 2017.

[6]Barabási, A.L., Albert, R.; Emergence of scaling in random networks, *Science* **286**, 509–512, 1999.

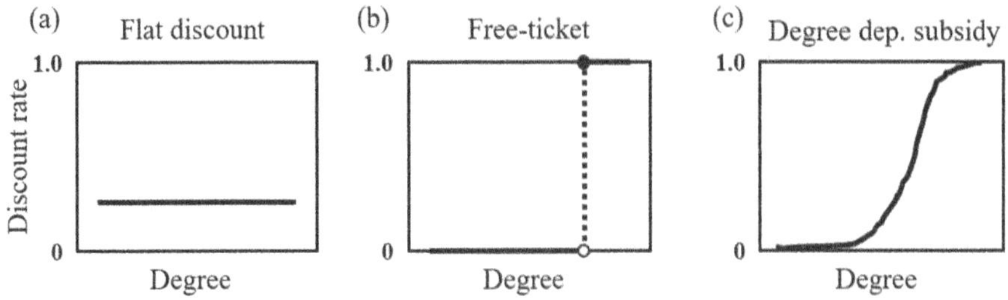

Fig. 11.1 A comparative schematic of three subsidy policies as functions of the degree of agents: (a) giving the same discount to all cooperative agents, i.e., flat discount; (b) giving free tickets only to hub agents (having higher degree); and (c) giving discounts according to the degree of agents (i.e., according to the degree-dependent subsidy)

the efficiencies, which are measured as the ASP of the newly introduced degree-dependent subsidy, flat discount, and free-ticket policy under different circumstances.

11.2 Model Design

As mentioned before, the present model considers a subsidy policy whereby incentives are given to cooperative agents (or C agents) according to their degree distribution on a scale-free network—that is, C agents with a greater degree will receive a higher discount, and vice versa. The main purpose of the degree-dependent subsidy is to investigate the efficiency of the current subsidy policy under different circumstances compared to the free-ticket and discount-ticket policies.

Another purpose of this new scheme is to explore whether a certain intermediate step between flat-discount and free-ticket policies is able to emerge as a more socially preferable result than either of those two. Note that the concept of a flat discount should be addressed in terms of social equality, and that of a free ticket might be important from a social-efficiency viewpoint. The present agent-based-modeling (MAS) approach follows our description in Sect. 3.3.

11.2.1 Vaccination Game on a Scale-Free Network

We consider a population of size N, who are connected through a scale-free network and decide whether to vaccinate before the onset of an epidemic. We further assume the spreading of a seasonal-influenza-like disease that circulates recurrently. Annual vaccination is recommended due to the waning immunity and antigenic drift of the virus strains. Therefore, individuals decide whether to vaccinate for the next season, perceiving their previous season's experience. The whole dynamical setup is then

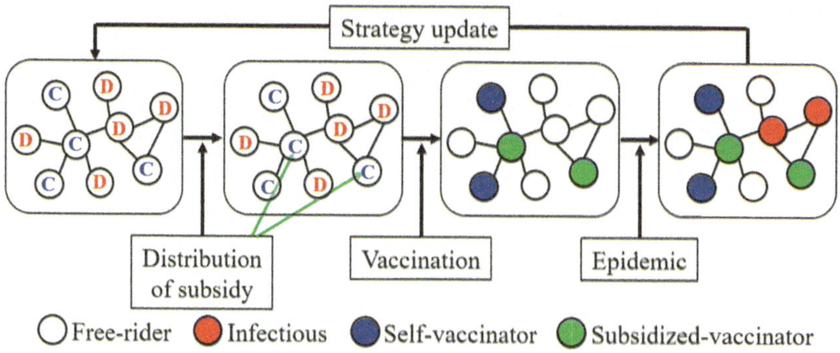

Fig. 11.2 The layout of the whole dynamical setup. The vaccination campaign is conducted prior to the onset of an epidemic. Some of the cooperative (C agent)-minded individuals are being subsidized (green colored), while some C agents vaccinate by self-financing (blue colored). Non-vaccinators can get infected (red colored) or remain healthy (successful free riders) during the epidemic. Revision of the strategies takes place before the onset of another epidemic. This process is repeated until we reach a steady state

divided into two: the circulation of an epidemic process and the evolution of the vaccination decision-making process. These two processes constitute one single period (i.e., a single season) in the model. To capture the equilibrium properties of the model, we iterate the procedure until we reach a steady state. The whole dynamical process is depicted in Fig. 11.2. At the steady state, we quantify the vaccination coverage (VC), final epidemic size (FES), and ASP.

Epidemic Spreading Consider a strain of influenza that is going to emerge among a population of size N. We start this process by assuming I_0 randomly selected initial infected individuals (we set $I_0 = 5$ throughout the study) among susceptible class. The disease spreads according to SIR/V dynamics, where transitions from the susceptible to the recovered stage occur via an infectious stage. The disease-transmission rate per day per person is assumed to be β [day^{-1} person^{-1}], while the recovery rate per day is denoted by γ [day^{-1}]. We choose $\beta = 0.833$ and $\gamma = 1/3$ by presuming the basic reproduction number $N_0 = \frac{\beta}{\gamma}$=2.5. Susceptible people who commit vaccination are presumed to have perfect immunity (for simplicity) throughout the epidemic season and to incur a vaccination cost, C_v. Those who avoid vaccination expose themselves to the possibility of infection. An infected individual possesses an infection cost C_i ($\geq C_v$); without loss of generality, one can choose $C_i = 1$ and normalize these costs by taking the relative cost of vaccination as $C_r = C_v/C_i$ $(0 \leq C_r \leq 1)$. However, a successful free rider who could avoid infection without vaccination incurs no cost at all.

Vaccination Campaign Since we are assuming the recurrent circulation of a seasonal influenza, individuals must decide whether to vaccinate before the onset of the next epidemic depending upon the perceived payoff attained in the previous season. Therefore, this stage is the "strategy-updating" stage. Furthermore, as vaccination is presumed to bestow perfect immunity, there is no possibility for a

Table 11.1 State of an individual along with their payoff

Strategy/State		Healthy	Infected
Vaccinator (C)	Self-financed	$-C_r$	
	Subsidized	$-C_r + r \cdot C_r$	
Non-vaccinator (D)		0	-1

vaccinator (denoted by V) to be infected, whereas a non-vaccinator (NV) may or may not be infected during the epidemic. Vaccinators can be of two types: self-financed vaccinators (i.e., those who commit vaccination voluntarily) and subsidized vaccinators. In the current study, subsidies are distributed only to cooperative (C) individuals, that is, to the people who intend to be vaccinated spontaneously. Defectors (D) are not offered any subsidy. At the end of the epidemic, a self-financed vaccinator attains a payoff $-C_r$, while a subsidized vaccinator receives a payoff $-C_r + r \cdot C_r$, where r is the rate or amount of discount given on the vaccination cost C_r. Obviously, $r = 1$ recovers the payoff for the free-ticket policy. Moreover, a NV who remains healthy during the epidemic achieves a payoff 0; however, if they become infected, they realize a payoff -1. All these payoffs, along with their strategies or states, are summarized in Table 11.1.

The update of the vaccination decision or strategy updating is performed using the pairwise Fermi rule. We presumed the individual-based risk assessment (IB-RA) introduced in Sect. 3.2.2. According to such a rule, an agent i having a strategy S_i randomly selects one of their neighbors, j with a strategy S_j, and decides whether to copy j's strategy with the probability

$$P(S_i \leftarrow S_j) = \frac{1}{1 + \exp\left[-(\pi_j - \pi_i)/\kappa\right]}, \tag{11.1}$$

where π_i denotes the payoff of agent i and $\kappa > 0$ signifies the strength of selection. A smaller κ corresponds to a higher sensitivity to the payoff difference. Here, we choose $\kappa = 0.1$.

11.2.2 Subsidy Policies

We will demonstrate the comparative effectivity of three different subsidy policies in terms of social cost and disease eradication: (i) distributing the same discount to all cooperative agents (flat discount); (ii) distributing free tickets to cooperative hub agents (free ticket); (iii) distributing discount tickets to C agents according to the degree distribution (degree-dependent subsidy). Suppose that σ is the fraction of the population to be subsidized. Thus, the total financial budget is $\sigma \cdot N \cdot C_r$. The parameter $\sigma \in [0, 1]$, in other words, depicts the budget size of the subsidy. A higher σ corresponds to a greater budget size.

(i) *Flat discount*: In this case, every cooperative agent gets the same discount, irrespective of their degree. The per capita discount to C agents can be defined by

$$d = \sigma N C_r / N_C, \tag{11.2}$$

where N_C is the number of cooperative agents in the population. Now, the rate of discount under this policy is

$$r = d/C_r = \sigma N / N_C. \tag{11.3}$$

Clearly, r is same for all cooperative agents for a fixed σ. If $d > C_r$, we set $r = 1$, and the remaining amount is returned to the taxpayer, i.e., *TAX* introduced in Eq. (11.7) is discounted. To avoid confusion, let us note that the discount-rate parameter r bears no relationship to the subscript of cost, C_r.

(ii) *Free ticket* (hub C-agent priority): In this policy, the whole vaccination cost, C_r, is subsidized for cooperative hub agents. Hence, the rate of discount under this free-ticket policy is $r = 1$. If $\sigma N C_r > N_C C_r$, the remaining budget is also returned, i.e., the *TAX* introduced in Eq. (11.7) is discounted.

(iii) *Degree-dependent subsidy*: In this subsidy policy, the discount ticket is distributed only to cooperators but according to the extent of their connection with other agents, meaning that individuals who are highly connected with others will receive a higher incentive. We first classify C agents by degree. Suppose that the degree (k) of agents on a scale-free network varies from k_{min} to k_{max}. As subsidies will be allocated according to the degree of C agents, it is plausible to give priority to those agents who are associated with a higher degree. In such a case, C agents (if any) having degree $k = k_{max}$ will be allocated the subsidy first, followed by less-connected individuals. That is, at the beginning, the whole budget $\sigma \cdot N \cdot C_r$ is available for these C agents. The amount of the budget for agents of degree k is allocated by dividing the available budget by the total number of degrees, ranging from k_{min} up to k, with which at least one cooperative agent is associated. If f_k is the available budget for degree k, then the maximum amount of the subsidy to be distributed to all C agents of degree k, denoted by g_k, is determined by the following formula:

$$g_k = \frac{f_k}{\sum_{l=k_{min}}^{l=k} \mathrm{H}\left[n_l^C\right]}, \tag{11.4}$$

where $\mathrm{H}\left[n_l^C\right]$ is the Heaviside function, yielding 1 if $n_l^C > 0$, otherwise 0. The notation n_l^C signifies the number of cooperators at degree l. The denominator in Eq. (11.4) counts the total degree ranging from k_{min} up to k if there is at least one C agent associated with degree l. Moreover, Eq. (11.4) confirms that no C agents will

Table 11.2 The allocation of the budget according to degree

k	f_k	g_k
k_{max}	σNC_r	$f_{k_{max}}/\sum_{l=k_{min}}^{k_{max}} \mathrm{H}[n_l^C]$
$k_{max}-1$	$\sigma NC_r - d_k^C \times n_{k_{max}-1}^C$	$f_{k_{max}-1}/\sum_{l=k_{min}}^{k_{max}-1} \mathrm{H}[n_l^C]$
$k_{max}-2$	$\sigma NC_r - \sum_{l=k_{max}-1}^{k_{max}} d_l^C \times n_l^C$	$f_{k_{max}-2}/\sum_{l=k_{min}}^{k_{max}-2} \mathrm{H}[n_l^C]$
⋮	⋮	⋮
k	$\sigma NC_r - \sum_{l=k+1}^{k_{max}} d_l^C \times n_l^C$	$f_k/\sum_{l=k_{min}}^{k} \mathrm{H}[n_l^C]$
⋮	⋮	⋮
k_{min}	$\sigma NC_r - \sum_{l=k_{min}+1}^{k_{max}} d_l^C \times n_l^C$	$f_{k_{min}}$

Notations: f_k= available budget at degree k, g_k= maximum amount of subsidy allocated for degree k, n_k^C = number of cooperators at degree k, $d_k^C = \min\{C_r, g_k/n_k^C\}$ = discounted amount at degree k, C_r= vaccination cost, H[*]= Heaviside function, σNC_r= total budget

be left without a subsidy. Thus, the discounted amount for each C agent with degree k can be calculated as

$$d_k^C = \mathrm{Min}\left[C_r, g_k/n_k^C\right]. \tag{11.5}$$

From this, we can define the discount relative to the cost C_r or discount rate r for the current subsidy policy as $r = C_r/d_k^C$. Eq. (11.5) ensures that the discounted amount does not exceed the vaccination cost, C_r. Note that every C agent with the same degree receives an equal amount of the subsidy. Therefore, the total amount allocated for degree k is $d_k^C \cdot n_k^C$. At this point, one can define the remaining budget at degree k as

$$f_k = \sigma NC_r - \sum_{l=k+1}^{l=k_{max}} d_l^C \cdot n_l^C. \tag{11.6}$$

Clearly, the second term on the right-hand side of Eq. (11.6) demonstrates the already allocated amount for C agents with a degree higher than k. We further note that the second term is zero for $l = k_{max}$, as the whole amount is available for agents having degree k_{max}. The process will be repeated if there is at least one C agent available to be subsidized. Table 11.2 summarizes the allocation of the budget to each degree according to the methodology discussed above. Figure 11.3 illustrates an instance (with $C_r = 0.5$ and $\sigma = 0.5$) of the distribution of subsidies as a function of degree according to this policy.

As we have mentioned, the efficiency of these three subsidy policies will be investigated by estimating the overall social gain, which can be characterized by the "average social payoff" index (*ASP*). Let us now evaluate *ASP* for the three subsidy policies. Suppose that *VC* is the fraction of vaccinators in the population. As $\sigma \leq VC$ is the fraction of the population to be subsidized, $VC - \sigma$ is the fraction of self-financed vaccinators. Due to the subsidy, the government must bear an extra

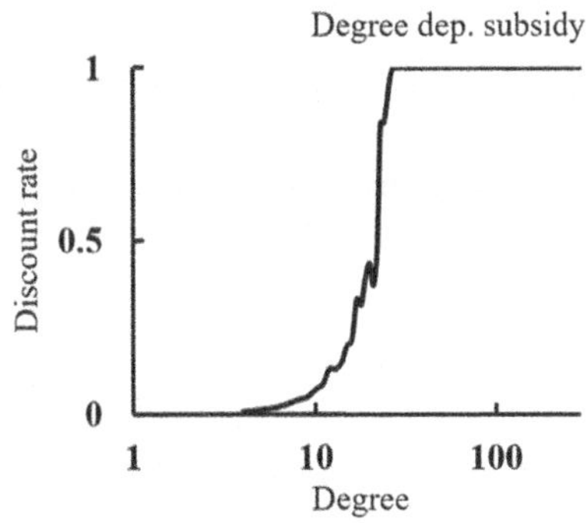

Fig. 11.3 The distribution of the subsidy as a function of degree according to the degree-dependent-subsidy policy. We choose an instance with $C_r = 0.5$ and $\sigma = 0.5$ to draw this figure

financial cost, which comes from the tax collected from citizens. As the whole budget for the subsidy is σNC_r, we can define the per capita tax as $\frac{\sigma NC_r}{N} = \sigma C_r$. However, as mentioned earlier, if the per capita discount, d, is higher than the cost C_r, i.e., if $r > 1$, then the remaining amount will be deducted from the tax. Considering this, we can define the per capita tax as

$$TAX = \sigma C_r - (d - C_r)\mathrm{H}[d - C_r], \tag{11.7}$$

where $\mathrm{H}[d - C_r]$ is the Heaviside function yielding $\mathrm{H}[d - C_r] = 1$ if $d - C_r > 0$; otherwise, $\mathrm{H}[d - C_r] = 0$. Therefore, we can estimate the ASP for the flat-discount policy as follows:

$$\begin{aligned} ASP_{Flat\ dis.} &= (-C_r + r \cdot C_r) \cdot VC - 1 \cdot FES - TAX \\ &= -C_r \cdot VC + r \cdot C_r \cdot VC - 1 \cdot FES - TAX, \end{aligned} \tag{11.8}$$

where FES depicts the final size of the epidemic. Conceptually, it is plausible to assume that the amount $r \cdot C_r \cdot VC$ and the TAX are equivalent because the extent of the subsidy distributed to vaccinators should be the same as that of the allocated budget. Therefore, Eq. (11.8) implies that

$$ASP_{Flat\ dis.} = -C_r \cdot VC - FES. \tag{11.9}$$

Unlike the "flat" discounting policy, the free-ticket policy (hub C-agent priority) does not subsidize all cooperative agents; accordingly, some C agents vaccinate by self-financing. In such a case, ASP can also be estimated by setting $r = 1$ in Table 11.1:

$$ASP_{Free\ tic.} = -C_r \cdot (VC - \sigma) - 1 \cdot FES - TAX = -C_r \cdot VC - 1 \cdot FES. \tag{11.10}$$

For the degree-dependent subsidy, all C agents are confirmed to receive subsidies, although their extent varies according to the degree of cooperative agents. Therefore, we estimate ASP for this policy as follows:

$$\begin{aligned} ASP_{dds} &= (-C_r + rC_r) \cdot VC - 1 \cdot FES - TAX \\ &= -C_r \cdot VC + r \cdot C_r \cdot VC - 1 \cdot FES - \sigma \cdot C_r. \end{aligned} \tag{11.11}$$

Using similar logic to that which held that $r \cdot C_r \cdot VC$ is equivalent to *TAX*, one can re-write the *ASP* for the degree-dependent subsidy as follows:

$$ASP_{dds} = -C_r \cdot VC - 1.FES. \tag{11.12}$$

Therefore, ASP is the same for all subsidy policies (i-iii). Eqs. (11.9), (11.10), and (11.12) demonstrate that *ASP* is independent of per capita tax, which is quite natural, since it is extremely difficult to distinguish the exact portion of one's taxes that eventually end up contributing to vaccination-subsidy policies. It is the government who decides the portion of revenue to allocate for subsidies. Therefore, a detailed breakdown of one's taxes remains invisible; however, subsidies via discount or free tickets are visible to a subsidized individual and impact their decision-making. Hence, subsidy policies work as a catalyst to enhance the vaccine uptake, which accordingly may grant a better state against the disease outbreak.

11.2.3 MAS Approach

We consider a finite population of size $N = 10^4$ on a Barabási and Albert scale-free (BA-SF) network with average degree $\langle k \rangle = 8$. We estimate the disease-transmission rate for MAS in such a way that the disease spreads over 90% of the population if no one commits to vaccination, which means that we use the effective transmission rate β_e introduced in Sect. 3.3.2. Each simulation result is estimated by taking an ensemble average of 100 independent realizations. Every simulation starts with presuming an equal extent of C and D agents in the population. The number of infected people at the beginning of each season is assumed to be 5. Furthermore, the SIR/V dynamics on this spatially structured population are simulated using the well-known Gillespie algorithm[7]. The performance of different subsidy policies is measured by estimating the FES, VC, and ASP as functions of vaccination cost C_r and subsidy size (σ) (i.e., the subsidized fraction of the population).

[7]Gillespie, D.T.J.; Exact stochastic simulation of coupled chemical reactions, *Journal of Physical Chemistry* **81**, 2340–2361, 1977.

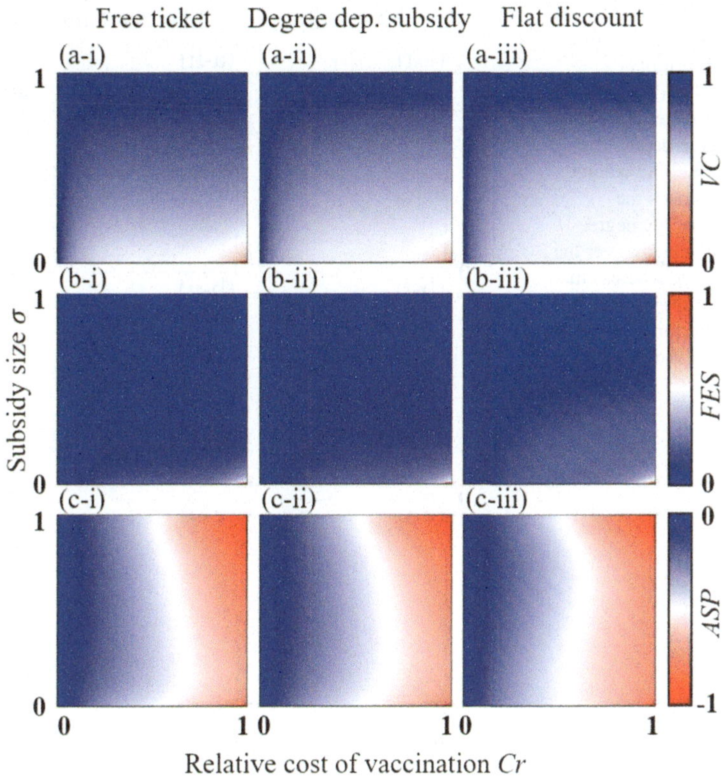

Fig. 11.4 Outcomes of three subsidy policies, namely the free-ticket policy (*-i), degree-dependent subsidy (*-ii), and flat discounting (*-iii), which are measured in terms of vaccination coverage (a-*), final epidemic size (b-*), and average social payoff (c-*) as a function of vaccination cost (C_r) and subsidy size (σ)

11.3 Result and Discussion

We begin our discussion by showing the VC, FES, and ASP under three different subsidy policies as a function of the relative cost of vaccination (C_r) and subsidy size (σ). Figure 11.4 presents the result. The outcomes of the three policies show almost the same tendency; however, to better understand the differences among subsidy policies, we next subtract the outcomes of the major two ones (i.e., free tickets and flat discounting) from that of the degree-dependent subsidy scheme. Figure 11.5 demonstrates that the free-ticket (hub C-agents priority) policy performs better than the degree-dependent subsidy in terms of VC and FES (especially when σ is low), although with respect to the ASP, either policy can outperform the other depending upon the cost and subsidy-size combinations. On the other hand, the degree-dependent subsidy seems to outweigh the flat-discount policy with regard to the three measurement indices (VC, FES, and ASP). With respect to disease

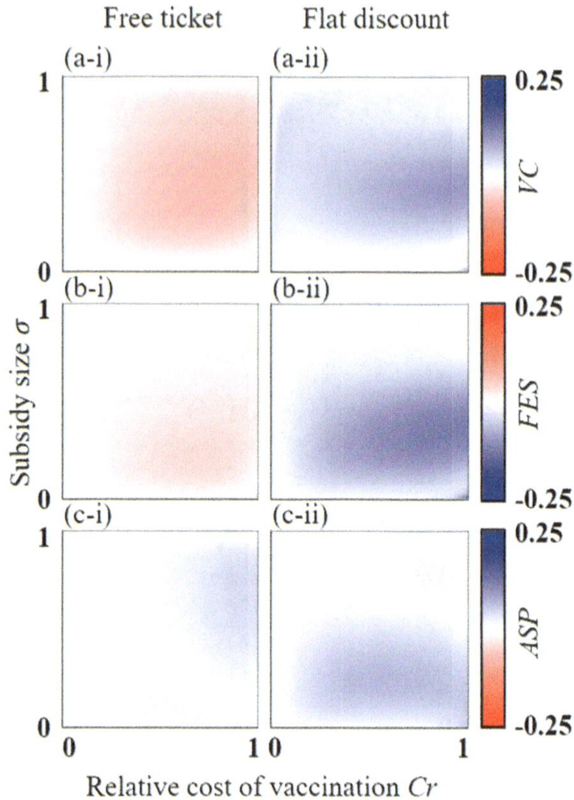

Fig. 11.5 Difference in the outcomes of the (*-i) free-ticket (hub-priority *C* agents) and (*-ii) flat-discount policies from that of the degree-dependent subsidy. The blue-colored region illustrates the superiority of the degree-dependent subsidy over the other policies, whereas the red-colored region indicates the better performance of the free-ticket/flat-discount policy

suppression, the free-ticket policy seems best. However, in terms of overall societal gain (i.e., ASP, which is measured by taking into account the cost of vaccination) as well as the cost of infection, a deeper understanding of which of the policies can efficiently suppress the disease with minimum social cost is required.

Figure 11.6 illustrates the best policy in terms of ASP under all possible combinations of C_r and σ. We note that the best policy has been determined such that each policy prevents the FES from being greater than 0.3, i.e., $FES \leq 0.3$, as depicted by the green line at the lower-right-hand corner of Fig. 11.6 (a). Below this line, $FES > 0.3$; consequently, the disease cannot be controlled by any subsidy policy. The best policy has been estimated via the following procedure: suppose that ASP_1, ASP_2, and ASP_3 are average social payoffs for three different policies at a certain (C_r, σ) point such that $ASP_1 > ASP_2 > ASP_3$. Note that ASP_i, $i = 1, 2, 3$ can be any of the three subsidy policies. Now, ASP_1 will yield the best social gain if it satisfies the following two conditions:

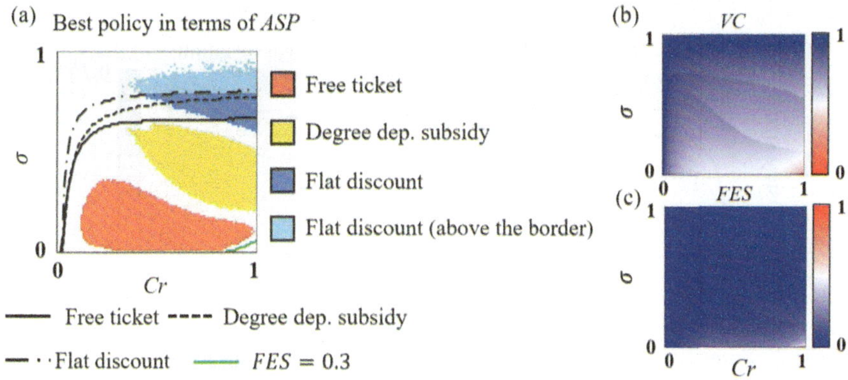

Fig. 11.6 (a) Estimation of the best policy among the three subsidy policies under different circumstances. The free-ticket policy performs best when the subsidy size (σ) or budget size is quite low. The flat – discount policy works well when the subsidy size is relatively high. The degree – dependent subsidy seems to be the best choice for a moderate budget. The solid, small – dashed, and long – dashed lines, respectively, demonstrate the boundaries above which the free – ticket, degree – dependent – subsidy, and flat – discount policies will result in a wasteful vaccination (i.e., allocating budgets above these lines will end up overusing vaccines). The green line in the bottom right – hand corner of panel (a) depicts the boundary above which $FES \leq 0.3$. Below this line, it is deemed a failure at controlling the disease. Panels (b, c) illustrate the corresponding outcomes for panel (a) in terms of vaccination coverage and final epidemic size

$$\left.\begin{aligned} ASP_1 - ASP_3 > 0.01 \\ ASP_1 - ASP_2 > 0.005 \end{aligned}\right\}. \tag{11.13}$$

The first inequality imposes the constraint that the best policy must outperform the third one by more than 1%, while the second inequality stipulates that the best policy must be superior to the second one by more than 0.5%.

With these assumptions, the free-ticket policy seems to perform best among all subsidy policies when the subsidy budget (σ) is quite low (red region in Fig. 11.6 (a)). The flat-discounting policy works well when the budget for the subsidy is relatively high and the cost for vaccination ranges from a moderate level up to the highest level (the deep-blue region in Fig. 11.6 (a)). On the other hand, the degree-dependent subsidy (yellow region in Fig. 11.6 (a)) outperforms the other policies for a moderate budget size. We further estimate the upper limit of the budget size corresponding to the vaccination cost, C_r for each subsidy policy above which subsidization would result in excessive use of vaccination. Fig. 11.6 (a) shows such upper limits for the three policies (represented by three black lines: solid, broken, and dotted). Interestingly, none of the policies except the flat-discount scheme (light blue regime in Fig. 11.6 (a)) crosses its upper limit while granting the best *ASP*. Such upper limits or threshold lines are estimated as follows: below an FES value of 0.01 (i.e., $FES < 0.01$ for each C_r), the disease is considered to have been eradicated; the *FES* is almost zero above the threshold line corresponding to

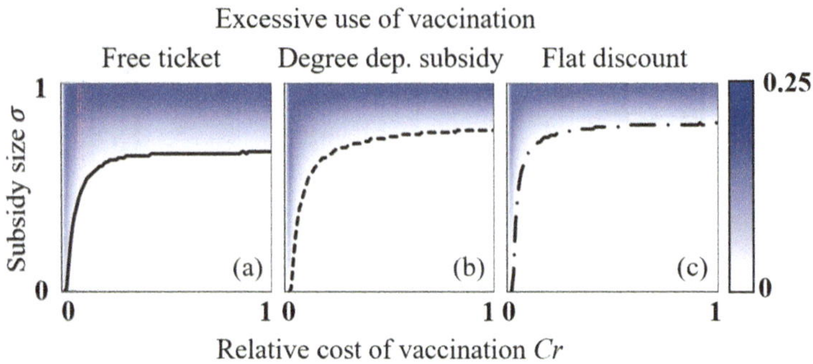

Fig. 11.7 The extent of excessive vaccination for the three policies as a function of C_r and σ. If the budget for a subsidy increases, the free-ticket (panel (a)) and flat-discount (panel (c)) policies result in the maximum and minimum excessive uses of vaccination, respectively. The degree-dependent subsidy (panel (b)) lies between the two extreme scenarios. Note that we only estimate the excessive use of vaccination above the border lines; therefore, the whiteout region below the lines does not correspond to the value of the color bar

each policy, and we call this the "quasi disease-free equilibrium" state, at which point it would be wasteful to spend more money on vaccination. We note that the conditions in Eq. (11.13) are not strictly followed in the whiteout region (Fig. 11.6 (a)) below the threshold lines. In a nutshell, in the whiteout region, we should say that the difference among those three policies is less significant. We draw the VC and FES heat maps (Fig. 11.6 (b, c)) corresponding to that in Fig. 11.6 (a). Observing panels (b) and (c) in comparison with panel (a) illustrates that it is possible to eradicate the disease without expenditures above the threshold limits.

Figure 11.7 illustrates the extent of wasteful vaccination (dV) for the three policies. This extent is determined by the formula, $dV = VC - VC_T$, where VC is the VC at each (C_r, σ) above the border line (where $FES < 0.01$), as in Fig. 11.6 (a), and VC_T is the VC at the threshold lines that confirms $FES < 0.01$ for the first time when the budget size σ increases from 0. In terms of the excessive use of vaccination, Fig. 11.7 suggests the following chronological order when the subsidy budget increases: Free ticket > Degree-dependent subsidy > Flat discount. This clearly demonstrates why the free-ticket policy outperforms the other two for a low-subsidy budget.

The flat-discount policy requires a higher budget to minimize infection because this scheme gives the same subsidy to all C agents, regardless of their degrees. Therefore, the enhancement effect of this policy is significantly lower than that of the other two policies. The free-ticket policy targets individuals (hub C agents) who are more connected with others. Hence, vaccinating hub agents with free tickets can diminish the disease prevalence more efficiently because vaccinated hub agents incur benefits not only to themselves, but also to connected individuals. Moreover, hub agents are also influential in encouraging others to vaccinate. Hence, if the budget size is low, then distributing free tickets to vaccination-intending hub agents can be more effective. The degree-dependent subsidy, on the other hand, intends to

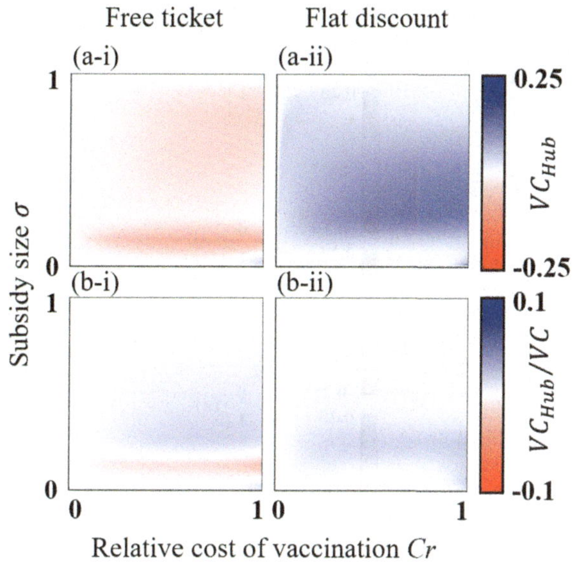

Fig. 11.8 Differences in the vaccination coverage of the hub agents (VC_{Hub} in upper panels) and the VC_{Hub} percentage relative to the total vaccine uptake (VC_{Hub}/VC in lower panels) under the free-ticket (*-i) and flat-discount (*-ii) policies compared to those under the degree-dependent subsidy. Although hub agents' vaccine-uptake level for the free-ticket policy seems superior (panel (a-i)) to the degree-dependent subsidy, the hub agents' vaccination-coverage percentage relative to the whole coverage is relatively better than that under the degree-dependent subsidy. However, in both cases, the degree-dependent subsidy outperforms the flat-discount policy. We note that the blue color illustrates the superiority of the degree-dependent subsidy

balance between the two other policies by subsidizing cooperative individuals according to their connectivity, thereby requiring a moderate subsidy budget unlike the free-ticket policy.

As hub agents are more vulnerable to disease infection, intensive vaccination of hub agents is needed to suppress disease contagion. We now proceed to investigate the hub agent's vaccine-uptake level in the degree-dependent subsidy, as compared to that of the other two policies. To this end, we subtract the hub agents' vaccine-uptake level (i.e., VC_{Hub}; Figure 11.8 (-*)) and the percentage of VC_{Hub} relative to the total vaccine uptake (i.e., VC_{Hub}/VC; Fig. 11.8 (b-*)) under the free-ticket and flat-discount policies from that of the degree-dependent subsidy. Although VC_{Hub} for the degree-dependent subsidy seems lower than that for the free-ticket policy, the VC_{Hub} percentage is better under the degree-dependent subsidy, especially when σ is not very low. The flat-discount policy is mostly surmounted by the degree-dependent subsidy for both indices. We further draw heat maps to explore the comparative performance of the degree-dependent subsidy for lower-degree (fringe), and middle-degree agents (see Fig. 11.9), depending on VC and infectious rate. In all cases, the degree-dependent subsidy outperforms the flat-discount policy, but is dominated by the free-ticket policy.

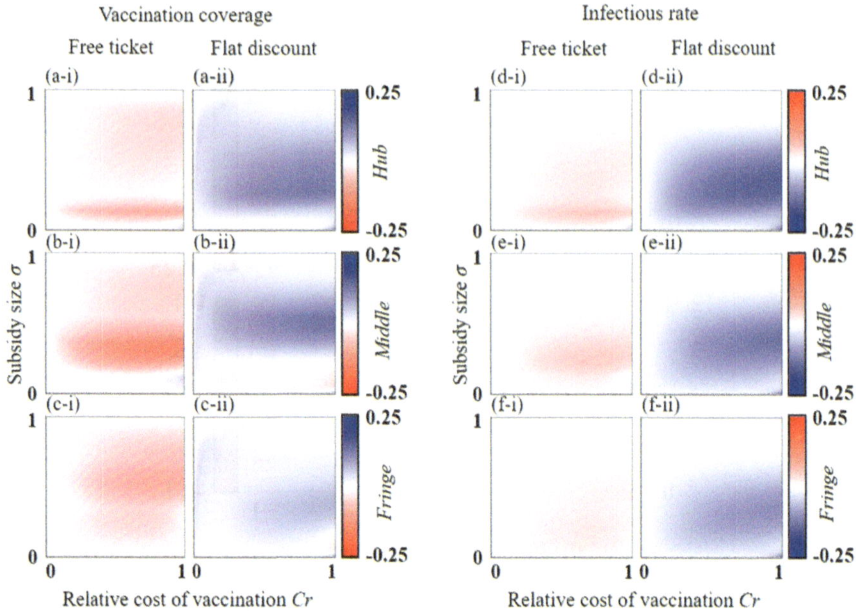

Fig. 11.9 The relative performances of the degree-dependent subsidy, as compared to that of the free-ticket and flat-discount policies in terms of vaccination coverage and the infectious rate for hub agents, agents of middle degree, and agents of lower-degree (fringe). In all cases, the degree-dependent subsidy is outperformed by the free-ticket policy while dominating the flat-discount policy

Let us conclude with a concise summary of what we have discussed, as below.

In this chapter, we introduced a new subsidy policy for vaccination against influenza-like diseases, the so-called degree-dependent subsidy, whereby agents on a scale-free network are subsidized according to their degree—i.e., agents having a higher connectivity or higher degree (hub agents) will receive a greater subsidy than lower-degree agents. This policy connects, to some extent, two major policies that have been previously presented, namely the free-ticket and flat-discount schemes, of which the former prioritizes only hub-cooperative agents while the latter policy distributes the same subsidy to every cooperative agent.

We have investigated the comparative performance of the new scheme with that of the other two policies, relying upon the evolutionary-vaccination game by undertaking the so-called MAS approach. While examining the average social gain resulting from the three policies, the free-ticket policy is seen to yield the best social gain when the subsidy budget is low. The flat-discount scheme outperforms the other two policies when the subsidy budget is quite high and the vaccination cost ranges from a moderate level to the highest level. The degree-dependent subsidy policy, on the other hand, seems to be the best choice for a moderate subsidy budget. Such findings might be meaningful to the public-health authority designing a subsidy framework.

As the aim of such policies is to eradicate the disease efficiently without overspending the money, we estimated the threshold limits (for each cost C_r) of the subsidy budget for each policy above which subsidies may result in unnecessary vaccination. The free-ticket policy, which works well for a lower budget, has the lowest threshold level among the three policies, while the flat-discount scheme possesses the highest threshold level. The threshold level for the degree-dependent subsidy policy lies between the two extreme cases.

Since hub agents are more vulnerable to disease contagion, intensive vaccination of hub agents can result in a more favorable state against the infection. Therefore, subsidizing hub agents who inherently intend to vaccinate with the highest priority or with the highest discount can help to eradicate disease. We compare the new policy with the other two schemes in terms of the hub agents' vaccine-uptake level (VC_{Hub}), as well as the percentage of VC_{Hub} relative to the total uptake level. Although the free-ticket policy seems to have a better uptake level for hub agents compared to the new policy, the percentage of their uptake level relative to the total coverage appears to be better under the new scheme. On the other hand, the flat-discount policy is found to be outperformed by two other policies. Hence, the performance of the degree-dependent subsidy scheme appears to lie, on the whole, between those of the other two policies.

Chapter 12
Flexible Modeling

This chapter shows another possible application of mathematical epidemiology. When considering either the vaccination or intervention-game framework, it is necessary to combine the mathematical-epidemiological model as a foundation—for instance, by combining the SIR process with evolutionary game theory to quantify the time evolution of an individual's decision about provisions for controlling an epidemic, including vaccination. Even relying upon the framework of the behavior model (introduced in Chap. 7), which is appropriate for considering the evolution of individuals' decision over the same timescale on which the disease spreads, another dynamical equation must be attached to the set describing the SIR process. In this chapter, we show that such individual-action dynamics can be implemented into the epidemiological dynamics. This is because the framework of mathematical epidemiology is so flexible that external events other than the spread of a disease can be also built into the epidemiological dynamics.

J. Tanimoto, *Sociophysics Approach to Epidemics*, Evolutionary Economics and Social Complexity Science 23, https://doi.org/10.1007/978-981-33-6481-3_12

12.1 Introduction and Background: A New Cyclic Epidemic-Vaccination Model: Embedding the Attitude of Individuals Toward Vaccination into the SVIS Dynamics Through Social Interactions

For the sake of a discussion template, we are concerned about how people's attitudes toward vaccination reflect their inherent recognition of the trade-off between vaccine acceptancy and the risk of infection, which has been one of the central themes in this book. Here, we model a new path for cyclic epidemic dynamics in vaccination by considering the vaccine acceptancy and fear of infection, whereby the social dynamics are embedded into the mathematical-epidemiological framework. We develop a model called the cyclic mean-field (CMF) model vis-à-vis the cyclic-behavioral (CBH) model that is based on Bauch's behavior model[1]. In the CMF model, we presume that, as with disease transmission, the propelling process for vaccination (i.e., the vaccine-acceptancy-disseminating process) also occurs through human contacts, and that deflating vaccination (low vaccination acceptancy) is also integrated into the social interactions of individuals. This concept enables the decision dynamics, originally depicted in the behavior model, to be buried within the SVIS dynamics.

Thanks to the flexibility of the mathematical-epidemiological model, wherein various compartments can be introduced to model an arbitrary scenario (as we have repeatedly discussed in the previous chapters), even individual-decision dynamics can be implemented. This is quite straightforward because a set of epidemic dynamics encompasses everything and an additional dynamical equation is required to describe the time evolution of human behavior.

Aiming at a more realistic situation in which immunity does not last for a long period, we intend to articulate the cyclic epidemic model by including imperfectly vaccinated compartments in an SIS-like model, whereby a portion of susceptible individuals can spontaneously be vaccinated at a given rate, or by relying on the decision-making process.

12.2 Model Depiction

As a general assumption, we presuppose a well-mixed and infinite population to ensure the mean-field approximation (MFA). The cyclic SVIS epidemic model, which considers both the disease spread and the imperfections in a vaccine, is considered to explore the human decision-making process in a well-mixed and infinitely large population. Each compartment can exist in one of three states:

[1]Bauch, C.T., Bhattacharyya, S.; Evolutionary game theory and social learning can determine how vaccine scares unfold, *PLOS Computational Biology* **8 (4)**, e1002452, 2012.

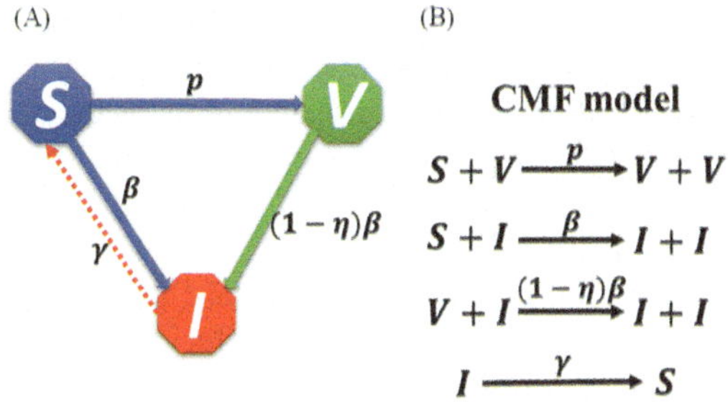

Fig. 12.1 (A) Schematic diagram of the SVIS epidemic model for both CBH and CMF epidemic models and (B) for only the CMF model. Suseptible individuals are infected by their infected neighbors at a transmission rate of β and then vaccinated at a rate of p. Vaccinated individuals become infected at a discount rate of $(1 - \eta)\beta$. Infected nodes become susceptible at a recovery rate γ

susceptible (S), infected (I), or vaccinated (V), as schematically shown in Fig. 12.1 (A). A susceptible individual acquires infection from their previously infected neighbors through social interactions at a transmission rate of β [day^{-1} person^{-1}]. A susceptible individual becomes a vaccinator at a rate of p. Note that this specific parameter p does not indicate vaccination coverage, but vaccination rate. According to the efficiency model, introduced in Sect. 3.2.1, a vaccinated individual can decrease the risk of infection by efficiency η, although the individual can still be infected at the rate $(1 - \eta)\beta$. Additionally, the rate at which a recovered individual acquires infection through social interactions is γ [day^{-1}].

Cyclic-Behavioral Model (CBH Model) The vaccination rate introduced above is treated as a time-dependent variable, $p(t)$. The following set of dynamical equations describe the spread of disease:

$$\frac{dS}{dt} = -p(t) \cdot S(t) - \beta \cdot S(t) \cdot I(t) + \gamma \cdot I(t), \tag{12.1}$$

$$\frac{dV}{dt} = p(t) \cdot S(t) - (1 - \eta) \cdot \beta \cdot V(t) \cdot I(t), \tag{12.2}$$

$$\frac{dI}{dt} = \beta \cdot S(t) \cdot I(t) + (1 - \eta) \cdot \beta \cdot V(t) \cdot I(t) - \gamma \cdot I(t). \tag{12.3}$$

Besides Eq. (12.1), we need to establish another dynamical equation to account for the time evolution of $p(t)$. Based on the behavior-model concept that was originally introduced by Bauch et al. as above, we can presume that

$$\frac{dp}{dt} = m \cdot p(t)[1 - p(t)][-C_v \cdot V(t) + q_{BH} \cdot I(t)]. \tag{12.4}$$

Here, C_v is the relative cost of vaccination, as compared with the disease cost that is set to 1 without loss of generality, and q_{BH} $(0 \leq q_{BH} \leq 1)$ indicates the relative recognition of disease risk. Thus, the term, $-C_v \cdot V(t) + q_{BH} \cdot I(t)$ indicates the benefit/risk trade-off for an individual regarding whether they accept or avoid vaccination. If the vaccination cost increases, the term becomes negative. Furthermore, if there is an affluent fraction of vaccinators, it would be negative, since an individual is affected by a defective mentality free riding on the possibility of herd immunity. By contrast, if the risk of individual infection increases due to either a large q_{BH} or a large fraction of infected individuals, the term would be positive, leading to an increase in the vaccination rate ($p(t)$). Incidentally, m is introduced as a proportionality constant to represent the inertial effect of these social dynamics.

Cyclic Mean-field Model (CMF Model) Below, we formulate a new framework, called the CMF model. The most important difference from the CBH model is that all state transfers S→V, S→I, and V→I (but not I→S, as depicted in Fig. 12.1 (B)) are caused by interactions. Our intriguing concept involves embedding the dynamics of an individual's state alterations into the epidemiological dynamical equation by adopting the SVIS dynamics. Thus, we build the following dynamics:

$$\frac{dS}{dt} = -p \cdot \mathrm{H}[C(S(t), V(t)) + q_{MF} \cdot S(t) \cdot I(t)] - \beta \cdot S(t) \cdot I(t) + \gamma \cdot I(t), \tag{12.5}$$

$$\frac{dV}{dt} = p \cdot \mathrm{H}[C(S(t), V(t)) + q_{MF} \cdot S(t) \cdot I(t)] - (1 - \eta) \cdot \beta \cdot V(t) \cdot I(\mathrm{t}), \tag{12.6}$$

$$\frac{dI}{dt} = \beta \cdot S(t) \cdot I(t) + (1 - \eta) \cdot \beta \cdot V(t) \cdot I(t) - \gamma . I(t), \tag{12.7}$$

where q_{MF} indicates the relative recognition of disease risk when comparing the effects of S–V and S–I interactions and H[∗] is the Heaviside-step function. Here, we introduce another parameter p, which denotes vaccine awareness as a time constant. Note that it does not indicate the vaccination rate ($p(t)$), as in the CBH model. We define $C(S(t), V(t))$ as below:

$$C(S(t), V(t)) = \mathrm{H}[V_\tau - V(t)] \cdot [S(t) \cdot V(t)] + (1 - \mathrm{H}[V_\tau - V(t)]) \times [-S(t) \cdot V(t)], \tag{12.8}$$

$$\mathrm{H}[V_\tau - V(t)] = \begin{cases} 1, & V_\tau \geq V(t) \\ 0, & V_\tau < V(t) \end{cases}. \tag{12.9}$$

Here, $\mathrm{H}[V_\tau - V(t)]$ is another Heaviside-step function corresponding to a relatively sharp transition at $V_\tau = V(t)$, which leads to either 0 or 1, as depicted in Eq. (12.9). V_τ represents the threshold of the vaccination rate over which interaction with vaccinators hampers the increase in the number of vaccinators in society due to the intention to free riding upon herd immunity. Hereinafter, V_τ is called the vaccination acceptancy; when the vaccination uptake, $V(t)$, exceeds the threshold vaccination level (denoted by V_τ), it results in $C(V(t), I(t)) = -S(t) \cdot V(t)$; thus, $-p \cdot \mathrm{H}[C(V(t), I(t)) + q_{MF} \cdot S(t) \cdot I(t)]$ becomes $-p \cdot [-S(t) \cdot V(t) + q_{MF} \cdot S(t) \cdot I(t)] < 0$, which implies that the rate of vaccination from S is discouraged by a relatively large fraction of vaccinators due to the free riding incentive despite the disease risk $q_{MF} \cdot S(t) \cdot I(t)$. By contrast, when $V_\tau \geq V(t)$, $-p \cdot \mathrm{H}[C(V(t), I(t)) + q_{MF} \cdot S(t) \cdot I(t)]$ becomes $-p \cdot [S(t) \cdot V(t) + q_{MF} \cdot S(t) \cdot I(t)] \ll 0$, which implies that the vaccination rate is significantly bolstered by both effects (i.e., a low fraction of vaccinators making individuals anxious and a large fraction of infected individuals working as well); thus, an individual has a greater incentive to be vaccinated.

12.3 Result and Discussion

In the below results with the exception of Figs. 12.3 and 12.4, we presume that $R_0 = 2.5$ and $\gamma = \frac{1}{3} = \frac{\beta}{R_0}$, as we would expect from an influenza-like epidemic, as was presumed in previous chapters.

Theoretical Analysis—Simplified Phase Diagram for the CMF Model First, let us discuss the CMF model in detail. The model given by Eqs. (12.3) and (12.4) has three equilibria, including the trivial ones $(S, V, I) = E_1(1, 0, 0)$, $E_2(0, 1, 0)$ and an internal point $E_3\left(s_1^*, v_1^*, i_1^*\right)$; these are schematically illustrated in Fig. 12.2 (A). Panels B and C in Fig. 12.2 show which equilibrium among the three becomes dominant for varying β (x-axis) and η (y-axis) when presuming (B) $V_\tau = 0$ and (C) $V_\tau = 1$. In both panels, as in the case of $\gamma = 0$ (all infected individuals must stay at I), the disease spread surges, which, with no hope of eradication, leads to the endemic phase $E_3\left(s_1^*, v_1^*, i_1^*\right)$.

Regarding Eq. (12.4), in the case of $V_\tau = 0$, the first term on the right-hand side of Eq. (12.5) always becomes $-p \cdot [-S(t) \cdot V(t) + q_{MF} \cdot S(t) \cdot I(t)] < 0$; thus, at equilibrium, a certain intermediate fraction of vaccinators can be achieved (realizing the coexistence of S, I, and V: $E_3\left(s_1^*, v_1^*, i_1^*\right)$) depending on the strength of a disease, which is primarily determined by β; otherwise, S dominates ($E_1(1, 0, 0)$) as long as β is less than some critical value.

By contrast, $V_\tau = 1$ always results in $-p \cdot [S(t) \cdot V(t) + q_{MF} \cdot S(t) \cdot I(t)] \ll 0$; thus, with a relatively high p and γ (implying difficulty in controlling a disease), an equilibrium presuming a low disease strength (β) or a high vaccine efficiency (η) is superseded by $E_2(0, 1, 0)$ instead of $E_1(1, 0, 0)$. Noting that $E_2(0, 1, 0)$ is more substantially swayed by the vaccine efficiency than $E_3\left(s_1^*, v_1^*, i_1^*\right)$, we naturally accept that the critical line between $E_2(0, 1, 0)$ and $E_3\left(s_1^*, v_1^*, i_1^*\right)$ becomes convex, unlike that between $E_1(1, 0, 0)$ and $E_3\left(s_1^*, v_1^*, i_1^*\right)$.

CBH Vs CMF Model Here, we conduct a qualitative comparison between the CBH (Fig. 12.3 panel A) and CMF (panel B) models in terms of time evolution. In Fig. 12.3, the green and red lines, respectively, indicate the vaccinated and infected fractions. We set the fixed parameters $\beta = 1.0$ and $\gamma = 0.1$. In the top row (*-i), the vaccine efficiency, which is the common parameter for the two models, is varied.

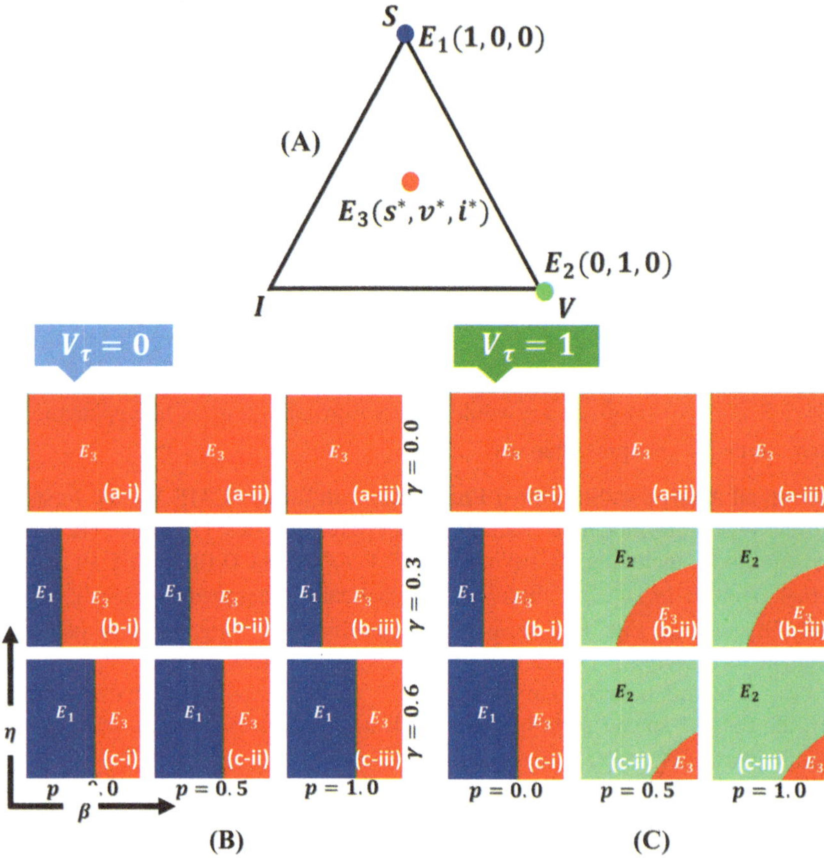

Fig. 12.2 (Panel A) $E_1(1,0,0)$, $E_2(0,1,0)$, and $E_3(s^*, v^*, i^*)$ for the CMF model. The 2D phase space along with $\beta - \eta$ of the CMF model is displayed for $V_\tau = 0$ (panel B) and $V_\tau = 1$ (panel C). The heat maps are drawn for the recovery rates of (a-*) $\gamma = 0$, (b-*) $\gamma = 0.3$, and (c-*) $\gamma = 0.6$. Additionally, sub-panels (*-i), (*-ii), and (*-iii) present the sensitivities resulting from vaccine awarenesses of $p = 0$, $p = 0.5$, and $p = 1.0$, respectively. We set $\gamma = 1/3$ and $\beta = \frac{R_0}{\gamma} = \frac{2.5}{\gamma} = 0.833$

Despite the difference in the final states of V and I, the two models share a general time-evolution tendency.

In the middle row (*-ii), m in CBH and p in CMF are varied, which are counterpart parameters representing the speed of behavior change in terms of vaccination attitude in the respective models. In the bottom row (*-iii), q_{BH} and q_{MF} are varied; these are counterpart parameters as well. Although there is difference in the detailed time-evolutionary tendency and final states of V and I depending on either CBH or CMF being applied, we judge that the general time-evolution tendency is analogously shared.

Now, we discuss the equilibrium comparison between the CBH and CMF models. Figure 12.4 shows the 2D heat maps of disease-transmission rate versus

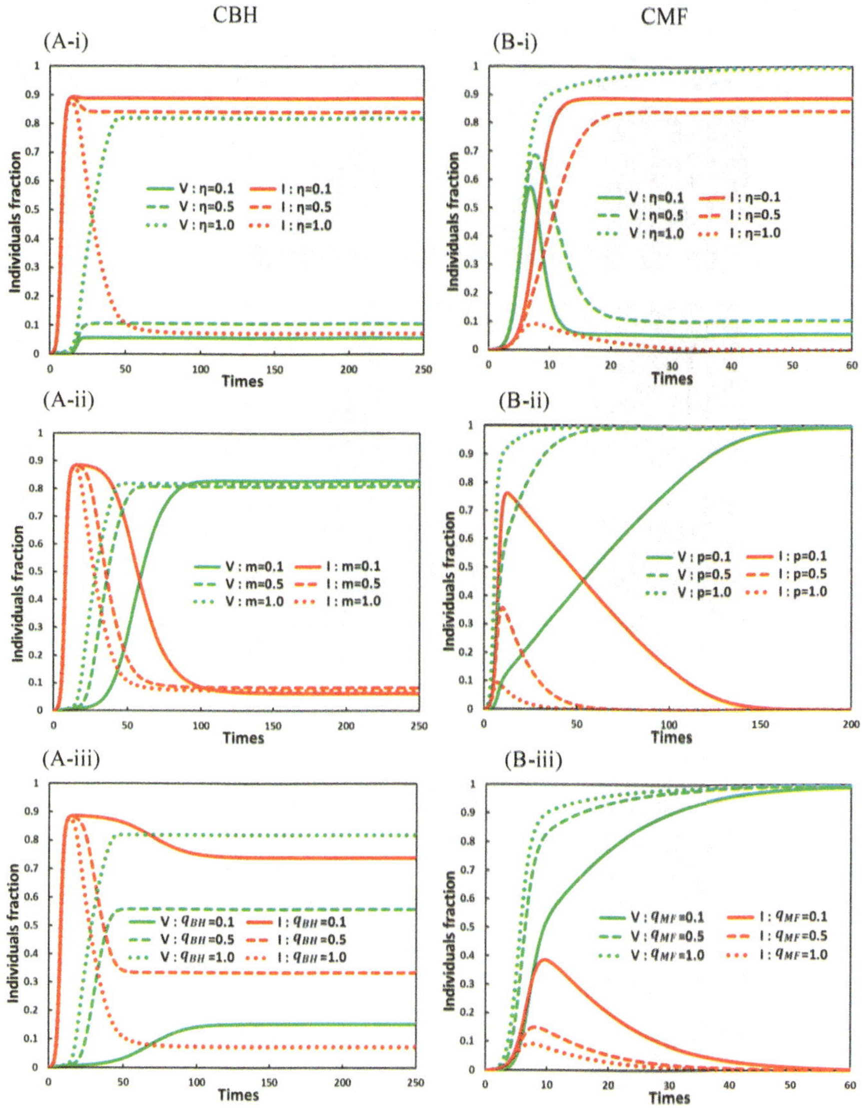

Fig. 12.3 The time-dependent sensitivity plots for the cyclic behavior (CBH) and cyclic mean-field (CMF) models are presented in Panels A and B, respectively. Here, sub-panel (*-i) presents the fraction of vaccinated (green) and infected (red) individuals with varying $\eta = 0.1$, $\eta = 0.5$,and $\eta = 1.0$. Sub-panel (A-ii) displays the transferring parameter, $m(0.1, 0.5, 1.0)$; sub-panel (B-ii) presents the vaccine awareness, $p(0.1, 0.5, 1.0)$; and sub-panels (A-iii) & (B-iii) show the relative recognitions of the disease-risk rate, $q_{BH} = 0.1, 0.5, 1.0$ and $q_{MF} = 0.1, 0.5, 1.0$, respectively. The other parameters are $\beta = 1.0$, $\gamma = 0.1$, and $V_\tau = 1.0$ (for the CMF model)

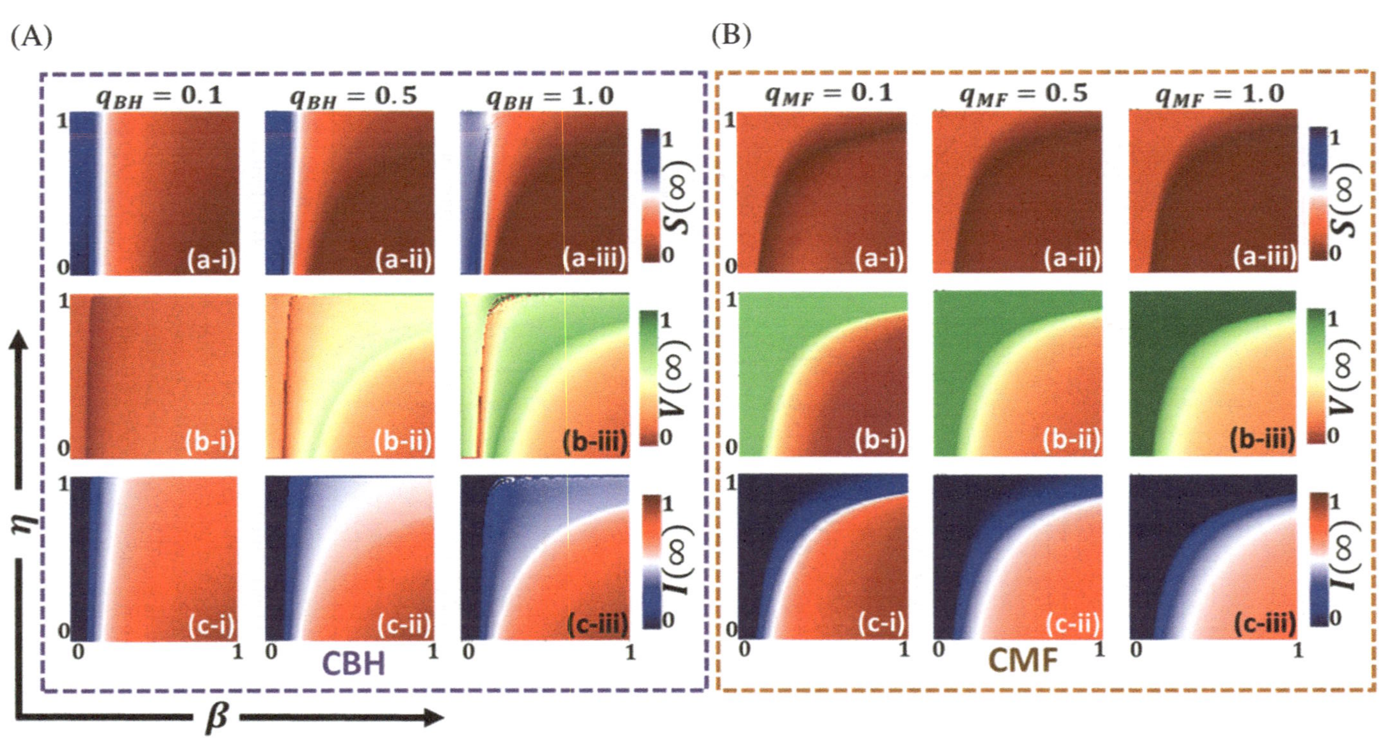

Fig. 12.4 2D phase diagram for the CBH (Panel A) and CMF (Panel B) models as functions of disease-transmission rate (β) and vaccine efficiency (η). Here, sub-panels (a-*), (b-*), and (c-*) present the results for the fraction of susceptible, vaccinated, and infected individuals at the steady state (equilibrium). In panel A, sub-panels (*-i), (*-ii), and (*-iii) display the results of varying the disease risk for CBH as $q_{BH} = 0.1$, $q_{BH} = 0.5$,and $q_{BH} = 1.0$, respectively. Here, $m = 1.0$ and $\gamma = 0.1$. In panel B, sub-panels (*-i), (*-ii), and (*-iii) display the results for varying the disease risk for CMF as $q_{MF} = 0.1$, $q_{MF} = 0.5$,and $q_{MF} = 1.0$, respectively. The other parameters are $p = 1.0$, $V_\tau = 1.0$, and $\gamma = 0.1$.

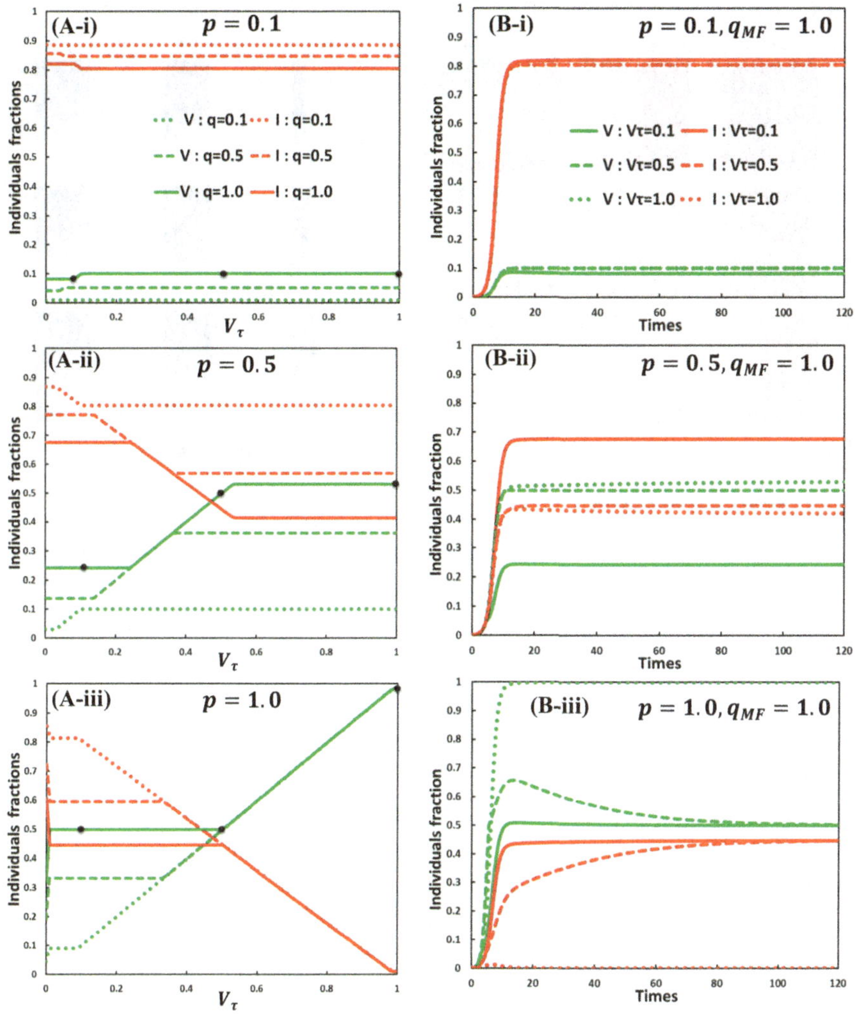

Fig. 12.5 (A) Impact of the vaccine acceptancy (V_τ) for the CMF model on the fraction of infected (red) and vaccinated (green) individuals at equilibrium for $q_{MF} = 0.1$, $q_{MF} = 0.5$,and $q_{MF} = 1.0$. (B) Based on the black circles in Panel A ($V_\tau = 0.1, 0.5, 1.0$), the time-series plots are presented in sub-panels (B-i), (B-ii), and (B-iii) for $p = 0.1$, $p = 0.5$, and $p = 1.0$, respectively. We set $\eta = 1.0$, $\gamma = 1/3$ and $\beta = \frac{R_0}{\gamma} = \frac{2.5}{\gamma} = 0.833$

vaccination efficiency, which are common parameters between the two models. Each of the three heat maps in the row direction adopt a different q_{BH} (panel A) and q_{MF} (panel B). Although the detailed features (e.g., the β (or η) values at which a phase transition occurs, or at which equilibrium is realized) differ, the same general tendency is shared between models. Thus, note that the newly introduced CMF

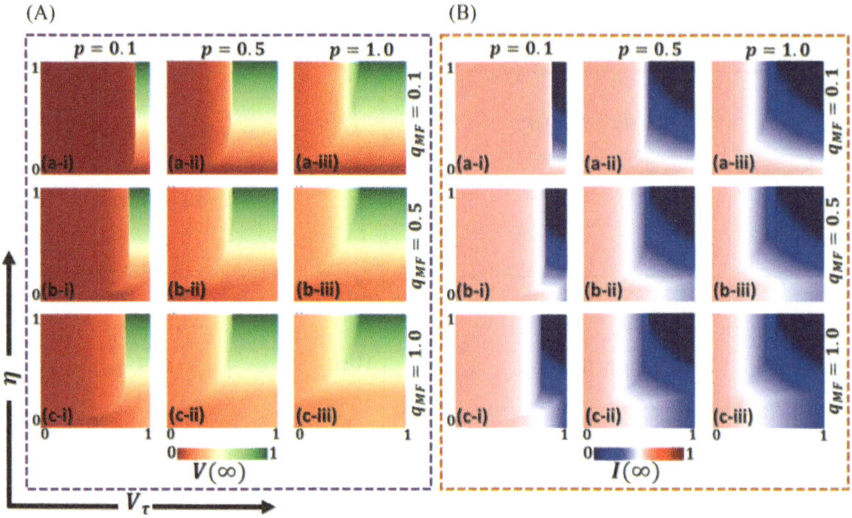

Fig. 12.6 Phase diagrams of the vaccinated (Panel A) and infected (Panel B) individuals for the CMF model as functions of vaccine efficiency (η) and vaccine acceptancy (V_τ). Here, sub-panels (*-i), (*-ii), and (*-iii) present the results of varying p as 0.1, 0.5, and 1.0, respectively. Furthermore, the outcomes of varying q as (a-*) $q_{MF} = 0.1$, (b-*) $q_{MF} = 0.5$, and (c-*) $q_{MF} = 1.0$ are presented. We presumed that $\gamma = 1/3$ and $\beta = \frac{R_0}{\gamma} = \frac{2.5}{\gamma} = 0.833$

model well-reproduces the CBH model based on the conventional behavior-model concept.

Vaccine Acceptancy Henceforth, we focus upon the CMF model. Figure 12.5 shows the sensitivity of vaccine acceptancy, V_τ, as the crucial parameter that significantly affects the model dynamics. With maximal consideration of the effect of the interaction with existing vaccinators as well as infected individuals ($p = 1$), if V in relation to V_τ is higher than a certain critical value (in the case of the solid line, around 0.5), this will lead to a monotonic increase in the fraction of vaccinators, which in turn leads to a decrease in the fraction of infected individuals. As discussed previously, when $V_\tau \to 1$, $-p \cdot \mathrm{H}[C(S(t), V(t)) + q_{MF} \cdot S(t) \cdot I(t)]$ is always negative in Eq. (12.5), leading to a monotonic increase in the fraction of vaccinators. Thus, at the highest p, all individuals are vaccinated and no individual is infected at equilibrium, i.e., $E_2(0, 1, 0)$ in Fig. 12.2 (C). With a relatively small p, it tends to be difficult to achieve full-scale vaccination equilibrium as $E_2(0, 1, 0)$. By contrast, $V_\tau \to 0$ allows the brackets in the formula above to have a zero point, which mathematically implies that the vaccinator fraction does not change after a certain moment in an evolutionary path; consequently, the sensitivity of V_τ has a certain minimal level, as observed here.

Efficiency Vs Acceptancy Figure 12.6 displays the fraction of vaccinated $V(\infty)$ (panel A) and infected $I(\infty)$ (panel B) individuals on the 2D heat map of vaccination acceptancy versus efficiency for different combinations of p and q_{MF}. If vaccine acceptancy and efficiency increase together, the vast majority of individuals will participate in vaccination, resulting in complete eradication of the disease, as is quite conceivable. With both p and q_{MF} (panel A c-iii) at maximum, even in the region of low acceptancy and efficiency, people are encouraged to be vaccinated to some extent. However, low acceptancy impedes complete eradication of the disease, even with a high efficiency (panel B c-iii). Therefore, acceptancy, rather than efficiency, determines whether a dynamic becomes endemic (but not a full wave of infection) or whether a disease-free situation is achieved.

Our discussion in this chapter has uncovered the possibility of a mathematical-epidemiological mode that consists of one of the two parts of the extended vaccination game (i.e., the intervention game). As we have learned, the mathematical-epidemiological model is able to incorporate the dynamics of human decision-making processes, which were originally dictated using the framework of evolutionary game theory.

Hence, the parts of the intervention game and its holistic framework are flexible and can be adjusted to various situations taking place during an epidemic event.

Postscript

The long journey of the present book is nearly over.

I highly appreciate the reader's patience in keeping company with me.

J. Tanimoto, *Sociophysics Approach to Epidemics*, Evolutionary Economics and Social Complexity Science 23, https://doi.org/10.1007/978-981-33-6481-3

Throughout this book, we have been concerned with how the spread of an epidemic on a complex human social network can be reproduced by mathematical-physics frameworks. Mathematical epidemiology, evolutionary game theory, and multi-agent simulation (MAS) are applied as major tools to approach the focal problem.

By coincidence, the time at which we wrote the manuscript happens to be synchronous with the COVID-19 pandemic. Never before in our history have we experienced so many harmful things—human and economic loss besides myriad tragedies, deep anger and distrust—at the same time, except during periods of global warfare.

Yet we still believe that science is powerful enough for us to combat against an unknown virus; not only by developing the vaccines and antiviral treatments to which the field of medical science has always paid great attention, but also by building up solid prediction frameworks concerning the spatiotemporal spread of an epidemic through our social systems. This is the domain of *Sociophysics*.

I would feel very honored if the present book were able to present a remedy, no matter how small, to ease the huge schism brought about by this odious virus.

As the present book has repeatedly addressed, mathematical prediction of a communicable disease on a complex human social network is one of the most difficult scientific challenges. This is because we must interpret it as a merging phenomenon of two different events with the same spatiotemporal structure: the spread of disease, which basically obeys the physics of a "diffusion phenomenon," and the dynamics of human decision making on a complex human social network. The diffusion property not only pertains to the "virus," anymore, but also to information that affects decision making, which work together to make the problem more difficult.

From the first chapter to the very last, this book has aimed to shed light on the concept of an advanced vaccination game, called the *intervention game*, whereby the mathematical-epidemiological framework is dovetailed with the evolutionary game theory. As the author, I definitely believe that the methodology presented by the book works as a useful "decryption key" for reading and understanding such complex phenomena, helping us to propose meaningful social provisions to win the battle with an invisible enemy: an unknown virus.

Putting down my pen, I would like to present some of my works of art, these being several images of Shinto shrines in Japan. Shintoism is Japan's traditional religion, blending indigenous belief and Buddhist faith into one belief system that has lasted thousands of years in our island society.

One of the religious missions of Shintoism over centuries has been healing and calming down the spread of epidemics over our land. Although our island's isolation has historically allowed it to avoid damage from pestilence like the Black Death suffered by Medieval Europe, we have experienced sporadic plagues. Then and still now, Shinto shrines have attracted people's prayers from the bottoms of their souls.

Kamo-mioya-jinja (Shimogamo-jinja); *Kyoto*

Kamo-wake-ikazuchi-jinja (Kamigamo-Jinja); *Kyoto*

Tsurugaoka Hachiman-gu Shrine, Kamakura; *Kanagawa*

Nakatsu-no-miya, Munakata Taisha Shrine; *Fukuoka*

Okitsugu Yohaijo, Munakata Taisha Shrine; *Fukuoka*

Izumo Taisha Shrine; *Simane*

Izumo Taisha Shrine; *Simane*

Dazaifu Tenmangu; *Fukuoka*

Dazaifu Tenmangu; *Fukuoka*

Index

J. Tanimoto, *Sociophysics Approach to Epidemics*, Evolutionary Economics and Social Complexity Science 23, https://doi.org/10.1007/978-981-33-6481-3

The manufacturer's authorised representative in the EU is Springer Nature Customer Service Centre GmbH, Europaplatz 3, 69115 Heidelberg, Germany. If you have any concerns regarding our products, please contact ProductSafety@springernature.com

Printed and bound by CPI Group (UK) Ltd, Croydon, CR0 4YY
30/09/2026
02225295-0008